Introducing 2-in-1: The Instant Cholesterol and Fat Control System

The revolutionary *Instant Cholesterol and Fat Control System* combines the cholesterol-producing effect of *both* saturated fat and dietary cholesterol into *one number* for more than six thousand food items. You will be getting two essential values combined into one "magic" score.

Here is all you have to do:

1. Determine how many calories you eat each day. (Follow the simple instructions in chapter 11).
2. Locate your personal calorie column in the 2-in-1 tables. (Choose the column nearest your daily calorie intake.)
3. So long as you consume no more than 100 points per day, you will be following the National Institutes of Health's guidelines for lowering serum cholesterol levels. That is the "magic" of the 2-in-1 tables—each calorie column is weighted for each food item, so that precisely 100 points are permitted regardless of which calorie column applies to you.
4. Look up *everything* you eat under your calorie column and add up the scores, being careful not to exceed your 100 points per day. It's easy!

Here's to good health and good eating!

2-IN-1:
The Instant Cholesterol and Fat Control System

F. M. Oppenheimer
and Dr. Herbert Pardell

PRENTICE
HALL
PRESS

New York   London   Toronto   Sydney   Tokyo   Singapore

The ideas, procedures, and suggestions in this book are not intended to replace the services of a trained professional. All matters regarding your health require medical supervision. You should consult your physician before adopting the procedures in this book. Any applications of the treatments set forth in this book are at the reader's discretion.

PRENTICE HALL PRESS
15 Columbus Circle
New York, NY 10023

Copyright © 1990 by Nutrition and Health Research, Inc.

PRENTICE HALL PRESS and colophons are registered trademarks of
Simon & Schuster, Inc.

Library of Congress Cataloging-in-Publication Data
2-in-1 : the instant cholesterol and fat control system / by. F.M.
Oppenheimer and Herbert Pardell.—1st Prentice Hall Press ed.
    p.   cm.
    ISBN 0-13-545104-3 (cl.)—ISBN 0-13-932088-1 (pbk.)
    1. Food—Cholesterol content—Tables.   I. Oppenheimer, F.M.
II. Pardell, Herbert.   III. Title: Two-in-one.
TX553.C43A12   1989
641.1—dc19   89-16349
                                                         CIP

Designed by Richard Oriolo

Manufactured in the United States of America

To Helene–
who has been there–with love and
encouragement–always.

We gratefully acknowledge the perceptive understanding, the diligence for detail, and the wise guidance of our editor, Toni Sciarra. This book would not be the same without her.

CONTENTS

PART III
Introducing the 2-in-1 Tables

The Tables

Contents

APPENDIXES

What This Book
Will Do for You

Good morning, and welcome to breakfast at your favorite fast-food restaurant. You are about to order scrambled eggs with sausage, hash browns, and a muffin. Real enjoyment at a reasonable price, or is it?

You know about the dangers of high cholesterol, but what does that have to do with your breakfast? Everything! If you order your usual breakfast, you will have consumed 45 grams—equivalent to 3 tablespoons—of pure fat, of which 18 grams were saturated fat. In addition, you will have eaten 575 milligrams of cholesterol.

You realize that you should reduce your consumption of saturated fat and cholesterol—and you want to—but how can you possibly analyze the nutritional content of every food you eat? *You no longer have to!* Instead, you take a quick glance at the 2-in-1 tables in this book. In only a few seconds you will know that if you order the scrambled eggs, sausage, hash browns, and a muffin, you will have used up 114 points of your daily allowance of 2-in-1 points just with breakfast—and you know that you are supposed to use only 100 points for the whole day! The 2-in-1 tables have alerted you to the fact that you should find an alternative.

You walk next door to your neighborhood coffee shop. After a quick check in the 2-in-1 tables, you order a glass of orange juice, a cold cereal with a banana and a cup of skim milk, an English muffin with a pat of butter and some marmalade, and a cup of coffee. A wonderfully filling and nutritious breakfast that totals only 16 points, instead of the 114 points you nearly consumed, and you still have 84 points left for the rest of the day.

Lunch is no problem. You have a date to have pizza with a few friends from your office. You have already checked the 2-in-1 tables, so you know that if you have two pieces of cheese, mushroom, and green pepper pizza with a soft drink, you will have used up just 12 points—well within reason, considering that you can use a total of 100 points per day.

When it's dinner time, you have a choice of the following:

1. An 8-ounce steak
   French-fried potatoes
   Onion rings
   A vegetable
   Coconut cream pie

Or

2. A bowl of Manhattan clam chowder, chicken gumbo, or onion soup
   (without cheese)
   6½ ounces of skinless chicken (1 drumstick, 1 thigh, ½ breast)
   Rice or noodles
   2 vegetables
   Fruit salad
   2 oatmeal cookies

Which do you choose? It will take you only a minute or two to check all the choices against the 2-in-1 tables. The result makes your choice clear: Dinner 1 has a score of 160; Dinner 2 has a score of 33.

That evening as you look back at your three meals, you see that you have consumed only a total of 61 2-in-1 points the entire day, including some butter for breakfast, cheese for lunch, and a large, satisfying dinner. You still have 32 points left and can certainly afford nearly any kind of snack before going to sleep.

What Makes the
2-in-1 System So Unusual?

The 2-in-1 system makes it simple, fast, and accurate for you to check the cholesterol impact on you of more than six thousand food items. The system does this by combining the cholesterol-producing effect of both the saturated fat and the cholesterol content of each individual food into just *one number*—something that has never been done before!

Still another part of the magic of the system is that it has been computed so that everyone has an exact limit of 100 points per day, regardless of daily calorie consumption. It was so designed because a 100-point daily limit does not involve long or cumbersome calculations, and a 100-point total is a number everyone can relate to and remember.

If you have been told that you have high cholesterol, or perhaps just that you should "watch" your cholesterol, the 2-in-1 tables will make it easy for you to know what you should (and should not) eat. They will do so without complicated formulas or the need to remember a multitude of numbers or equations. All you will do is look up the score for each food that you eat, and try never to exceed 100 points per day. It's as simple as that!

The tables will also show you why some products labeled "cholesterol free" may really be more harmful to your cholesterol level than other products that actually contain some cholesterol. For example, they will show you which three oils—even though they contain no cholesterol— are the most dangerous for your cholesterol levels and yet are used in thousands of products without your realizing it.

The Most-of-the-Time/ Once-in-a-While Rule

One more thing the 2-in-1 system will do for you: It will let you have fun, because if you always plan your menus under the guidance of the 2-in-1 tables and never exceed 100 points, then you can indulge yourself once in a while and eat anything you want. Your cholesterol level does not change from meal to meal. It changes for the better or the worse over a period of one to two months, depending on your total consumption of foods during that period. This book will show you that if you eat carefully *most of the time*, then you can still enjoy many of your favorite foods—foods that you might have thought were forever forbidden to you.

Therefore, if you chose the second breakfast and the second dinner above, and continue to follow the 2-in-1 system carefully each day, then go to that great party, have that special birthday dinner, and enjoy yourself. Not only will the food taste better than ever before, but you know that you can eat whatever you like *once in a while*, because you have learned to control your cholesterol the rest of the time.

How the 2-in-1 System Began

The 2-in-1 system got its start because one day one of the authors, F. M. Oppenheimer, went to see his personal friend and physician (and now coauthor) Dr. Herbert Pardell, for an annual checkup. The results of blood tests indicated that Oppenheimer had very high cholesterol and triglyceride levels, and Dr. Pardell recommended a special diet low in dietary cholesterol and fats.

Oppenheimer wanted to know more about the subject and about the types of foods he could eat. He went to the major book stores in his area and to his local public library. He read dozens of books that attempted to explain in great detail what causes "high cholesterol" and what can be done to reduce it. Most of these books were so packed with medical and chemical facts, names, and charts, however, that after the first few chapters Oppenheimer found that there were many aspects of the subject that he could not understand until he discussed them at length with Dr. Pardell. He realized then that most people don't have the opportunity to have in-depth conversations with their physicians. They need simple explanations that define exactly what saturated fat and cholesterol are, how much of each is contained in the foods we eat, and how we can effectively control how much of each we consume in our daily diets.

This book contains a clear, easy-to-follow explanation of what high cholesterol is, what causes it, and what you can do about it. The chapters are short on purpose, because you don't need material intended for the scientific world. Instead, we have tried to simplify and clarify the material, so that all of it can be readily understood without a medical or nutrition background.

In researching our subject we also found that sometimes recipes comprised more than half of other books. This, we felt, was fine for chefs, but it seems unrealistic to ask people to change their cooking habits dramatically and start baking special types of cholesterol-reducing muffins or breads. We therefore decided not to include any recipes in this book.

Then we discovered that some books required the reader to do complicated mathematical calculations in order to determine the cholesterol effect of a given food. Readers were told to determine the number of calories in a particular item, such as a fried chicken drumstick, which contains 193 calories. Then they were told to find the amount of fat in the drumstick, which came to 11.3 grams. The next task was to find the percentage of fat that was saturated, which equalled 26.4 percent. This figure then had to be related to the total fat by multiplying 11.3 by 0.264, resulting in 2.98 grams of saturated fat. As each gram of fat contains 9 calories, this result had to be multiplied by 9, which came to 26.8, to arrive finally at the calories of saturated fat in the drumstick. Readers then had to compare the 26.8 calories of saturated fat to the drumstick's total of 193 calories, to determine whether they exceeded 10 percent of that number.

They did, but by now, if you were the reader, you were so confused that you lost your appetite for the drumstick or anything else—or you had gotten so hungry that you had already eaten it! And even if you successfully completed all these calculations for determining the saturated fat content of the drumstick, you still had to assess its cholesterol content, which came to 62 milligrams. Perhaps you were told that you were allowed 300 milligrams of cholesterol per day, so you think it is all right to eat the drumstick after all—or is it?

When we realized how extremely difficult and complicated it was to use this method, we decided to use our computer to list the fat, saturated fat, and cholesterol contents of the more than six thousand food items for which we could find the relevant nutrient information. Then, using scientifically and medically accepted formulas, we designed a unique group of equations that combined the cholesterol-producing effect of *both saturated fat and dietary cholesterol* on a weighted basis for each food item and for eight different calorie intake levels. These equations were then programmed into the computer, making it possible to arrive at *one single score* for each food product, representing its total serum cholesterol value.

Using our system, your decision of whether to eat that drumstick has been made easy: For instance, if you consume an average of 2,400 calories per day, all you have to do is look up the drumstick entry under "Poultry, Chicken" in the tables in this book, and in the column for 2,400 calories you will find the 2-in-1 score of "15." You know that you can consume up to a total of 100 points per day, and the "15" immediately tells you exactly how many points are used up by that drumstick.

There are eight different calorie columns, but you need only look for the score in the calorie column that applies to you. The tables in chapter 11 will show you how many calories you should consume for your height, weight, and lifestyle. The magic of the 2-in-1 system is that each food item has been indexed for each calorie level so that regardless of each individual's calorie intake, everyone has a daily limit of exactly 100 points.

Appendix B explains how we developed our formula, but you can skip that part if you wish. Our goal was to simplify the concept of dietary cholesterol and fats so that you could immediately recognize the values of all the foods you like to eat.

We think we accomplished our objective. Using this system, you should find it easy to control your cholesterol while still enjoying good food and eating well!

Does the 2-in-1 system work? It did for Oppenheimer! In chapter 13 we will show you how he lowered his cholesterol level by 36 percent—from 306 milligrams to 197 milligrams—in just a few short months.

We hope this book will do as much for you!

HERE'S TO GOOD HEALTH— AND BON APPÉTIT!

PART I

Cholesterol Made Easy— All You Need to Know for a Healthier Life

I

Cholesterol–The *Real* Definition

You have undoubtedly heard and read much about the dangers of "high cholesterol." You may be worried—or perhaps your doctor has told you—that your own cholesterol level is too high. Stop worrying and relax! In most cases cholesterol levels can be reduced substantially in a brief period, through the use of one or more methods—all of which are analyzed in this book.

Besides being worried, you may be puzzled. What is cholesterol, anyway? You can't grapple with something you don't really understand. Again we say: Relax! Cholesterol is not complex or difficult to understand. Here is what it's all about:

Cholesterol is a fatty substance found only in animals and never in plants. This includes *all* animals—not just beef, but also lamb, pork, and veal, as well as chicken, turkey, and other poultry, and all fish from salmon, snapper, lobster, and herring, down to the littlest sardine. In short, all animals produce the fatty cholesterol substance, including the most developed animal there is: the human animal. All of us—men, women, and even small children—have cholesterol in our bodies. Not only is cholesterol found in all animals, it is also found in all animal products, such as milk, cheese, butter, lard, and eggs.

Whenever we eat meat, poultry, or fish (or their products), we cannot help but eat cholesterol in that food. This cholesterol that we eat is

3

called *dietary cholesterol*, which differentiates the cholesterol we eat from the cholesterol that we ourselves carry.

The cholesterol in our blood is called *serum cholesterol*, "serum" being the name for the liquid portion of our blood. Serum cholesterol is made up of dietary cholesterol and cholesterol manufactured in our liver.

Some Cholesterol in Your Blood Is Good for You

There is absolutely nothing wrong with having cholesterol in your blood—everyone has it and everyone needs it, because cholesterol is an essential building block in the production of hormones and cell membranes. One important idea to understand at this point: You do not need to consume dietary cholesterol to be healthy. If you never again ate any cholesterol, your body would still have more than enough of it because your liver manufactures all the cholesterol you need.

What can be wrong is that you may have too much cholestrol in your blood. Many scientific studies over several decades seem to prove that an elevated cholesterol level may have a serious effect on your health. The well-known Framingham scientific study found that the risk of heart disease is four times as great if your serum cholesterol level is 300 than if it is only 180.

It seems that some of the fatty substance that we call cholesterol has a tendency to stick to the walls of your arteries as it moves through your bloodstream. (The medical term for this condition is *atherosclerosis*, a term your physician might use in discussing your cholesterol level with you.)

As the years go by, more and more of the cholesterol (along with other substances) adheres to your arteries. The opening through which your blood must pass gets smaller and smaller. One day, the plaques (as the cholesterol-containing material that is attached to your arteries is called) may become so dense that they momentarily completely stop the flow of blood. And you have just had a heart attack or a stroke!

What Can You Do about High Cholesterol?

The solution to the problem seems easy: Just eat less cholestrol. That *is* an important part of the answer. There is more to it than that, though, and what you can do about it is more fully described in the following chapters.

Is There a Good As
Well As a Bad Cholesterol?

Until recently, when your doctor checked your serum cholesterol level, he did just that: He drew a little blood and had his lab tell him how many milligrams of cholesterol were found in your blood. This, however, is no longer considered sufficient. Research has proven that it is now imperative that all the parts that make up the total cholesterol count be individually scored.

While we said before that the cholesterol in your body moves through the bloodstream, it actually cannot do this unassisted. As it is just a fatty substance, it would lie heavily in the arteries unless some mode of transportation, something like a water taxi, were provided to move it about.

Our body, being the incredibly sophisticated machine it is, provides not just one but three different "water taxis." These "taxis" are actually tiny capsules of water-soluble protein, which encompass even tinier drops of the cholesterol in your blood.

These tiny capsules come with different densities. Some of them are of high density, while others are of low density. The high density ones are called HDL, (high density lipoprotein), while the low density ones are called LDL (low density lipoprotein). Lipoprotein is the term for the combination of fat and protein that moves through the bloodstream.

The cholestrol being transported by HDL capsules is considered "good cholesterol," because the high density capsules take the cholesterol and carry it straight to the liver, where it is eventually converted to harmless, but necessary, body chemicals.

The LDL capsules, on the other hand, carry "bad cholesterol" because, if the liver does not have sufficient receptor sites to accept and assimilate the low density capsules, they will be forced back into arterial circulation. After being rejected by the liver, these LDL capsules may get stuck along artery walls, and might eventually totally block one or more of these arteries.

This is why one of the most important facts to remember about the cholesterol in your blood is this: It is not enough for you to know that you have a certain count of serum cholesterol—it is equally important for you to know how much of your cholesterol is "good" (HDL), and how much of it is "bad" (LDL).

It has been shown that the higher your HDL score is, the less likely it is that you will have a heart attack. HDL is the only value in the whole cholesterol count that you want to have high rather than low. On average, men have an HDL count of 45 milligrams, while women usually score higher, with a 55 milligram average. It has also been shown that HDL levels often decrease with advancing age—all the more reason that the older you are, and the more at risk, the more you want to keep track of your HDL count.

Again, the solution seems simple. If we could just increase the HDL capsules and reduce the LDL ones, we would make real progress. Used correctly, this book will help you achieve exactly that.

Cholesterol-HDL Ratio

We cannot emphasize the importance of your HDL level too much. The higher the HDL count is, compared to the total cholesterol count, the more cholesterol is actually being removed from the bloodstream. Conversely, the lower the HDL count, the less cholesterol is being removed.

Therefore, if your HDL level is high, you can afford to have a more elevated total cholesterol level, because the *ratio* between your total cholesterol and your HDL values is perhaps an even more important measuring tool than your cholesterol score by itself.

The *Cholesterol-HDL Ratio* has been called one of the most reliable indicators of impending heart disease. The ratio is determined by dividing your total cholesterol by your HDL count. Say, for instance, that your cholesterol count is 225 and your HDL score is 45. By dividing 225 by 45, you arrive at the cholesterol-HDL ratio of 5, which is about average for men. (Women have a lower average of about 4.4.)

It is important to understand that this ratio is one more number that you want to have *as low as possible*, because the lower your cholesterol count and the higher your HDL score, the lower will be the ratio. For example, let's suppose that in the readings above of 225 milligrams cholesterol and 45 milligrams HDL, the cholesterol could be lowered to 180. Then dividing 180 by 45 results in a lower ratio of 4 instead of the previous 5. Now suppose you could also raise your HDL score to 54. You would then divide 180 by 54 and get an even lower ratio of 3.33, compared to the original 5. While this improvement may not seem like much, you would now have *less than half* the average risk of having a heart attack.

The famous Framingham study of cardiovascular disease has followed the risk factors affecting diseases of the heart in thousands of the residents of Framingham, Massachusetts, for nearly forty years. It was the Framingham study that developed the following table of heart attack risk based on the cholesterol-HDL ratio:

Table 1. How a Lower Cholesterol-HDL Ratio Can Sharply Reduce Your Risk of Heart Disease

RATIO	
2.6	Vegetarians
3.4	Half average risk (Marathon runner)
4.4	Average risk for women
5.0	Average risk for men
4.6–6.4	Average victim of heart disease—women
5.5–6.1	Average victim of heart disease—men
7.1	Twice average risk—women
9.6	Twice average risk—men
11.0	Triple average risk—women
23.4	Triple average risk—men

Triglycerides

There is one more important component to the cholesterol story: triglycerides. These make up another fat in the bloodstream, similar to cholesterol, but with one major difference: While dietary cholesterol is acquired only through the consumption of animals and their products, nearly every fat found in both animals *and* vegetables contains some triglycerides.

Of greater importance than the actual consumption of dietary triglycerides, however, is the production of triglycerides by the liver. This organ converts excess calories and some foods, such as fats, sugar, and alcohol, into triglycerides. Again the point is this: You don't need dietary triglycerides. Your liver produces all you need—and far too much if you eat an excess of the wrong foods.

Your body needs triglycerides because they are essential in the formation of all cell walls. However, too high a level can cause problems similar to those caused by too high a cholesterol level. Furthermore, there seems to be a correlation between triglycerides and HDL: usually the higher the triglyceride count, the lower the HDL score—and the greater your risk of heart disease.

Therefore, you want to lower your triglyceride level, just as you want to lower your LDL and your total cholesterol score.

Triglycerides are transported through the bloodstream by a protein "taxi" called VLDL (very low density lipoprotein). While triglycerides are measured and reported separately in your blood workup, they are in effect also included in your total cholesterol count. That is because every triglyceride-carrying VLDL capsule will also carry one-fifth as much cholesterol as it is transporting triglycerides. Therefore the lab simply divides your triglyceride score by five and thus determines the amount of VLDL cholesterol in your blood.

Your Total Cholesterol Count

You may know what your cholesterol count is, either from your physician or from an independent screening service, but you should also know the values of the three components comprising your total cholesterol score. Your cholesterol count is calculated by adding together the individual scores for LDL, HDL, and VLDL. Using arbitrary figures as an example, the totals may look like this:

155 mg LDL

45 mg HDL

25 mg VLDL (⅕ of 125 triglycerides)

225 mg TOTAL CHOLESTEROL

In each case the "mg" (sometimes also shown as mg/dl) stands for milligrams per deciliter (equal to one-tenth of a liter). This is the standard measurement for serum cholesterol.

A Quick "High-Low" Cholesterol Summary

You should try to have low levels of everything involved with cholesterol *except* your HDL score:

HDL	should be high
LDL	should be low
VLDL	should be low
TOTAL CHOLESTEROL	should be low
CHOLESTEROL-HDL RATIO	should be low

Average Cholesterol Levels

Table 2 shows the "average" cholesterol, triglyceride, LDL, and HDL levels for American men and women. Sometimes, these figures are even called "normal," because they represent the average score. It must be pointed out, however, that these levels are just that: "average." That does not mean they are good. The National Institutes of Health (NIH) has now called for a *maximum of 200 milligrams* of serum cholesterol (and 130 milligrams of LDL), regardless of your age or sex. Their findings are more fully discussed in chapter 3.

As a matter of fact, while the "average" cholesterol level is indicated below as being not much in excess of 200 milligrams, it means, in essence, that more than 50 percent of all men over age 40 (and more than 50 percent of all women over age 45) have serum cholesterol levels higher than 200 milligrams. Further, the facts are that only 10 percent of all men under age 40 have levels lower than 170, and only 10 percent of women under age 40 have levels lower than 180. Again we must repeat: "Normal," meaning "average," is not necessarily good or healthy.

We, therefore, present these figures only because they may be of interest to the reader, not because they should be used as guidelines.

Table 2. Average Cholesterol Levels

Age	Choles- terol	LDL	HDL	Triglyc- erides
		MEN		
10–14	155	95	55	65
15–19	155	95	45	80
20–24	165	105	45	100
25–29	180	115	45	115
30–34	190	125	45	130
35–39	200	135	45	145
40–44	205	135	45	150
45–54	215	145	50	140
54–69	215	145	50	140
70+	205	145	50	135

Age	Choles- terol	LDL	HDL	Triglyc- erides
WOMEN				
10–19	160	100	55	75
20–24	170	105	55	90
25–34	175	110	55	90
35–39	185	120	55	95
40–44	195	125	60	105
45–49	205	130	60	110
50–54	220	140	60	120
55–64	230	150	60	125
65 +	230	150	60	130

Source: Adapted from Lipid Research Clinics Program, U.S. Department of Health and Human Services, National Institutes of Health

As pointed out earlier, these are purely "average" figures. Your goal should be to achieve levels that are considerably better than the numbers shown. The 2-in-1 System will show you how you can accomplish that goal quickly and easily.

2

Dietary Fats–
Some Are Actually *Good*
for You!

As important as it is to limit your cholesterol intake, it is perhaps even more important to watch the fats you eat. Recent research shows that *excess consumption of the wrong fats will raise your serum cholestrol level even more than dietary cholesterol.*

In October 1987, the National Heart, Lung, and Blood Institute (which is part of the NIH), joined by all the major national health organizations, recommended that Americans limit their consumption of fat to a maximum of 30 percent of total calories. It was further recommended that not more than one third of all fat calories be made up of saturated fat, meaning that saturated fat intake should never exceed 10 percent of the total calories consumed daily.

Practically every food we eat contains some fat or oil. As you will see, however, some foods have far more fat than other foods, and not all fats are bad for your serum cholesterol. Some fats may actually be helpful in lowering your cholesterol count.

The fats we eat fall primarily into two main categories: *saturated* and *unsaturated*. Unsaturated fats are of two distinct types: *poly*unsaturated and *mono*unsaturated. The chemical differences are as follows:

All fats consist of long chains of fatty acids containing carbon and hydrogen atoms. Fats are called *saturated* when all available spaces between the carbon atoms in the chain are taken up with hydrogen molecules. If there is *one* double space free of hydrogen atoms, then the fat is *monounsaturated*. If there are *two or more* double spaces free of hydrogen atoms, then the fat is *polyunsaturated*.

Saturated Fats

Many books have stated that you can recognize saturated fat easily, because it is usually solid at room temperature. This is not true, because many oils are far more highly saturated than some solid fats. The U.S. government has defined any fat that is solid as a "fat," and any fat that is liquid as an "oil." The degree of saturation actually has no bearing at all on whether it is a solid fat or a liquid oil.

Saturated fats are "bad" for you, since the consumption of these fats will raise your cholesterol count. They are found primarily in animals—that is, all meats, poultry, and fish, and all animal products, including all dairy items such as milk, ice cream, cheese, and butter.

As you will soon see, some types of meat and poultry have far less saturated fat than others. And some fish, because of the unique fats they contain, can actually be beneficial to you, even though they are part of the animal kingdom and may be quite fatty. Therefore, you must take great care not just to limit your consumption of animal foods but to differentiate between those animal foods that may be worse for you than others. The 2-in-1 system is designed to help you do just that. (Make sure, also, that you read chapter 8, Foods That Can Actually Lower Your Cholesterol.)

Let us give you an example of the difference in saturated fat content between foods (based on a daily intake of 2,400 calories). Suppose you feel like eating a 6-ounce serving of short ribs. These contain 2.5 ounces of fat, of which more than 1 ounce is saturated. As each ounce is approximately 28 grams, and as each gram of fat has 9 calories, this 1 ounce of saturated fat is equivalent to 252 calories of saturated fat, or 10.5 percent of the 2,400 total calorie intake. According to the federal guidelines, this one item alone will exceed the 10 percent limit of saturated fat that you should consume in one day.

If, instead, you eat 6 ounces of steamed or poached salmon, you would consume a total of only about one quarter of an ounce of fat, of which less than 20 percent, or a total of one-twentieth of an ounce, is saturated. Thus the salmon, with only one-twentieth of the saturated fat of the ribs, would permit you far more leeway in the other foods you can eat the rest of the day.

The 2-in-1 tables, of course, save you from all this cumbersome arithmetic. If you look at the short ribs entry in the 2,400-calorie column, as an example, the tables show you that the ribs have a 2-in-1 score of 90, while the salmon has 14. It makes it so simple to see which foods are better for you! (Don't forget that the score also takes cholesterol into account, not just saturated fat.)

Here is another example, this time with your dessert (and again based on a daily intake of 2,400 calories). If you have a piece of chocolate cheesecake after dinner, you will consume nearly 60 percent of the saturated fat that you are allowed for the day. If, instead, you have a similar size piece of coffee cake, you will consume only about 4 percent of the allowable daily saturated fat.

Again, the 2-in-1 tables make it simple for you: The chocolate cheesecake has a score of 53, while the coffee cake rates only a 5. Isn't it smarter to choose the coffee cake, and enjoy the meal that precedes it, because you know that your dessert will not materially affect your 2-in-1 limitations for the day?

There seems to be only one exception to the rule that all saturated fats raise your cholesterol score. This exception is a fat called stearic acid, which for more than twenty years has been accepted as being neutral with respect to producing cholesterol. This fatty acid can be found as part of the saturated fats in practically every food that contains such fats (including beef, lamb, pork, poultry, and cheese), and usually represents only about 10 to 15 percent of the total fats in these foods.

Some recent tests indicate that stearic acid may actually lower serum cholesterol under certain conditions, but results so far are inconclusive because of the limited number of tests undertaken and the special circumstances under which they were administered.

Furthermore, as nearly all foods containing saturated fat have a similar amount of stearic fat, it in effect becomes a common denominator and has no particular bearing on any particular food or food group. Undoubtedly, this is the reason why the National Institute recommended that the saturated fat calories represent not more than 10 percent of total calories consumed, regardless of the makeup of the saturated fat.

Unsaturated fats

In contrast to saturated fats, which come mostly from animals, *unsaturated fats* are usually derived from vegetables in the form of oils. They are either soft or liquid at room temperature. (The four major exceptions are coconut, palm and palm kernel oils, and cocoa butter. See the Caution on the following page.)

As mentioned on the previous page, unsaturated fats fall into two categories: polyunsaturated and monounsaturated.

Polyunsaturated fats

Polyunsaturated fats may be good for you. In the last thirty years, many studies seemed to prove that polyunsaturated fats actually help to lower serum cholesterol, and a formula was developed stating that a given quantity of polyunsaturated fat will lower cholesterol by half as much as the same quantity of saturated fat will raise it. (Other studies seem to indicate that while polyunsaturated fats will indeed decrease LDL levels, there is a tendency in some people for the important HDL levels to decline as well.)

Polyunsaturated fats are found in most animal and vegetable products in varying degree. The best sources for polyunsaturated fats are fish, about which we talk more in chapter 8, and most vegetable oils, such as corn, soybean, safflower, and sunflower oils.

Monounsaturated fats

Monounsaturated fats may also be good for you. Recently, much research has been done with foods containing large amounts of mono-unsaturated fats. These tests were undertaken primarily because it was discovered that among people in Greece and Italy, high serum cholesterol was largely unknown. Research seems to indicate that the main dietary reason is olive oil, a food very rich in monounsaturated fat, and consumed in large quantities in those two countries. Many scientists now believe that monounsaturated oils will also decrease cholesterol and LDL levels, but are not yet certain whether it is in the same proportion as polyunsaturated fats. They have found, however, that while polyunsaturated fats tend to decrease "good" HDL levels while decreasing "bad" LDL, monounsaturated fats will not only lower LDLs but may also actually *raise* HDL scores.

While there is still debate as to whether monounsaturated fats will actually lower serum cholesterol levels, all major health organizations now recommend that about one-third of your total fat intake be monounsaturated. Besides olive oil, some other good sources of mono-unsaturated fats are almond and rapeseed (canola) oils.

Caution: Three vegetable oils are extraordinarily high in saturated fats: coconut, palm, and palm kernel oils. Beware of them in all packaged foods.

These so-called tropical fats are used extensively by manufacturers because they extend the shelf life of foods, particularly of baked goods, in supermarkets. Further, they are produced in such vast quantities in Asia that they are much cheaper than other fats or oils.

One other vegetable fat not ordinarily used in cooking but which is also highly saturated is cocoa butter. Therefore, you must be very prudent in the quantity of chocolate you eat. Cocoa powder usually has most of its fat removed and can be used much more freely.

You may already have reduced the amount of animal fat you eat. Perhaps you pass up that 12-ounce prime steak or that goose drumstick with the skin. However, if you then eat cookies, cakes, crackers, or any of the other products that list one or more of the tropical oils as an ingredient, you will probably have consumed more saturated fat than if you had eaten that steak or the goose drumstick and skipped the dessert altogether. Read all labels, be careful, and check the 2-in-1 tables!

Hydrogenated Oils

Because of continuing bad publicity surrounding the tropical oils, many food manufacturers have recently undertaken to eliminate these offensive fats. They usually substitute other vegetable oils, such as soybean or corn oil. Because they need a more solid product, however, they "partially hydrogenate" the oil by adding hydrogen ions to the oil. (If they were to "fully" hydrogenate the oil, it would become a solid block of saturated fat.)

You must be aware of two problems with "partially hydrogenated" oil. In the first place, you cannot know how much of the oils is hydrogenated from the word "partially." Is it 25, 50, or 75 percent? The difference is that if only a small percentage of hydrogenation is added, then some of the polyunsaturated fats are turned into mono-unsaturated fats. If more hydrogenation is used, then some of the monounsaturated fats can become saturated, and if a great deal of hydrogenation is used, then you could end up with mostly saturated fats.

The second problem stems from the hydrogenation process, which produces an abnormal type of fat (called a *trans* fat) that is not found in a natural state. While some medical and scientific authorities do not believe that trans fats are harmful, many physicians and biochemists are not that certain. They believe that these trans fats can be detrimental to the cell structure of your body, may negatively influence the immune system, and may disrupt normal heart function.

Given all the evidence, it seems that if you have a choice between a product containing saturated fat and one containing partially hydrogenated unsaturated fat, you must certainly choose the latter. But if you have a third choice—a product that uses vegetable oils that are not hydrogenated—then that should be the product you buy. Again, read all labels and check the 2-in-1 tables.

The Predominant Fat Contents of Selected Fats and Oils

While the 2-in-1 tables list the scores for every fat and oil, giving the appropriate value to their saturated fat and cholesterol content, they cannot show which *unsaturated* fats are prevalent in various products.

Table 3 outlines which type of fat (saturated, polyunsaturated, or monounsaturated) is dominant in the most widely used fats and oils. The favorable values for safflower, sunflower, and walnut oils (polyunsaturated), as well as for olive, almond, and canola oils (monounsaturated) are immediately obvious.

Equally obvious are the highly unfavorable saturated fat values for the tropical oils (coconut, palm, and palm kernel), and the poor values for all fats derived from animals. And don't forget that the animal fats also contain dietary cholesterol (31 milligrams for a tablespoon of butter, 13 milligrams for either tallow or lard, and 7 milligrams per tablespoon for shortening).

It is certainly better to substitute one of the margarines for butter if you have been using butter in cooking rather than olive, safflower, or one of the other good oils. If you use butter sparingly only once or twice a week on toast or on a roll, however, then by all means continue doing so, especially if you use a whipped butter. A half pound tub should last two people about four weeks, and at that rate your butter consumption should not affect your cholesterol level.

Of particular benefit is *rapeseed oil*, with a very low amount of saturated fat and about twice as much mono- as polyunsaturated fat. It also has the largest percentage of the important Omega-3 fat (explained in detail in chapter 8). Produced primarily in Canada, rapeseed oil has just recently become available in the United States. Puritan Oil, one brand now being sold, consists solely of "canola oil," the name under which rapeseed is now being marketed. Like safflower or sunflower, it is a superb oil to use in any type of cooking or baking.

Of the nut oils, *walnut oil* is similarly interesting. It is not only high in polyunsaturated fats but it also has a higher percentage of Omega-3 fats than any other nut oil.

Table 3. Fat Contents of Selected Fats and Oils

	SATURATED
Coconut oil	87%
Palm kernel oil	82
Butter	62
Cocoa butter	60
Tallow, edible	50
Palm oil	49
Shortening (animal and vegetable fat)	41
Lard	39

	Poly	Mono
	UNSATURATED	
Safflower oil	74%	12%
Sunflower oil	65	29
Walnut oil	63	23
Corn oil	59	24
Cottonseed oil	51	18

	Mono	Poly
	UNSATURATED	
Almond oil	70%	18%
Olive oil	73	8
Rapeseed (canola) oil	61	27
Margarine (average stick or tub)	46	34
Peanut oil	46	32
Soybean oil	43	38

Source: U.S. Department of Agriculture

Two fat substitutes are currently being considered for approval by the Food and Drug Administration. One is Simplesse, to be used only with cold items such as salad dressings, ice cream, mayonnaise, yogurt, and cheese spreads. The other is Olestra, which can be used for other purposes such as frying. Neither item has as yet been approved. No one knows for certain what, if any, side effects will develop from these products until they have been used for many years by many thousands of people. And, remember that such fat substitutes have no nutritional value. You might be tempted to fill up with such "harmless" products as no-fat ice cream or yogurt, but you will not receive any nutritional benefits. You would be better off enjoying a piece of fresh fruit or simply using a regular low-fat product, such as low-fat yogurt. These would give you the nutritional values you need and still not increase your serum cholesterol.

3

What the Nation's Health Experts Recommend

During the last two or three decades there have been many different opinions of what constitutes healthy cholesterol levels. Recommendations have varied not only in the total acceptable score for each individual but also in differentiating between the sexes and various age groups. There was also much debate about exactly how fats and dietary cholesterol influenced our serum cholesterol levels.

Finally, in October 1987, the National Heart, Lung, and Blood Institute announced the U.S. government's official guidelines for recommended maximum cholesterol levels, as well as for maximum consumption of dietary cholesterol and fat. Twenty-three major medical associations and health organizations (including the American Medical Association, the American Heart Association, and the American College of Cardiology) joined in the National Institute's findings.

The panel also made recommendations regarding the use of cholesterol-reducing drugs, for those cases in which dietary restrictions by themselves did not reduce serum cholesterol levels sufficiently.

Listed on the following pages are the key features of these guidelines:

The National Institutes of
Health Recommended Guidelines for
Serum Cholesterol Levels

Desirable serum cholesterol level	less than 200 mg
Borderline-high serum cholesterol	200–239 mg
High serum cholesterol	more than 240 mg
Desirable LDL level	less than 130 mg
Borderline-high risk LDL	130–159 mg
High risk LDL	more than 160 mg

Additional heart disease risk factors to be taken into account by physicians in treating patients with "borderline-high" levels include the following:

- Gender—male greater than female
- Family history of early heart disease
- Cigarette smoking
- High blood pressure
- HDL levels under 35 milligrams
- Diabetes
- Obesity
- History of arteriosclerosis (hardening of arteries)
- Blockage of blood vessels to the head, extremities, or brain due to atherosclerosis

Below are the dietary guidelines, as set forth by the NIH, for the maximum daily consumption of fats, saturated fat, and dietary cholesterol. These suggested limitations are for all adult Americans, regardless of age or sex.

The National Institutes of
Health Dietary Guidelines for
Fat and Cholesterol

Total fat intake	Less than 30% of total calories daily
Saturated fat intake	Less than 10% of total calories daily
Dietary cholesterol	Less than 300 milligrams daily

Now that we know the goals, let us look at the facts that challenge those goals.

Facts about Serum Cholesterol

FACT. More than half of all men and women in the United States have more than 200 milligrams serum cholesterol, placing them in the "borderline-high" category.

FACT. Approximately 40 percent of all men and women between the ages of 45 and 55 have levels of more than 240 milligrams of cholesterol, defined as "high." Many people have cholesterol levels of 300 to 400 milligrams; some even exceed those numbers.

FACT. At 240 milligrams, the risk of heart disease is *twice* as great as it is with a 200-milligram cholesterol level.

FACT. At 300 milligrams of serum cholesterol, the risk is *four times* as great as it is at 180 milligrams.

FACT. An overall reduction of just 10 percent in cholesterol levels by all Americans could save 100,000 lives a year in the United States.

FACT. A 15 percent reduction in your cholesterol level should reduce your risk of heart disease by 30 percent!

Facts about Diet

FACT. The average American consumes more than 40 percent of all calories in fats, whereas the maximum should be 30 percent.

FACT. The maximum intake of saturated fat should be 10 percent of total calories, but the average American consumes about 15 percent, and many consume as much as 20 percent of their total calories in saturated fat.

FACT. The maximum daily dietary cholesterol intake should be 300 milligrams, but the average American consumes between 350 and 450 milligrams daily, with many exceeding 500 milligrams per day.

Can You Beat These Facts?

Yes, you can! And the 2-in-1 system will help you do so. The NIH is convinced that a diet low in fat, saturated fat, and dietary cholesterol will substantially reduce the serum cholesterol level of most people. Many people can achieve this reduction through what the National

Institute calls their "Step 1 diet," in which total fats do not exceed 30 percent of daily calories, saturated fats do not exceed 10 percent of total calories, and dietary cholesterol is limited to a maximum of 300 milligrams per day.

If satisfactory progress in lowering your serum cholesterol with the Step 1 diet has not been achieved in about three months, the NIH recommends that you reduce your daily consumption of saturated fat and dietary cholesterol to what they call the "Step 2 diet." This diet calls for saturated fat not to exceed 7 percent of total calories (instead of 10 percent) and a maximum daily intake of 200 milligrams of dietary cholesterol (instead of 300 milligrams).

You can easily follow the Step 2 diet with the 2-in-1 tables simply by limiting yourself to a daily score of 70 (instead of 100). This holds true because of the special way the 2-in-1 tables have been designed to accommodate any level of calorie consumption.

If you have been on the Step 2 diet for three months and your cholesterol levels are still not in the recommended ranges, there may be genetic or other physical factors that are causing your high cholesterol score.

The NIH report suggests that under those circumstances niacin or one or more prescription drugs may be called for. Your physician may recommend such drug intervention, combined with a dietary program.

The next chapter more fully describes the effects of niacin and some of the prescription drugs currently on the market. Remember, though, that a proper dietary regimen should always be the first, and perhaps the most important, remedy.

4

What About Niacin or Prescription Drugs?

Even though a dietary regimen should always be first and foremost in the approach to treating high serum cholesterol levels, certain cases may call for the use of drug therapy. One case might be those people who have such a severely high cholesterol score that it could be indicative of a genetic predisposition to dangerously high cholesterol. Another case might be advanced atherosclerosis, especially in those people who have had bypass surgery. A third case might be high-risk patients for heart disease, such as those with high blood pressure and people who refuse to stop smoking. The final candidates would be those people who do not change their ingrained eating habits sufficiently, and those who cannot achieve a satisfactory reduction in their serum cholesterol levels even after changing their diet and lifestyle.

In the latter case, as in the first, your physician will investigate whether the reason is genetic, which can be determined through various tests. A genetic predisposition might be indicated when even reducing your 2-in-1 score to 70 points for several months does not achieve satisfactory reduction in serum cholesterol. Genetic predisposition is divided into five distinct types:

- Type I—cholesterol normal, triglycerides greatly elevated
- Type II—cholesterol and LDL unduly high, triglycerides normal
- Type III—all elements (cholesterol, LDL, and triglycerides) elevated
- Type IV—triglycerides slightly high, cholesterol normal
- Type V—cholesterol elevated, triglycerides extremely elevated, and a diminished, but present, LDL count

If you suspect or have been told that your cholesterol level is too high, your first step is to get a complete reading of your blood values, including triglycerides, LDL, and HDL. If it then appears that your LDL level is too high (or HDL too low), or if your cholesterol/HDL ratio is too high, your next step is to reduce your saturated fat and dietary cholesterol intake. Use the 2-in-1 tables to check each food you eat. Try not to exceed a total daily consumption score of 100 points.

If you have not made satisfactory progress after two or three months, discuss with your physician the efficacy of reducing your daily 2-in-1 score to 70.

If, after three months of this more stringent program, you still do not show substantial improvement in your serum cholesterol level, you may wish to discuss the use of niacin with your physician.

Niacin

You should consult with your physician before taking niacin, because even though it is a nonprescription vitamin, it can be considered a drug when taken in large quantities. Recent research has shown that niacin may substantially reduce serum cholesterol levels (including triglycerides and LDL), while substantially raising HDL levels.

Your own physician should recommend the specific dosage of niacin that may be suitable for you. Dr. Pardell suggested that his coauthor F. M. Oppenheimer start out by taking one 50-milligram tablet three times a day, with meals. After a few days, the dosage was increased to 100 milligrams three times a day, followed by additional gradual increases until Oppenheimer was taking 500 milligrams three times a day, for a total of 1,500 milligrams daily (1.5 grams).

In its October 1987 report, the National Heart, Lung, and Blood Institute suggested that the daily dose could be as much as 3 full grams, and that with some patients with marked serum cholesterol elevation, a daily dose of 6 grams is even occasionally warranted. As stated before, however, you must consult with your own physician regarding whether niacin is suitable for you, and in what quantities.

You may experience a flushing sensation shortly after taking niacin. This effect is an expected and natural reaction because niacin relaxes the muscle tone of the arteries, causing a dilation of your blood vessels. It is perfectly normal, and not at all dangerous. Pharmaceutical authorities recommend that you take niacin three times daily, with meals, to reduce the flushing sensation. Taking an aspirin about a half hour before taking the niacin will also reduce the flushed feeling and avert the rare headache that might accompany the taking of niacin. (You should, of course, not take aspirin if you are sensitive to it or if your physician has advised against your use of it.)

You may also take a simple antacid tablet regularly with your niacin to prevent any gastrointestinal distress as a side effect. Your doctor may periodically order a blood test to make certain that the niacin has not elevated your liver enzymes, which could indicate liver cell damage. He may also ask you about any gouty symptoms, which can occur.

Niacin is also available in sustained-release form (which will reduce the flushing sensation) but this is considerably more expensive than regular niacin.

Prescription Drugs

Several drugs on the market can help lower serum cholesterol levels. Your physician may prescribe one of these if neither improved diet nor niacin has worked for you. Some of the drugs are onerous to take (causing nausea, vomiting, and/or diarrhea and other side effects), some have only been tested for a relatively short period, and some are very expensive. However, in those cases where other methods have been unsuccessful in reducing serum cholesterol, some of these drugs have often proved helpful.

Cholestyramine and Colestipol are known as bile acid sequestrants. By removing bile acids from the bowel, these two drugs cause the cholesterol in the liver to be converted into bile acids, thus lowering the cholesterol in the blood. Both of these drugs are powders which, when mixed with water or fruit juice, are taken two or three times daily with meals. Side effects include gastrointestinal distress, constipation, interference with absorption of other medication you may be taking, and an occasional increase of triglycerides in the blood.

Many physicians do not consider an older drug formula, clofibrate, a safe drug, as it has been reported to cause long-term toxicity and liver cell damage. It has not been shown to reduce heart disease.

The long-term safety of two other drugs, probucol and Gemfibrozil, has not yet been established, but Gemfibrozil has been proven to reduce the risk of heart disease. Both drugs are sometimes prescribed in extreme cases of elevated serum cholesterol, and then often in combination with one of the other drugs.

Finally, there is Lovastatin, the latest drug approved by the government for sale in the United States. It comes in tablet form, is taken once or twice daily with meals, and has been shown to substantially reduce serum cholesterol levels. The long-term effects of this drug are, as yet, not certain. Side effects may include abnormal enzyme changes in the liver, cataracts in the eyes, as well as gastrointestinal distress.

You may have realized by now that no one drug treatment is right for every person. A proper diet is the mainstay of treatment and should be the first line of therapy in all cases. However, if diet alone does not reduce your cholesterol satisfactorily, then you and your physician will work together to formulate a more aggressive plan, which might include drug therapy to effectively reduce your serum cholesterol.

5

Children and
the 2-in-1 System

The American Academy of Pediatrics has recommended that physicians check the cholesterol, triglyceride, and HDL levels of any child over the age of 2 whose family has a history of high serum cholesterol or early coronary heart disease.

A person has a "family history" of high cholesterol if high cholesterol is found in either parent, a sibling, a grandparent, an aunt, or an uncle. "Early heart disease" is defined as occurring under age 50 in men or age 60 in women.

The Academy also recommended that any child with a serum cholesterol level exceeding 176 should be considered for dietary counseling.

(This, of course, means that the child's parents must become familiar with the saturated fat and dietary cholesterol values in all the foods their child eats. The 2-in-1 tables seem to be the best and simplest answer!)

Lastly, the Academy stated that, since the long-term efficacy of drug intervention has not been evaluated in children, medication should be considered only for children with extremely high cholesterol levels (over 200 milligrams) and only after dietary modification alone is not successful.

Regardless of family history, it is not suggested that any child under the age of 2 be put on a fat-restricting diet. Youngsters, especially those under 2 (and, particularly infants), have a far greater need for fats than adults do. While eating cookies, ice cream, rich pastries, and candy may be close to suicide for an adult, their fat content might not be harmful at all to very young children. Malnutrition is a greater danger than cholesterol when considering the diets of very young children.

In recent years many studies have been undertaken to measure cholesterol levels in children. Foremost among these is probably the research done over a fifteen-year period by Louisiana State University. Other recent studies have been conducted by the University of Michigan and the University of California. From these studies it has become apparent that approximately 30 percent of all children in the United States have cholesterol levels in excess of the 176 level recommended by the American Academy of Pediatrics.

There seems to be very little doubt among most physicians and scientists that too high a level of serum cholesterol in children will usually lead to elevated cholesterol levels in adulthood, and that it is therefore imperative that parents be aware of their children's cholesterol levels.

Although the Academy suggested that only children over the age of 2 who have a familial history of heart disease or high cholesterol be tested for their cholesterol level, we do suggest that *all* children over the age of 10 (even if there is no family history of early heart disease or elevated cholesterol) should be tested for cholesterol, triglycerides, LDL, and HDL. At that age, every child's parent should be aware of the child's blood values, so that if remedial dietary intervention is necessary, it can be started early.

Under no circumstances, however, should your child be put on a fat-limiting regimen unless your pediatrician recommends it. If your physician does feel that your child should limit his or her fat and cholesterol consumption, then by all means use the 2-in-1 system for your boy or girl, just as you would use it for yourself. Be sure to check with your child's physician to determine the appropriate number of calories your child should be consuming each day, and refer to that particular calorie column when checking your child's foods in the 2-in-1 tables. You will find the 2-in-1 system to be just as effective in lowering your child's serum cholesterol as it is in lowering your own.

PART II

Cholesterol and Food–
The Good, the Bad,
and the Best!

6

You Can Enjoy Great
Meals While Cutting Down on
Your Fat and Cholesterol

Eating a low-fat, low-cholesterol diet need not be arduous, especially if you let the 2-in-1 tables do the analyses and calculations for you.

As you check each food against the 2-in-1 tables, you will soon become so familiar with the values of your favorite foods that you will immediately know how much of a particular food you can eat.

Beware of Invisible Fats

It's easy to realize that butter, margarine, or any other fat or oil is pure fat. It's also easy to see that bacon has a lot of fat, or that there is half an inch of fat surrounding a tenderloin steak. These and other obvious fats are easy to recognize and eliminate.

The problem is that you can't know how much fat there really is in some food items, because you don't actually see it. That is why we call it "invisible fat." Four ounces of tenderloin, for instance, with fat attached, contain 360 calories, of which 66 percent (239 calories) is fat. But 4 ounces of *trimmed* tenderloin, coming in at 263 calories, *still* have 126 calories (48 percent) of fat hidden in the meat itself.

Not only meat is loaded with invisible fat—so is cheese, cake, ice cream, french fries, and dozens of other items. This chapter will expose both the obvious and the invisible fats, help you eliminate the worst offenders, and suggest delicious and healthy alternatives.

Remember that:

- Your daily intake of fat should not exceed 30 percent of your total calories.
- Your saturated fat intake should be less than 10 percent of your total calories.

To maintain these guidelines, the simplest first rule to follow is never to exceed those limits with *any one* food. For example, 70 percent of the calories of your favorite cheese comes from fat, and 60 percent of those is saturated fat. This means that 42 percent (60 percent of 70 percent) of the calories in the cheeses come from saturated fat. Seventy-five to 80 percent of the calories in luncheon meats and sausages comes from fat, and 40 percent of those comes from saturated fat, meaning that most of these products carry about 30 percent of saturated fat, or three times the amount to which you should limit yourself.

While you should check each food product against the 2-in-1 tables, you can use the broad guidelines below to help you achieve an immediate reduction in fat and cholesterol intake.

Fatty Meats

Is there anything wrong with beef as such? Not really, but 8 ounces of heavily marbled steak a few times a week can be a killer! This warning includes *all* meats—not just beef—and not just steaks and chops.

"Prime" grade meats have more marbling (and therefore more fat) than "choice," and "choice" meats have more fat than meats graded as "good." Lean cuts are better for you than fatty cuts in beef, lamb, veal, and pork.

Some people mistakenly believe that veal is better for them than beef, but an untrimmed veal cutlet (with fat attached) has more total fat and more saturated fat than a lean T-bone steak—and more cholesterol, too! This 4-ounce veal cutlet scores 29 on the 2-in-1 tables (based on a 2,100 daily calorie intake). Even if you switch to an all-lean veal cutlet, the score is still 24. And what about the 4-ounce T-bone steak? It also comes in at 24, as good as the lean veal cutlet, but still, of course, quite high. Four ounces of ground veal have a 2-in-1 score of 32. Not much different from the score of 31 for 4 ounces of extra lean ground beef.

The fact is that none of them is ideal for you! (By the way, regular ground beef scores 37—20 percent more than the extra lean ground beef, making it an item that is best avoided.)

And how does lamb compare? Four ounces of a roasted lean lamb leg score 22 on the 2-in-1 tables, while a lean loin chop scores 24. But when you switch to an untrimmed shoulder lamb chop, the score becomes a terrifying 52. The same fat story is true about pork. A lean, roasted 4-ounce loin pork chop has a score of 26, but if you choose a pan-fried untrimmed loin chop, your score nearly doubles to 48.

The lesson is clear: There is nothing wrong with eating meat, as long as it is the right kind, prepared the right way, and not eaten more than twice a week in small quantities. Keep in mind these basic rules:

- "Good" meats have the least fat.
- "Choice" meats have more fat than "good" meats.
- "Prime" meats have the most fat.

- Broiling produces the least fat and cholesterol.
- Roasting is nearly as good as broiling.
- Stewing creates more fat and cholesterol.
- Frying scores worst on the 2-in-1 tables.

- Always buy the leanest cut available.
- Always trim off all visible fat before cooking.

Organ Meats

Organ meats do not have much fat, but they are so high in cholesterol that you should avoid them if you want to lower your serum cholesterol level. If you like liver, kidneys, brains, or heart, check the 2-in-1 tables before eating them. You will see that just 4 ounces of fried calves liver contain more dietary cholesterol than you should consume in a whole day. Even with its low fat content, the liver scores a whopping 75, far in excess of what any single food item should be.

Luncheon Meats

Practically all luncheon meats are very high in fat, including salami, bologna, bacon, sausages, frankfurters, pastrami, corned beef, liver-wurst, and all mixed-meat luncheon loaves. Stay away from them if you can!

For example, three slices of bacon score 11 on the 2-in-1 tables. One frankfurter has 16 points (so two frankfurters alone would make up a full one-third of your 100 daily points), one 1-ounce slice of salami is worth 7 points, while a similar slice of bologna has a score of 11 (a sandwich with four slices of bologna would be equal to nearly half your daily limit!). On the other hand, a 1-ounce slice of pastrami made of turkey weighs in at only 3 points. Equally low are turkey frankfurters and other meats. (Be careful, though, of the nitrites and salts often used in processing these turkey products.)

Generally speaking, all red meats, as well as pork and veal, have a higher saturated fat content than poultry or fish. Therefore, you will want to monitor your intake of these meats more carefully than may be necessary with most poultry or fish.

Poultry

Poultry, in general, is good for you. While it often has about the same cholesterol content as most meat ounce for ounce, its fat content is usually considerably lower—unless you eat the skin. Two examples: Four ounces of a roasted, skinless chicken drumstick have a 2-in-1 score of 18, and 4 similar ounces of chicken breast have a score of 16 (about one-third less than the leanest cut of any meat).

Chopped turkey tastes great as hamburgers or meatballs, and 4 ounces of ground turkey register only 20 points on the 2-in-1 tables compared to 31 for extra lean ground beef.

Remembering the following simple rules:

- Chicken and turkey have less fat than duck or goose.
- Light meat has less fat than dark meat.
- Never eat the skin of any bird. (The 4-ounce skinless drumstick that scored 18 points on the 2-in-1 tables scores 22 if you eat the skin.)

All poultry organs, like those of the other meats, are very high in cholesterol: Four ounces of fried chicken livers have nearly 700 milligrams of cholesterol (more than twice what your daily maximum should be). Despite their relatively low fat content, they score a total of 98 on the 2-in-1 tables, using up nearly all of your total daily allowance of 100 points. Giblets, gizzard, heart, and liver should therefore rarely, if ever, be eaten.

Fish

Fish and shellfish are, without question, the best main course food you can eat. Most fish have about the same cholesterol content as meat and poultry, but not only is fish usually considerably lower in total fat, it is practically always much lower in saturated fat. Two examples: Four ounces of broiled flounder fillet have a 2-in-1 score of only 11 and a similarly prepared perch has a score of 12, both of which are about half the score of the leanest cuts of meat.

Not only is fish low in saturated fat, but its unsaturated fats often include large amounts of the important Omega-3 fats. This fat has been found to be very helpful in actually lowering serum cholesterol levels. (See chapter 8, Foods That Can Actually Lower Your Cholesterol, and table 5, The Great Fish Oils.)

For many years it was thought that shellfish might be harmful, because some shellfish have a relatively high cholesterol content. However, since the saturated fat content of shellfish is usually so low, it is now accepted that it can be eaten as freely as any other main course meal—with the usual discretion in the quantites you consume and in how it is prepared. Broiling or steaming is always best, followed by baking or stir-frying in one of the "good" oils, such as olive or safflower oil. Never use butter, and never fry or deep-fry your fish or shellfish: Frying immediately negates the good values of the food.

Among shellfish, we should perhaps mention lobster and shrimp. Both of them have a small-to-insignificant total fat and saturated fat content. (One of the least fatty meats, chicken breast, has *five times* as much saturated fat as lobster and both have the same cholesterol content.)

While the cholesterol level of lobster is average, shrimp has more than twice as much dietary cholesterol, so you should watch your consumption of shrimp carefully. (Four ounces of steamed lobster have a score of only 13, which is very good, but the same quantity of shrimp has a total of 30 points, putting shrimp into the same category as a lean hamburger.) Both clams and crab are low in saturated fat and only slightly higher in cholesterol than lobster. A good example of the point value of a mollusk would be 4 ounces of scallops, which have a score of only 7 when steamed or boiled, or 9 if they are broiled or baked.

All in all, whether you eat fish or shellfish, you will usually score much lower than you will with meat or poultry, but do check the 2-in-1 tables for the score of each individual item.

Milk and Milk Products

While most dairy products are extremely high in saturated fats, this is not a blanket indictment against all these products. Beware of all cream—heavy, whipping, medium, half-and-half, and sour. Just one tablespoon of medium cream in your coffee costs 8 points. If you drink this combination four times in a day, you will have used up nearly one-third of your allowable 100 points on just four cups of coffee!

Whole milk has nearly double the saturated fat content of 2 percent milk, which in turn has about double the content of 1 percent milk. But even that has about five times the saturated fat content of skim milk. Initially, skim milk may not taste the same as whole milk, but once you realize that it is as nutritious and has as much calcium as whole milk but only one-twentieth of the saturated fat, you will probably never drink or cook with whole milk again. (Just remember this: One cup of whole milk has a 2-in-1 score of 19; one cup of skim milk costs a mere 1 point!)

Buttermilk, surprisingly, has a comparatively low fat content, only about as much as 1 percent low-fat milk, so enjoy it once in a while.

Butter is 100 percent fat, 62 percent of which is saturated. One tablespoon of butter is worth 23 points, so be careful of the quantities you use. Watch out for hidden butter in cakes, pastries, and other products usually made with butter. If you like butter once in a while on your bread or rolls, use whipped butter—a small quantity can go a long way and won't add too many points to your score.

A tablespoon of regular margarine has a 2-in-1 score of only 5, while diet margarine scores just a 3. You should, therefore, give serious consideration to using margarine as a substitute for butter. We believe, however, that for cooking purposes you would be better off using one of the good, pure vegetable oils, rather than a hydrogenated fat.

Ice Cream and Yogurt

Ice cream is similar to regular cream: The higher the fat content (as in some of the "designer name" ice creams), the worse it is for you. One cup of a regular commercial ice cream scores a whopping 31 points on the 2-in-1 tables. As if that weren't enough, look at the score of 1 cup of French ice cream: a withering 55—more than half of your allowable daily intake! Try sherbets instead: They come in at just 8 points a cup and are delicious.

Yogurt may be made from whole milk, low-fat milk, and now even from nonfat milk. It is an excellent substitute for sour cream in sauces, toppings, and dressings. Whole-milk yogurt has twice as much saturated fat as low-fat yogurt, and *eighteen times* as much saturated fat as nonfat yogurt. It isn't hard to decide which one is best for you, is it?

Cheese

Most cheese is extremely high in fat, with about 50 percent of its calories coming from saturated fat. Cheese should, therefore, be consumed rarely and only in small quantities. There are some exceptions: Low-fat (1 percent) cottage cheese has about a quarter of the saturated fat of regular cottage cheese, and only about an eighth as much saturated fat as most other cheeses. If you like Parmesan cheese on your pasta, go right ahead and enjoy it. Though it, too, is high in fat, if you use it grated, a tablespoon will go a long way. Low-fat mozzarella has about half as much saturated fat as most other cheeses, and low-fat ricotta has only about a quarter as much. If you wish to eat cheese, then certainly mozzarella, ricotta, and cottage cheese are preferable.

You can also purchase low-fat processed cheeses, such as American or Swiss. These usually have about one-quarter the saturated fat of their regular counterparts, but be sure to check their labels carefully!

When you combine cheese with another product that by itself is high on the 2-in-1 scale, look out! That is why cheeseburgers or chocolate cheesecake should really disappear from our vocabulary.

Cookies and Candies

Always look at the ingredients of packaged cookies and candies. Even though some of these products have been reformulated to eliminate the tropical oils, many of them (especially candies) are still made with a combination of partially hydrogenated oils (see chapter 2,) and/or coconut, palm oil, and animal shortening. Read the ingredients carefully. Don't buy it if any of these are listed: coconut, palm, palm kernel, lard, shortening, or hydrogenated oil (without specifying "partially").

If you want a cookie or two once in a while, then why not have two oatmeal cookies with a 2-in-1 score of only 4, or even two chocolate-chip cookies with the same 4-point total. Stay away from that brownie, though! Just one, with icing, will set you back 12 points.

Cakes, Pies, and Pastries

Most cakes, pies, and pastries are made with butter, shortening, and eggs. They are, therefore, rarely acceptable according to the 2-in-1 guidelines. There are some baked goods, though, that can be eaten without worry, as long as they are consumed in limited quantities. We mentioned chocolate cheesecake before—one piece scores 58 on the 2-in-1 scale, whereas a piece of marble cake has a value of only 7! Stay away from coconut cream pie: With its combination of cream, eggs, and coconut, it scores a staggering 73 points! A single eclair has 29 points, but if you have a piece of coffee cake, instead, you will have used up only 5 points. If you want some cake, a piece of pie, or a pastry for dessert—check the 2-in-1 tables first!

Salad Dressings

Most salad dressings are very high in saturated fat. Regular salad dressings have from 9 to 14 saturated fat calories per tablespoon, with a score of 3 to 4 for just that one tablespoon. Diet dressings, on the other hand, have only about 2 saturated fat calories for the same quantity, so that they have only about one-sixth the point value. If you must use mayonnaise or Miracle Whip, use the light versions (without cholesterol)—they have only about half the saturated fat of the regular product. Better yet, stick to olive oil and vinegar.

Eggs

Eggs are a wonderful food. They don't contain too much fat (unless they are scrambled or fried), and they taste so good! But are they ever loaded with cholesterol! One egg, hard- or soft-boiled, has 213 milligrams of cholesterol, just about your total limit for the day.

There are already so many eggs hidden in our daily diet—in breads, bagels, croissants, cakes, sauces, and in so many home- or restaurant-prepared dishes, that you really should not eat any eggs by themselves, except very occasionally.

The cholesterol in eggs is contained only in the yolk. Many good cooks have discovered that they can substitute egg whites for whole eggs and still have their recipes taste great. They have also made scrambled eggs using just one yolk with three or four egg whites, and found that it serves two people well.

By the way, the high cholesterol content of eggs also holds true for the most expensive eggs of them all: caviar. If you just have a little bit as a garnish, enjoy. But don't celebrate that special occasion with 2 or 3 ounces, because you will have used up nearly a day's allowance of cholesterol!

Fried Foods

Stay away from *anything* fried, because even "good" oils used in frying become saturated when heated. Try any other method of cooking instead, such as broiling, roasting, baking, stewing, and even stir-frying with minute quantities of safflower or canola oil.

The Tropical Fats

Coconut, palm, and palm kernel oils are the worst saturated fat ingredients you can find. Watch out for them in cookies, crackers, coffee creamers, whipped toppings, potato chips, cereals, frozen dinners, cake mixes and frostings, canned soups, breading and stuffing, vegetable shortening, and a myriad of other products. *Always read the label!*

Fruits and Vegetables

Eat and enjoy fruits and vegetables. They have no cholesterol and practically no fat. The only exception is avocados, which are very high in fat and should therefore only be consumed in limited quantities. (The avocado's fat is primarily unsaturated, but it still registers a 12 on the 2-in-1 tables.) And do watch out for any vegetables prepared with a cream or cheese sauce!

7

Fast-Food Facts
at Your Fingertips

Most fast foods are so high in saturated fats and cholesterol that the easiest suggestion would be to recommend that you never set foot in another fast-food restaurant. But sometimes you may want to go or it's just convenient to go. The simplest advice we can give you is to check the Fast Food section in the 2-in-1 tables before you go.

We should point out here that while some fast-food chains were very cooperative in supplying the data we needed for the 2-in-1 tables, others gave us very little information, or none at all. Those that did supply the required data are included in the tables. If you are looking for your favorite fast food in the 2-in-1 tables and cannot find it, that may be because the chain selling that particular food may not have supplied sufficient information to calculate the score and list the food in the 2-in-1 tables.

As a fast-food rule of thumb, stay away from hamburgers, cheese-burgers, french fries, fried chicken or fish, or anything else fried. The salad bar is probably your best bet at any fast-food establishment.

Why is most of the food in the fast-food restaurants so bad for your health? First, a great deal of the food is fried. By now you know well that frying is probably the worst kind of cooking for you because of the large amount of oil or fat used in the process.

Second, it isn't just that the foods are fried, but what they are fried in. You might expect that your hamburger will be fried in high-fat beef tallow, but could you really be expected to know that the fish and the chicken, and yes, even the french fries, are often fried in beef tallow? And could you be expected to know that, even if beef tallow is not used, the vegetable oils used are so heavily hydrogenated and saturated that these foods may be *worse* for you than if beef tallow had been used? And could you know that the worst oils of all—the tropical coconut and palm oils—are freely used on other fast-food items?

Let us give you some examples, all taken straight from the 2-in-1 tables:

ITEM. Do you remember the fast-food breakfast we offered as one of the choices in the introduction to this book? That little eye-opener cost 114 points out of an allowable 100 for the *whole day!*

ITEM. For a fast-food lunch, how does a Monterey Burger, french fries, and a chocolate milk shake sound? Probably not very good if you looked up the items in the 2-in-1 tables and found that they total 117 points, again, more than your total for the whole day.

ITEM. Lets say you opted instead for a Taco Salad with Ranch Dressing. That, too, is 118 points.

ITEM. That's too much, so you select the Taco Light Platter instead, but even the "light" platter has 90 points.

ITEM. For a change of pace at dinner, let's have some fast-food chicken. How about an Extra Crispy wing, breast, and thigh, some french fries, and two biscuits? Really filling—and it should be at 94 points.

ITEM. You don't like chicken? Then how about some healthy fish. A Fillet of Fish, french fries, apple pie, and a milk shake will cost you 57 points—still too much for just one meal!

We don't mean to say that you should never again eat in a fast-food restaurant, but check the 2-in-1 tables first. For instance, you can get a Seafood Platter totaling 33 points, or a Chicken Plank Dinner with a total of 25 points. But decide against a dessert of pecan pie, which alone has 26 points. Wouldn't you be much better off with lemon meringue pie? Delicious, and only 6 points!

As we said before, your best choice is the salad bar. But here, also, be careful. Avoid the bacon bits, the potato salad, and all the other salads that may be made with mayonnaise and eggs. Ignore the herring if it's made with a cream sauce. Forgo the Russian, French, blue cheese, or thousand island dressing. Stick to olive oil and vinegar—it's much better for you!

If you want something other than a salad, stay away from red meat products, and order those chicken or fish dishes that rate best in the 2-in-1 tables. Say "no thanks" to rich cakes or pastries, ice cream, and milk shakes. Have some fresh fruit, or bring a packet of raisins, figs, or dates for a nutritious, sweet treat.

Your best bet is to eat at home or at a regular restaurant, where you have far more control over what you eat and which ingredients are used in the preparation of your favorite foods. You should have the right to know exactly what you are eating and how it is prepared, for it is *your* health that will be affected by the food you eat.

8

Foods That
Can Actually Lower
Your Cholesterol

Great news! Many individual foods can help you meet your goal of lower cholesterol (although one food by itself can never accomplish that goal for you).

Fiber

You probably have heard that fiber is good for you because it is roughage, and that is true. But there are two kinds of fiber, both of which will give you roughage. One type, known as insoluble because it is not digestible, will greatly help your elimination but does not lower serum cholesterol. The other type of fiber, which *is* soluble, has been demonstrated to be effective in lowering cholesterol levels. This type of fiber is found in all grains as well as in legumes and most other vegetables and fruits.

Oat bran contains more soluble fiber than any other part of the oat grain, and it also has more soluble fiber than the bran from any other grain. Regardless of whether it is eaten as a hot or cold cereal or baked into muffins, it has been found to be helpful not just in lowering serum cholesterol levels but also in raising HDL levels.

Oat bran, however, vividly demonstrates why you cannot rely on just one food to lower your cholesterol. Recent research conducted by the Northwestern University Medical School in Chicago showed that people who consumed approximately 40 grams of oat bran for forty days reduced their cholesterol level by only 3 percent. While higher intakes of oat bran have resulted in greater cholesterol reductions (another study showed that a daily consumption of 100 grams of oat bran led to a reduction of 19 percent in serum cholesterol levels), it is clear that you cannot rely on one food alone to lower your cholesterol level substantially. Rather, your total diet should conform to the NIH guidelines, which stipulate that saturated fat calories do not exceed 10 percent of total calories and that dietary cholesterol intake be limited to 300 milligrams per day. The 2-in-1 System will help you do exactly that!

Some reports indicate that rice bran may also be helpful in reducing serum cholesterol. While this has not yet been proven, and while rice contains very little bran, the best way to obtain rice bran is by eating brown rice. It certainly can't hurt, because brown rice (with the bran and with all the vitamins and minerals) is inherently healthier than white rice.

Legumes such as pinto, kidney, or navy beans, or dried lentils or peas have also had a decisive effect on lowering cholesterol levels. The form in which you consume these legumes is not particularly important: You may like bean, split pea, or lentil soup, or you may like baked beans or a three-bean salad. Just remember to make these soups or baked beans without the usual ham hock or sausage. Instead, use spices and herbs for a great change in taste. Eat legumes as often as possible, in whichever way you like them best.

Table 4 lists the grains, grain products, legumes, vegetables, and fruits with the highest soluble fiber content. All of these will help reduce your serum cholesterol, and you should try to enjoy at least three of these items as part of your diet every day.

Table 4. Foods with High
Soluble Fiber Content

	Grams of Soluble Fiber
GRAINS	
Oat bran, raw—4 oz (113 gm)	7.1
Millet, hulled raw—4 oz (113 gm)	6.4
Oats, rolled, or oatmeal, dry—4 oz (113 gm)	5.8
Wheat bran, crude—4 oz (113 gm)	3.4
Wild rice, raw—4 oz (113 gm)	2.9
CEREALS (ready-to-eat)	
Granola—½ c (56 gm)	2.5
Oat cereal—1 c (48 gm)	2.4
Bran flakes—1 c (45 gm)	1.1
BREAD & PASTA	
Tortilla, corn—4 tortillas (60 gm)	1.5
Tortilla, wheat flour—3 tortillas (60 gm)	1.3
Spaghetti or macaroni—1 c cooked (140 gm)	1.1
Pumpernickel—2 slices (52 gm)	0.9
Cracked wheat bread—2 slices (52 gm)	0.8
Rye bread—2 slices (52 gm)	0.8
LEGUMES	
Baked beans, sweet or tomato sauce—½ c (128 gm)	5.8
Kidney beans—½ c cooked (88 gm)	4.1
Pinto beans—½ c cooked (85 gm)	3.8
Cowpeas—½ c cooked (86 gm)	3.0
VEGETABLES	
Brussels sprouts—1 c boiled (156 gm)	3.9
Cabbage, common—2 c raw (140 gm)	2.0
Broccoli—1 c cooked (156 gm)	1.9
Carrots, raw—2 medium (122 gm)	1.7
Onions, raw—1 c chopped (160 gm)	1.4
Turnips—1 c boiled (156 gm)	1.4
Potatoes, boiled—1 medium (142 gm)	1.3
Cauliflower—1 c cooked (124 gm)	1.2
Celery, raw—4 medium stalks (160 gm)	1.1

	Grams of Soluble Fiber
FRUIT (fresh)	
Oranges, raw—1 orange (151 gm)	2.3
Strawberries, raw—1 c (149 gm)	1.8
Peaches, raw—2 peaches (174 gm)	1.4
Kiwifruit, raw—2 kiwifruit (152 gm)	1.1
Apples, raw—1 apple (138 gm)	0.8
FRUIT (dried)	
Prunes—4 prunes (34 gm)	1.5
Peaches—3 halves (39 gm)	1.3
Apricots—10 halves (35 gm)	1.0
Raisins—¼ c (36 gm)	1.0

Source: U.S. Department of Agriculture

Garlic and Onions

Scientists have reported that both garlic and onions may be useful in controlling cholesterol levels. Here again, the lower cholesterol levels of individuals in some Mediterranean countries, such as Italy and Greece, gave the first clue that garlic and onions might be beneficial. While no exact measurements have as yet been established, several studies seem to confirm that both garlic and onions may lower serum cholesterol levels somewhat.

Olive Oil

Whereas many oils are high in polyunsaturated fat, olive oil is highest in *mono*unsaturated fat. As discussed in chapter 2, polyunsaturated fat helps to decrease your overall cholesterol level, but too much polyunsaturated oil might lower your HDL score as well, which could lead to the development of cancer. Between the two types of unsaturated fats, you should consume about an equal amount of poly- and monounsaturated. There is no better monounsaturated oil than olive oil—and it tastes so good in salads and other foods!

Safflower Oil

Among all the polyunsaturated oils, safflower oil ranks number one, with the highest concentration of polyunsaturates. You most certainly should consider using it any time that you want to use a substitute for the heavily saturated dairy or animal fats. Unsaturated oils should be about equally split between mono- and polyunsaturates, and we don't think you could find a better combination than using olive oil for half your needs and safflower oil for the other half.

Should you prefer one of the other good polyunsaturated oils, such as corn, sunflower, or walnut, by all means use them. The difference is not that substantial. The important thing is that you start making the "good" mono and poly oils part of your daily diet.

Fish Oils

At one time it was thought that while some fish are good for you, others may contain too much fat. Once again, however, statistical evidence from other countries changed our understanding. The famous "Seven Country Study" of coronary heart disease, conducted by Dr. Ancel Keys of the University of Minnesota, showed that the Japanese living in Japan had the lowest serum cholesterol count and the lowest incidence of heart disease. They also happened to have a diet very low in saturated fat and cholesterol. In contrast, the Japanese living in the United States, who were eating an average American diet, had considerably higher cholesterol scores and far greater incidence of heart disease than their counterparts in Japan. Statistics disclosed that the average Japanese man's serum cholesterol level was 30 percent lower than that of his American counterpart.

In the same study, the people in Finland were found to have the most saturated fat in their diet. They were also found to have more cholesterol in their blood and more heart attacks than individuals in any of the other countries. The average Finn's risk of heart disease was shown to be nine times greater than that of the average Japanese.

Dietary differences between the Finns and the Japanese had to be responsible, in large part, for the differences in their health and longevity. Researchers found that the Finnish people ate large amounts of cheese and other milk products, while the Japanese, especially along the coastal region, ate a preponderance of seafood, as well as other foods naturally low in fats and cholesterol.

Another group of people in this study provided a fascinating fact: The Eskimos living in Greenland had an extremely high intake of total fats, but very low serum cholesterol and only minor incidence of heart attacks. The reason for this remarkable paradox was that their fat intake came primarily from fish and seal.

Researchers concluded that the fats in fatty fish are mostly unsaturated and that this unsaturated fat tends to lower serum cholesterol. It was also found that cold-water fish are especially effective in lowering cholesterol, and that the fattier the fish, the more it seems to lower cholesterol.

From those facts came the exciting discovery that a certain unsaturated fat, found in most fish and other seafood, is a distinctive type of fatty acid that scientists called Omega-3. This particular unsaturated fat seems to have the ability to reduce serum cholesterol and thus reduce heart disease.

One component of the Omega-3 fats is *eicosapentaenoic acid*, which has been abbreviated to EPA. Much research now seems to indicate that foods containing Omega-3 and EPA are among the most important cholesterol-lowering products you can eat.

The fish with the highest concentration of Omega-3 (per 100-gram weight of fish) are:

Fish	Grams of Omega-3
Mackerel, Atlantic	2.6
Mackerel, chub	2.2
Mackerel, king	2.2
Dogfish	2.0
Trout, lake	2.0
Herring, Pacific	1.8
Herring, Atlantic	1.7
Tuna, bluefin	1.6
Sablefish	1.5
Salmon, chinook	1.5
Sturgeon, Atlantic	1.5
Tuna, albacore	1.5
Whitefish, lake	1.5
Anchovy, European	1.4
Salmon, Atlantic	1.4

The expanded table below shows the major sources of the valuable Omega-3 polyunsaturated fish oil. The table also shows that the saturated fat content in most of these fish is well below one-third of their total fat, and that there is usually about as much Omega-3 fat as there is saturated fat in each item.

All entries are based on 3.5 ounce servings (100 grams) raw, without shells or bones.

Table 5. The Great Fish Oils

	Total Fat (grams)	Saturated Fat (grams)	Omega-3 (grams)
FINFISH			
Anchovy, European	4.8	1.3	1.4
Bass, freshwater	2.0	0.4	0.3
Bass, striped	2.3	0.5	0.8
Bluefish	6.5	1.4	1.2
Burbot	0.8	0.2	0.2
Carp	5.6	1.1	0.6
Cafish, brown bullhead	2.7	0.6	0.5
Catfish, channel	4.3	1.0	0.3
Cisco	1.9	0.4	0.5
Cod, Atlantic	0.7	0.1	0.3
Cod, Pacific	0.6	0.1	0.2
Croaker, Atlantic	3.2	1.1	0.2
Dogfish, spiny	10.2	2.2	2.0
Dolphinfish	0.7	0.2	0.1
Drum, black	2.5	0.7	0.2
Drum, freshwater	4.9	1.1	0.6
Eel, European	18.8	3.5	0.9
Flounder, unspecified	1.0	0.2	0.2
Flounder, yellowtail	1.2	0.3	0.2
Grouper, unspecified	1.3	0.3	0.3
Grouper, red	0.8	0.2	0.2
Haddock	0.7	0.1	0.2
Hake, Atlantic	0.6	0.2	Tr.
Hake, Pacific	1.6	0.3	0.4
Hake, red	0.9	0.2	0.2
Hake, silver	2.6	0.5	0.6
Halibut, Greenland	13.8	2.4	0.9
Halibut, Pacific	2.3	0.3	0.5
Herring, Atlantic	9.0	2.0	1.7

	Total Fat (grams)	Saturated Fat (grams)	Omega-3 (grams)
Herring, Pacific	13.9	3.3	1.8
Herring, round	4.4	1.3	1.3
Mackerel, Atlantic	13.9	3.6	2.6
Mackerel, chub	11.5	3.0	2.2
Mackerel, king	13.0	2.5	2.2
Mullet, striped	3.7	1.2	0.6
Perch, ocean	1.6	0.3	0.2
Perch, white	2.5	0.6	0.4
Perch, yellow	0.9	0.2	0.3
Pike, northern	0.7	0.1	0.1
Pike, walleye	1.2	0.2	0.3
Plaice, European	1.5	0.3	0.2
Pollock	1.0	0.1	0.5
Pompano, Florida	9.5	3.5	0.6
Rockfish, brown	3.3	0.8	0.7
Rockfish, canary	1.8	0.4	0.5
Sablefish	15.3	3.2	1.5
Salmon, Atlantic	5.4	0.8	1.4
Salmon, chinook	10.4	2.5	1.5
Salmon, chum	6.6	1.5	1.1
Salmon, coho	6.0	1.1	1.0
Salmon, pink	3.4	0.6	1.0
Salmon, sockeye	8.6	1.5	1.3
Sea bass, Japanese	1.5	0.4	0.4
Sea trout, sand	2.3	0.7	0.3
Sea trout, spotted	1.7	0.5	0.2
Shark	1.9	0.3	0.5
Sheepshead	2.4	0.6	0.2
Smelt, pond	0.7	0.2	0.3
Smelt, rainbow	2.6	0.5	0.8
Smelt, sweet	4.6	1.6	0.6
Snapper, red	1.2	0.2	0.2
Sole, European	1.2	0.3	0.1
Sprat	5.8	1.4	1.3
Sturgeon, Atlantic	6.0	1.2	1.5
Sturgeon, common	3.3	0.8	0.4
Sunfish, pumpkinseed	0.7	0.1	0.1
Swordfish	2.1	0.6	0.2
Trout, arctic	7.7	1.6	0.6
Trout, brook	2.7	0.7	0.6
Trout, lake	9.7	1.7	2.0
Trout, rainbow	3.4	0.6	0.6

	Total Fat (grams)	Saturated Fat (grams)	Omega-3 (grams)
Tuna, albacore	4.9	1.2	1.5
Tuna, bluefin	6.6	1.7	1.6
Tuna, skipjack	1.9	0.7	0.4
Whitefish, lake	6.0	0.9	1.5
Whiting, European	0.5	0.1	0.1

CRUSTACEANS

Crab, Alaska king	0.8	0.1	0.3
Crab, blue	1.3	0.2	0.4
Crab, Dungeness	1.0	0.1	0.3
Crab, queen	1.1	0.1	0.3
Crayfish	1.4	0.3	0.1
Lobster, European	0.8	0.1	0.2
Lobster, northern	0.9	0.2	0.2
Lobster, spiny Caribbean	1.4	0.2	0.3
Shrimp, Atlantic brown	1.5	0.3	0.3
Shrimp, Atlantic white	1.5	0.2	0.4
Shrimp, Japanese prawn	2.5	0.5	0.5
Shrimp, northern	1.5	0.2	0.5
Shrimp, other	1.2	0.3	0.3

MOLLUSKS

Abalone, New Zealand	1.0	0.2	Tr.
Abalone, South Africa	1.1	0.3	Tr.
Clam, hard-shell	0.6	Tr.	Tr.
Clam, hen	0.7	0.2	Tr.
Clam, Japanese hard-shell	0.8	0.1	0.2
Clam, littleneck	0.8	0.1	Tr.
Clam, soft-shell	2.0	0.3	0.4
Clam, surf	0.8	0.1	0.2
Conch	2.7	0.6	1.0
Cuttlefish	0.6	0.1	Tr.
Mussel, blue	2.2	0.4	0.5
Mussel, Mediterranean	1.5	0.4	0.2
Octopus, common	1.0	0.3	0.2
Oyster, eastern	2.5	0.6	0.4
Oyster, European	2.0	0.4	0.6
Oyster, Pacific	2.3	0.5	0.6
Periwinkle, common	3.3	0.6	0.7
Scallop, Atlantic deep-sea	0.8	0.1	0.2

	Total Fat (grams)	Saturated Fat (grams)	Omega-3 (grams)
Scallop, calico	0.7	0.1	0.2
Squid, Atlantic	1.2	0.3	0.4
Squid, short-finned	2.0	0.4	0.6

FISH OILS

	Total Fat (grams)	Saturated Fat (grams)	Omega-3 (grams)
Cod liver oil	100	17.6	19.2
Herring oil	100	19.2	12.0
Menhaden oil	100	33.6	21.7
MaxEPA, concentrated fish body oils	100	25.4	29.4
Salmon oil	100	23.8	20.9

Source: Compiled from U.S. Department of Agriculture statistics

Note: Tr. = Trace

9

Six Simple Strategies for Lowering Your Cholesterol

In addition to lowering serum cholesterol through proper diet, six additional strategies are useful in lowering serum cholesterol levels.

1. *Maintain the weight that is healthiest for you.* If you eat more than you should, not only are you eating too many calories, but you probably are also consuming too much saturated fat and cholesterol. The odds are that the excess fat in your diet is one major cause of your being overweight.

   Excess consumption of fat can lead to a larger number of fat cells in your body, as well as to an increase in the size of individual fat cells. This can increase triglyceride levels and lower the HDL count. There are, however, some people whose excess weight may be a medically correctable problem, such as an endocrine or metabolic dysfunction. Therefore, before blaming obesity on diet alone, a thorough evaluation by your family physician is advisable. (To determine what your proper weight should be, based on your height, frame, and lifestyle, see tables 6, 7, 8, and 9 in chapter 11.)

2. *Reduce stress.* There have been indications that the liver will produce more serum cholesterol when the body is under stress. People

under stress also tend to eat more and are less aware of what they are eating. Thus, stress itself often leads to increased consumption of just the foods that should be avoided: chocolates, candy, cookies, potato chips, and all the other snack foods that bring a sense of comfort, and in some cases a quick boost of energy, to the body.

Instead of snacking, substitute other, healthy activities whenever you are confronted with a stressful situation: Take a brisk walk for ten minutes, listen to a tape of your favorite music, call a friend and gossip for a few minutes, go shopping, have an early, healthful lunch—in short, do anything that will take your mind off whatever it was that caused the stress. It's practically impossible to eliminate all the stress that affects us daily, but we can change our reactions to that stress.

3. *Free yourself of smoking.* The relationship between cigarette smoking, heart disease, and cancer has been well established. Studies have also indicated that smoking will reduce HDL levels, which, of course, could be a contributory factor to the smoker's increased risk of heart problems.

4. *Monitor your sugar consumption.* Refined sugars—white, brown, crystallized or in any other form—do not benefit you. You get much more sugar than you need on a daily basis from the sugar added to practically every manufactured product. If you ate no other sugar besides that which comes naturally in fruit, you would be far better off.

The consumption of refined sugar has been shown to reduce HDL levels. Moreover, sugar consumption results in an additional, unexpected consequence: It causes the saturated fats and cholesterol that you consume in other foods to result in an even higher level of serum cholesterol (and triglycerides) than would result *without* such refined sugar intake.

5. *Reduce your alcohol intake.* Limited consumption of alcoholic beverages (at the most the equivalent of two drinks a day) has often been recommended by physicians. From the cholesterol-reducing point of view, more than a very moderate consumption of alcohol has been shown to increase triglyceride levels, which may lower your HDL. Remember that the liver converts excess refined sugars and alcohol primarily into triglycerides.

6. *Take up enjoyable exercise.* Insufficient exercise for your age and physical condition may well be one of the causes of your elevated cholesterol levels. There seems to be no question that proper and

sufficient exercise, always in consultation with your physician, will lower cholesterol levels, while raising the all-important HDL score. We are not suggesting hard, strenuous physical exercise if you are not used to it. But a program of brisk walking, healthful swimming, gentle calisthenics, or any other activity that gets your body moving, should help lower your cholesterol and will help you look good and feel great!

10

Easy Tips for Eating
Well with the 2-in-1 System

Much has been written about reducing saturated fat and cholesterol through complicated or hard-to-maintain diets. We feel the time has come to offer some simple solutions. None of the tips below involves your learning to cook, bake, or learn new recipes. They are simply some basic strategies that we have found effective in limiting our fat and cholesterol intake while using the 2-in-1 tables.

BREAKFAST

Pass Up

Eggs
Bacon, sausage, chicken livers, ham, or other breakfast meats
White bread or rolls, croissants, or egg bagels
Cream, whole milk, or butter
Pancakes or waffles (made with butter and/or eggs)

Choose

Hot cereal (preferably oatmeal)

Cold cereal (such as oat bran or whole grain)

Scrambled eggs made with three egg whites and one yolk (for two people—not more than twice a week)

Herring, smoked salmon, or other fish (Limit these if you have to watch your sodium intake.)

Pumpernickel or other whole grain breads, water bagels

Skim milk

Soft margarine

Fruit (fresh or canned in its own juice)

Jam, marmalade, or other preserves

LUNCH

Pass Up

Salami, bologna, ham, pastrami, or other meat sandwiches

Omelets, egg salad, egg sandwiches, or other egg dishes

Cheese on sandwiches, pizzas, or other cheese dishes (except low-fat processed cheeses)

Hamburgers or frankfurters

Fried foods

Salads made with mayonnaise (even if they are tuna or chicken)

Crackers made with palm, palm kernel, coconut, or hydrogenated oil

Ice cream

Pies or cakes made with butter or eggs

Cookies made with palm, palm kernel, coconut or hydrogenated oil

Choose

Homemade soups (except those made with whole milk, cream, eggs, or meat)

Salads prepared with olive oil

Three-bean or other legume salads

Sardines, tuna, or other fish salads or sandwiches (but not if they are made with mayonnaise)

Frankfurters or pastrami made from turkey

Broiled fish

Chicken or turkey sandwiches

Pita bread, melba toast, or bread sticks

Roasted or broiled chicken or turkey (skinless)
Pasta with tomato sauce
Low-fat cottage, mozzarella, ricotta, or processed cheese
Low-fat yogurt
Buttermilk
Sherbet
Dried or fresh fruit

DINNER

Pass Up

Pâtés (except those made of vegetables) and caviar
Salad dressings (except low-fat ones)
Prime grade steak or chops
More than 6 ounces (170 grams) of any beef, lamb, pork, or veal
Sauces made with cream, butter, and/or eggs (such as hollandaise or
    béarnaise)
Skin of chicken or turkey
Duck or goose
Liver, kidneys, or other organ meats
Cold cuts, such as corned beef, tongue, or pastrami
Any fried foods
Hamburgers, pizza or other fast foods
Spareribs
Pastries, pies, mousse, or ice cream

Choose

Broiled or steamed fish
Shellfish such as lobster, clams, mussels, crabs, and scallops
Roasted, stewed, or broiled chicken, Cornish hen, or turkey without the
    skin
Meat balls made from fresh chopped turkey (without skin)
Up to 6 ounces of beef, lamb, pork, or veal, broiled or roasted, with all
    fat trimmed away (not more than twice a week, for a total of not more
    than 12 ounces per week)
Spaghetti, macaroni, or other pasta with tomato or clam sauce and
    grated Parmesan cheese
Rice dishes, in any variety, except those made with whole milk

Chinese, Japanese, or Thai food (except dishes that are deep fried or made with meat or eggs. Wok-prepared food uses hardly any oil and is highly recommended)

Mexican food—especialy beans, chicken, tortillas, and avocado (limited amounts), but nothing deep fried, or prepared with cheese

Cuban food—especially beans, rice, onions, plantains, lamb, and chicken

Mediterranean cuisine—such as Greek, Italian, Spanish or Moroccan— particularly olives, figs, grains, legumes, onion, garlic, pasta, fish, lamb, and chicken

Salads made with olive oil

All vegetables, especially bean dishes

Sherbets or ice milk

Frozen yogurt

Puddings prepared with skim milk

Meringues, soufflés, or other dishes made with egg whites

Dried and fresh fruit

When deciding on your meals, keep the following information in mind:

All spices and herbs can be used freely. Therefore, make your foods more delicious by using mustard, horseradish, vinegar, Worcestershire sauce, tarragon, oregano, basil, curry, chives, parsley, and all other spices and herbs.

Enjoy unlimited amounts of celery and radishes, all lettuces, cucumber, pickles, and salsa.

Avoid regular mayonnaise. Use light or cholesterol-free varieties instead.

Use only diet salad dressings and then not more than 1 tablespoon per meal. Never use regular salad dressings; they are loaded with fat. Better yet, use olive oil and vinegar, or lemon juice and Parmesan, or low-fat yogurt spiced with dill.

Don't use nondairy creamers and imitation toppings. They are extremely high in the saturated tropical fats.

Say no to crackers made with coconut or palm oil, as well as lard. Instead, use the Norwegian and Swedish flat breads, such as Rye Crisp, or melba toast. They are much better for you!

If you must sauté, use butter or margarine very sparingly. But why not steam or stir-fry in a wok instead? Delicious—and do use olive or safflower oil.

Vegetables and fruit are great for you. They never contain cholesterol and rarely much more than a trace of fat. (As mentioned in chapter 6, the only exception is the avocado, which is loaded with fat. Use discretion in how much you eat.) One other idea bears repeating: Watch those cream and cheese sauces on vegetables. Better yet, watch them on someone else's plate.

Eat nuts only occasionally as snacks. All nuts contain substantial amounts of fat and a high calorie content. From a cholesterol point of view, the preponderance of the fat in all nuts is unsaturated. Therefore, eaten in limited quantity, they are an excellent food. The nuts scoring lowest on the 2-in-1 tables are walnuts, pecans, almonds, pistachios, and hazelnuts. The worst nuts for you (having the highest 2-in-1 scores) are cashews, macadamia, and brazil nuts. Coconuts (the source of coconut oil) rate the worst of all the nuts.

Dried fruit is a satisfying alternative to candies and chocolate. Try raisins, dates, figs, banana chips, apricots, apples, and all the others that are available in their pure form, without any additives. They are tasty and good for you. They contain neither cholesterol nor saturated fat, but they do contain a lot of calories. If you eat too many of them you will gain weight, so moderation, as always, is important.

Enjoy homemade soups. Homemade soups are just wonderful for you, especially bean or split pea soup (made without ham bones or sausage), minestrone or other vegetable soups, onion soup, and clam chowder, even chicken soup (if you remove the fat gathered on the top once the soup has cooled). Avoid soups that contain heavy or other cream, whole milk, eggs, or meat.

When dining out, be sure to check the dishes against the 2-in-1 tables. If the waiter can't tell you how a dish is prepared, stay away from it. When a dish is sautéed, ask if the chef is willing to use olive oil instead of butter. Go to the restaurants where you know that the chef prepares delicious poultry, fish, or pasta dishes. If there is a salad bar, enjoy it, but watch those salad dressings! Have a lobster, but don't have the butter sauce. Enjoy that baked potato, but don't have the sour cream with it. Don't give in to the temptation of that wonderful-sounding cream sauce with the Dover sole; have it broiled, baked, or poached. Eat those shrimp you like so much, but not prepared the "scampi" way, which floats them in a pure-fat butter sauce. Veto the duck: The chicken will be fine, especially after you remove the skin. Have a dessert, but instead of ordering cheesecake, ice cream, or a napoleon, opt for some fresh fruit, a small piece of coffee cake, a meringue, or sherbet. *Don't ever be reluctant to ask what is in a dish and how it is prepared—it's your money and your health.*

If, once in a while, you decide not to follow these guidelines, that's fine, but why waste your daily 100 points on inconsequential or ordinary foods? Is that hamburger or those sausages with scrambled eggs really worth it?

Aren't you better off living within the 2-in-1 guidelines most of the time, and then occasionally splurging on some really worthwhile occasion—a super party, a great dinner at a famous restaurant, family cookouts with steak and pie à la mode, or once in a while even savoring caviar or chopped liver? Enjoy it, but after you have had your fun, come back to the 2-in-1 system until the next celebration!

11

How Many Calories Should You Consume Each Day?

The National Institutes of Health's guidelines state that saturated fat calories should not exceed 10 percent of your total caloric intake. Your total calorie consumption, therefore, determines how much saturated fat you can consume. This is why the 2-in-1 tables were designed with eight different calorie columns, each one of which is weighted to permit exactly 100 points per day. You will have to use just one of these columns—the one that is nearest to your own calorie intake—to total up your own daily 2-in-1 score.

In order to use the 2-in-1 tables, all you need to know is the number of calories that you should eat each day.

If you are not certain how many calories you should eat, either to maintain your present weight or perhaps to lose a few pounds, read on.

Calories are units of energy, supplied by food. The more strenuous our work is, the more calories we must eat to supply the energy that our body needs. Similarly, if we engage in strenuous sports, such as tennis or football, we will need more energy (and thus more calories) than if we play gin rummy with a few friends.

If you consume more calories than you expend in energy, you will gain weight. If you weigh more now than you would like to, then you will lose some of your weight if you reduce your caloric intake. It really is as simple as that.

As a rule of thumb, you can figure that 3,500 calories, in excess of the calories you need for the energy you expend, will add one pound to your weight. Conversely, we can say that you will reduce your weight by one pound for every reduction of 3,500 calories from your present diet. Thus, if you want to take off two pounds per week, it means that you must reduce your caloric intake by 7,000 calories per week, or 1,000 calories per day.

Your Personal Calorie Profile

In order to find out how many calories you should consume each day, you need just a few basic facts which, combined with tables 6, 7, 8, and 9, will give you the number of calories recommended for you.

STEP 1. Determine whether your build is small, medium, or large. If you don't know, there is a simple way to judge the size of your frame. Just encircle either wrist, just below the hand, with the thumb and middle finger from the other hand. If you cannot make your finger and thumb touch each other, then you have a large frame. If they just touch, your frame is medium, and if they comfortably overlap, you have a small frame.

Another way to determine your frame size, while a little more difficult, is usually considered to be more accurate. Extend one of your arms and bend your forearm up, so that it is parallel to your body. There are two prominent bones on either side of your elbow. Determine the distance between them by placing two fingers against these bones and then carefully measuring the distance between your two fingers with a ruler. (This measurement can be done more easily and accurately if you have a large caliper available with which to measure the distance between these two bones.)

Now locate your measurement on table 6 (page 64), which lists the measurements for medium-framed men and women. If your measurement is less than the range shown for your height, it means that you have a small frame. If your measurement is greater than the range shown, it indicates that you have a large frame.

Table 6. Medium-Frame Measurements

MEN		WOMEN	
Height	Elbow Breadth	Height	Elbow Breadth
5′1″–5′2″	2½–2⅞″	4′9″–4′10″	2¼–2½″
5′3″–5′6″	2⅝–2⅞″	4′11″–5′2″	2¼–2½″
5′7″–5′10″	2¾–3″	5′3″–5′6″	2⅜–2⅝″
5′11″–6′2″	2¾–3⅛″	5′7″–5′10″	2⅜–2⅝″
6′3″	2⅞–3¼″	5′11″	2½–2¾″

Source: Metropolitan Life Insurance Company.
Note: Height is in stocking feet.

STEP 2. Now look at table 7 under Men or Women. Go down the left column until you find your height. Then move to the right on the table to the appropriate frame column (small, medium, or large). There you will find the recommended range of weights, for your sex, build, and height. If your weight falls between these two guides, then use your actual weight when you come to step 4. If your weight does not fall within the recommended range but you agree that you would like to lose enough weight (or gain enough) to meet the guidelines, then select a weight that is halfway between the upper and lower limit of your range when you come to step 4. You can, of course, also use your actual weight, even if it does not fall within the range, if you are satisfied with your weight as it is.

Table 7. Weight Determination
from Height and Type of Frame

Height is in stocking feet. Weight is without clothes. If you are wearing clothing, adjust your weight by deducting three pounds for women and five pounds for men.

Height		Frame Size		
Feet	Inches	Small	Medium	Large
MEN				
5	1	123–129	126–136	133–145
5	2	125–131	128–138	135–148
5	3	127–133	130–140	137–151
5	4	129–135	132–143	139–155
5	5	131–137	134–146	141–159
5	6	133–140	137–149	144–163
5	7	135–143	140–152	147–167
5	8	137–146	143–155	150–171
5	9	139–149	146–158	153–175
5	10	141–152	149–161	156–179
5	11	144–155	152–165	159–183
6	0	147–159	155–169	163–187
6	1	150–163	159–173	167–192
6	2	153–167	162–177	171–197
6	3	157–171	166–182	176–202
WOMEN				
4	9	99–108	106–118	115–128
4	10	100–110	108–120	117–131
4	11	101–112	110–123	119–134
5	0	103–115	112–126	122–137
5	1	105–118	115–129	125–140
5	2	108–121	118–132	128–144
5	3	111–124	121–135	131–148

Height		Frame Size		
Feet	Inches	Small	Medium	Large
5	4	114–127	124–138	134–152
5	5	117–130	127–141	137–156
5	6	120–133	130–144	140–160
5	7	123–136	133–147	143–164
5	8	126–139	136–150	146–167
5	9	129–142	139–153	149–170
5	10	132–145	142–156	152–173
5	11	135–148	145–159	155–176

Source: Based on data from *Metropolitan Life Insurance Company Statistical Bulletin,* Jan.–Jun. 1983.

STEP 3. Look at the five alternative lifestyles in table 8, and determine which lifestyle (from inactive to strenuous) best fits your daily activity schedule.

Table 8. Determining Your Lifestyle

Inactive	Light Activities	Moderate Activities	Vigorous	Strenuous
Sitting while at work	Carpentry or similar work	Plastering Painting	Heavy construction work	Boxing Wrestling
Reading	Food preparation	Sweeping or washing floors	Logging	Team basketball or football
Writing	Dusting	Gardening	Competitive sports (tennis, swimming)	Competitive running or cycling at least six hours per week
Typing	Ironing	Swimming		
Sewing	Washing dishes or clothes	Skating	Running at least three hours per week	
Painting	Shopping	Table tennis		
Driving	Walking at moderate speed	Light skiing		
Watching TV	Bowling	Golf	Cross-country skiing	
Minimal slow walking	Hunting Fishing	Doubles tennis		
		Dancing		
		Jogging at least one hour per week		

STEP 4. Now look at table 9 under "Men" or "Women." In the left-hand column find your desirable weight, which you determined from step 2. Move to the right and stop under the appropriate column for your lifestyle, as determined from table 8. There you will find the total daily calorie intake commensurate with your recommended weight. (If you eat more, you will probably gain weight, and if you eat less, you will probably lose weight.)

If your actual weight is different from your desirable weight, you will have to decide whether to use that or the weight that is recommended for you, to determine your daily caloric need. It depends solely on whether you want to lose weight if you are overweight, or if you want to gain weight if you are underweight. In either case, you will come up with the number of calories that you should consume daily. This calorie number is the one you will always use in conjunction with the 2-in-1 tables.

Table 9. Recommended Daily Calorie Intake Based on Weight and Lifestyle

Recommended Weight (lb.)	Calories				
	Inactive	Light Activities	Moderate Activities	Vigorous	Strenuous
			MEN		
110	1,430	1,540	1,690	1,870	2,090
115	1,490	1,610	1,760	1,950	2,180
120	1,560	1,680	1,840	2,040	2,280
125	1,620	1,750	1,920	2,120	2,370
130	1,690	1,820	1,990	2,210	2,470
135	1,750	1,890	2,070	2,290	2,560
140	1,820	1,960	2,150	2,380	2,660
145	1,880	2,030	2,220	2,460	2,750
150	1,950	2,100	2,300	2,550	2,850
155	2,010	2,170	2,380	2,630	2,940
160	2,080	2,240	2,450	2,720	3,040
165	2,140	2,310	2,530	2,800	3,130

	Calories				
Recommended Weight (lb.)	Inactive	Light Activities	Moderate Activities	Vigorous	Strenuous
170	2,210	2,380	2,610	2,890	3,230
175	2,270	2,450	2,680	2,970	3,320
180	2,340	2,520	2,760	3,060	3,420
185	2,400	2,590	2,840	3,140	3,510
190	2,470	2,660	2,910	3,230	3,610
195	2,530	2,730	2,990	3,310	3,700
200	2,600	2,800	3,070	3,400	3,800
205	2,660	2,870	3,140	3,480	3,890
210	2,730	2,940	3,220	3,570	3,990
215	2,790	3,010	3,300	3,650	4,080
220	2,860	3,080	3,370	3,740	4,180

WOMEN

90	1,040	1,130	1,240	1,400	1,580
95	1,090	1,190	1,310	1,470	1,660
100	1,150	1,250	1,380	1,550	1,750
105	1,210	1,310	1,450	1,630	1,840
110	1,270	1,380	1,520	1,710	1,930
115	1,320	1,440	1,590	1,780	2,010
120	1,380	1,500	1,660	1,860	2,100
125	1,440	1,560	1,730	1,940	2,190
130	1,500	1,630	1,800	2,020	2,280
135	1,550	1,690	1,870	2,090	2,360
140	1,610	1,750	1,940	2,170	2,450
145	1,670	1,810	2,010	2,250	2,540
150	1,730	1,880	2,080	2,330	2,630
155	1,780	1,940	2,150	2,400	2,710
160	1,840	2,000	2,220	2,480	2,800
165	1,900	2,060	2,290	2,560	2,890
170	1,960	2,130	2,360	2,640	2,980
175	2,010	2,190	2.430	2,710	3,060
180	2,070	2,250	2,500	2,790	3,150
185	2,130	2,310	2,570	2,870	3,240
190	2,190	2,380	2,640	2,950	3,330
195	2,240	2,440	2,710	3,020	3,410
200	2,300	2,500	2,780	3,100	3,500

Sample Calorie Calculation

Here is an example that illustrates how to calculate an appropriate level of calorie intake. Blank space is provided as well for you to use in calculating your own calorie needs. (In this example, the tables designated for men were used.)

	Example	You
Frame Size (from p. 64)	Medium	_____
Height	5'10"	_____
Weight Determination (Median from table 7)	155	_____
Lifestyle (from table 8)	Moderate	_____
Recommended Calorie Intake (from table 9)	2,380	_____

Another Calorie Calculation Method

There is, of course, one other way to calculate how many calories you actually consume each day. That method is to accurately count, for at least four days, the actual calories you consume each day. The 2-in-1 tables list the calories for each food item, so that this calculation is made easy for you.

But you must remember to count everything—from the orange juice in the morning to the saltines and the pat of butter at lunch; from the soup for dinner to the nuts and fruit and ice cream that you snack on after dinner.

Do this really accurately for at least four days of what you would consider "typical" consumption days. Then find the average number of calories consumed by adding all the calories for the four days and dividing by 4. This will give you the actual number of calories you consume on an average day.

The difference between this score and the number you found on table 9, if there is one, is the number of calories you may want to eliminate if you want to reach your desired weight. You would maintain your existing weight by continuing with your actual present calorie consumption.

It does not matter whether you use the calculations from the preceding tables to determine your desired calorie intake or your actual calorie consumption: Either one will let you use the 2-in-1 tables accurately and effectively.

PART III

Introducing the 2-in-1 Tables

12

The 2-in-1 System
in a Nutshell

The 2-in-1 system, using scientifically and medically accepted formulas, combines the amount of serum cholesterol produced when you eat saturated fats with the serum cholesterol produced when you eat dietary cholesterol, weighted into eight different calorie levels. Through this method, the 2-in-1 tables provide just one number that denotes the serum cholesterol value for each of more than six thousand food items. *

Here is all you have to do to start using the 2-in-1 tables:

1. Because the NIH has recommended that your saturated fat intake be limited to 10 percent of your total caloric intake, you must know how many calories you consume each day.

   If you have not done so already, use the tables in chapter 11 to determine how many calories per day you are now consuming. You need to do this step only once, unless your weight or lifestyle changes. These tables take into account your height, weight, bone structure, and activity level.

---

* If you are interested in a technical explanation of the mathematics involved in producing the 2-in-1 tables, we have provided full details in appendix B, The Mathematics of the 2-in-1 System.

2. Once you know what your daily caloric intake should be, you need only look up each food you eat under that particular calorie column in the 2-in-1 tables. Thus, if you consume 2,400 calories per day, just look up the score for the food item you are interested in under the 2,400 calorie column.

   Now you will understand how we arrived at the 2-in-1 scores of 114 and 16 for the two breakfasts mentioned in the introduction of this book. There we also showed a lunch with a score of 12, and two sample dinners, the first with 160 points, and the second with 33. Just for practice, check the 2-in-1 tables for the individual items on each menu, and you will see how easy it was to arrive at the final figures. You will see that we have provided eight different columns for caloric intake, ranging from 1,200 to 3,600. If your daily calorie intake does not fall precisely into one of these columns, select the one that is closest to your figure. Thus, if you consume 2,000 calories daily, which puts you between the 1,800 and 2,100 columns, we suggest that you use the 2,100 column—it's simple and will be very accurate.

3. Regardless of your caloric intake, the 2-in-1 system is designed so that everyone has the same daily limit of 100 points. The numbers in the tables change as the caloric intake changes. For instance, 6 ounces of a lean, broiled T-bone steak have a score of 44 if you consume 1,500 calories daily, but a score of only 32 if your daily intake is 2,800 calories. This occurs because your daily limit of 10 percent permits you only 150 calories of saturated fat if your total calories amount to 1,500, but your 10 percent limit amounts to 280 calories of saturated fat if you consume 2,800 calories daily. (Also factored into the 2-in-1 numbers, of course, is the daily dietary cholesterol limit.)

   Look up the score of each food you eat under your respective calorie column. Soon you will be able to see quickly whether you are surpassing the 100-point limit each day. You also will soon recognize which foods are adding an undue amount to your total score each day, thereby limiting your consumption of other foods. For example, if you consume 1,500 calories daily and are trying to decide whether to eat a 6-ounce T-bone steak (with a score of 44), 6 ounces of chicken breast (with a score of 27), or 6 ounces of scallops (with a score of only 12), it is easy to see that scallops are the best choice for you. They are not only delicious, but also leave 88 points still available for other foods that you like.

A *Special Note:* In its definitive report of October 1987, the National Heart, Lung, and Blood Institute recommended that if satisfactory reductions in serum cholesterol have not been achieved in about three months, saturated fat intake should be further reduced from ten percent to seven percent of total calories. At the same time, dietary cholesterol consumption should be reduced from 300 to 200 milligrams daily. This is what they call the "Step 2 diet." If you—with your physician's approval—feel that you should follow the "Step 2 diet" you can do so simply by reducing your total 2-in-1 point limit from 100 to 70.

What Else Can You Learn From the 2-in-1 Tables?

Besides the 2-in-1 scores, our tables include a few important nutrient details for each food item. (See chapter 14 for a more detailed explanation of every feature in the tables.)

The total fat, saturated fat, and cholesterol contents of each food item have been included for those readers who may be interested in the specific values for some products, but we want to emphasize again that you need none of these other values: The 2-in-1 tables have been designed specifically to do all the work for you.

Polyunsaturated and monounsaturated breakdowns were not provided in the tables for two reasons. First, the emphasis has to be on saturated fat—that is the one that causes the problems and should be watched. Second, the good benefits of monounsaturated fats have not yet been conclusively established. Until they are, the best advice is the recommendation from the NIH to consume approximately equal amounts of polyunsaturated and monounsaturated fats and oils. A normal diet will give you about half of each. As we suggested earlier, if you use equal amounts of olive and safflower oil in your salads and for cooking, you will have achieved a healthy balance.

13

What Can
You Expect?

You may choose to follow the guidelines in this book meticulously, always limiting your intake to 100 points. Or you may decide to "cheat" once in a while. You may be invited to a great party, with lots of hamburgers, hot dogs, and fried chicken—and why shouldn't you have a good time! Or perhaps you are having house guests and they bring a little caviar, some pâté de foie gras, and delicious little butter cookies. It would be downright impolite not to enjoy these fine gifts with your guests!

There is absolutely nothing wrong with cheating once in a while! When you make a habit of healthy eating, having fun occasionally is not going to affect your cholesterol level forever. However, he who rarely cheats will have consistently better results, overall, than those of us who break the rules too often.

Your present cholesterol level also has an important bearing on the results you might expect. For instance, if your count right now is 400, you should expect a larger reduction in pure numbers than if your count right now is only 250. You might achieve the same percentage of decrease in both cases, but twenty percent of 400 is a reduction of 80 milligrams, while the same twenty percent from an original count of 250 is only a 50 milligram decrease.

Another major factor is the amount of saturated fat and cholesterol you now consume. For example, let's say that you now consume 600 milligrams of cholesterol daily, and 20 percent of the calories you consume are from saturated fat. If you reduce your intake of cholesterol to the recommended 300 milligrams and your intake of saturated fats to 10 percent of your total calories, then you will have reduced both the dietary cholesterol and the saturated fat intake by a very substantial 50 percent.

If, however, you now consume about 400 milligrams of cholesterol daily, and your saturated fat intake is approximately 13 percent of your total calories, you would reduce your intake of both by only about 25 percent (100 milligrams of cholesterol and 3 percent saturated fat) to achieve the same level of recommended consumption. Obviously, your serum cholesterol score will decline more if you reduce your consumption of dietary cholesterol and saturated fat by 50 percent than if you reduce it by only 25 percent.

Finally, your success rate also depends on other changes you may have made in your lifestyle besides those made in your diet. For instance, if you are overweight, have you reduced? Are you exercising more now? Are you still smoking? Have you reduced your sugar and alcohol intake? Are you using olive oil and safflower oil whenever possible? Are you eating enough seafood containing the great fish oils? Are you eating oat bran, beans, and the other foods containing soluble fiber? All of these factors are important, and if you do these as well as watch your daily diet, you will optimize your chances for substantially lowering your serum cholesterol score.

We believe that if you follow the 2-in-1 system carefully, you should be able to reduce your total cholesterol, LDL, and triglyceride levels by about 15 percent in the first four to six weeks. If you also follow the suggestions in the previous paragraph, chances are you will reduce these levels by an additional 5 percent, for a total of 20 percent. At the same time, your all-important HDL level should increase by about 10 percent if you use the 2-in-1 tables, and by an additional 5 percent (for a total of 15 percent) if you follow the previous recommendations.

After ten to twelve weeks, just by following the 2-in-1 system, your total serum cholesterol, LDL, and triglyceride scores should be about 25 percent lower than when you started the program, and 33 percent lower if you also followed the suggestions above. Your HDL levels by this point should have increased by about 20 percent from when you started, or by about 30 percent if you also followed the other recommendations.

If your results are considerably less than this, and you are still smoking, eating the wrong foods, and not exercising, don't you think you should give yourself the best chance possible to succeed? But if you are doing the right things, including reducing your daily limit to only 70 points for a few weeks, yet you still can't get satisfactory reductions in your serum cholesterol level, you might be one of those people whose liver tends to manufacture too much cholesterol. If your doctor diagnoses this as the problem, he may prescribe medication to help lower your cholesterol to a satisfactory level (see chapter 4, What about Niacin or Prescription Drugs?).

A Real Success Story

You might recall that one of the authors of this book, F. M. Oppenheimer, had very high cholesterol levels. Here is his actual record of success in reducing those levels:

	Initial Reading	After 7 Weeks	After 14 Weeks	After 28 Weeks
Total Cholesterol	306	279	216	197
Triglycerides	218	171	68	70
LDL	220	203	141	105
HDL	42	42	61	78
VLDL (⅕ of Triglycerides)	44	34	14	14
Cholesterol/HDL Ratio	7.29	6.64	3.54	2.53

Both Oppenheimer and Dr. Pardell were somewhat disappointed in the readings after seven weeks. They therefore decided to try 1.5 grams of niacin daily, in addition to the 100-point program. As you can see, the readings after just an additional seven weeks seem nearly miraculous, but they were validated with the next readings, an additional fourteen weeks later (a total of twenty-eight weeks since the initial reading).

The combination of the 100-point program and 1.5 grams of niacin certainly worked for Oppenheimer! Not only was his cholesterol reduced from 306 to 197 but his LDL and triglycerides showed even more astounding declines. Especially significant was the increase in Oppenheimer's HDL readings from 42 to 78, and a reduction in his cholesterol/HDL ratio from 7.29 to an amazingly low 2.53.

You should be able to achieve the reductions mentioned earlier in this chapter and perhaps even do much better. You may see results that will truly astound you. When you do get these results, by all means go out and celebrate and eat anything you want to—for one day!

After that, return to the 2-in-1 system. You will want to, anyway, because besides a lower cholesterol level, you will have achieved an additional important benefit—that of feeling better than ever before— now that you are finally eating the proper way for lifelong health and well-being!

14

Important Tips
for Using the 2-in-1 Tables

With the exception of fast foods, the foods in the 2-in-1 tables, as well as their nutritional values, were compiled and consolidated from data supplied by the United States Department of Agriculture (USDA). The data for fast foods were supplied directly by the individual companies.

All of the foods are grouped by food categories, such as Meat, Poultry, Milk and Milk Products, and so forth. Some of these categories then have subgroups, such as Beef, Game, Lamb, Pork, and Veal under the Meat Group, and Cream, Milk, Ice Cream, and Yogurt under the Milk Group.

For quick reference, the contents lists these various groups and subgroups and the pages where they are located. All the food items within each group or subgroup are alphabetized, making it possible for you to see instantly which types or cuts of a certain product may be better for you than a different cut, or which method of preparation may be preferable. You can also easily compare different products, such as veal and beef, or yogurt and ice cream.

There are two special categories we want to call to your attention. One is Fast Foods, where we have listed the products of eight different fast-food companies that supplied us with nutritional breakdowns for their products. (See also chapter 7, Fast-Food Facts at Your Fingertips.)

The other special section is called Complete Dishes—Home Cooked or Restaurant. It lists alphabetically many different meat, seafood, poultry, pasta, and other dishes that you might eat at home or order in a restaurant.

What the Tables Contain

Food Description. Every item is described as to whether it is in its raw state or, if cooked, what method of preparation is used. Many items are shown with two or more different cooking methods, which makes it easy for you to compare the value of broiling, for instance, against the value of deep-frying. Many meat items are designated as "lean and fat" or "lean only." In the first case, some fat remains around the meat, and both the meat and fat are assumed to be eaten. In the second case, the fat has been trimmed from the meat, and only the meat is eaten. In some cases, we also indicate if an item is frozen, or, if it is canned, the type of fluid used in canning. Make sure that the entry you are using corresponds as closely as possible to the actual item you are eating.

Quantity. The food description also shows the weight in ounces of each item (if applicable), or the volume measurement (such as fluid ounces, cups, or tablespoons), or a description (such as a can, a jar, a drumstick, a lamb chop, or a slice). You can easily adjust the 2-in-1 score to the actual quantity of the item you consume. For example, if you eat 8 ounces of a product, and we give you the score for 4 ounces, be sure to multiply the score by 2 in order to evaluate the product accurately. Likewise, if the score for a drink is given for half a cup, and you drink three-quarters of a cup, you should increase the score by one-half. Be aware of this as well when using canned or packaged goods—make sure you use similar weights or volume, or adjust the 2-in-1 score to the amount you use.

Weight. The weight of every item is shown in grams. The weight for cooked foods is given *after* cooking. In some instances, the weight both before and after preparation is shown.

Calories. Every item shown lists the number of calories in the given quantity. You will notice that in most cases the higher the 2-in-1 score, the higher the calories. The reason, of course, is that the fat content in the food is responsible for a major part of its calories.

Total Fat. Total fat is shown in calories so that you can easily see how many of an item's calories come from fat. If you want to know the weight of the fat, just divide the fat calories by 9. This will give you the weight of the fat in grams, but we provided this measurement in calories, instead of grams, on purpose, because the most important aspect of the fat content is its relationship (in calories) with total calories. Avoid those products where the total fat exceeds 30 percent of total calories.

Saturated Fat. Saturated fat is also given in calories so that it can easily be related to total fat and calories. Whenever you see an item where the saturated fat calories are more than one-third of its total fat content, or more than one-tenth of its total calories—beware!

Cholesterol. Cholesterol is shown in milligrams so that you can compare the dietary cholesterol value of all food items.

Measurement Equivalents. See appendix A for measurement equivalents.

There was just a few foods for which one or more of the important nutrients could not be determined. In those cases, a dash (—) is used for the missing nutrient. As it was impossible to compute an accurate 2-in-1 score for those few items, a dash (—) was also used in the tables to indicate that a proper value could not be calculated.

Here They Are—The 2-in-1 Tables

For the first time the cholesterol-producing effect from both the saturated fat and from the cholesterol that you eat are combined into *one number* for more than six thousand food items.

By now you know exactly what the 2-in-1 System is, how it works, and what it can do for you! For easy reference, here, once more, is a short summary that boils it all down to four easy steps:

Here Is All You Have To Do:

1. Determine how many calories you eat each day. (Follow the simple instructions in chapter 11.)
2. Locate your personal calorie column in the 2-in-1 tables. (Choose the column nearest your daily calorie intake.)
3. You are allowed to consume exactly 100 points per day. That is the "magic" of the 2-in-1 system—each calorie column is weighted for each food item, so that precisely 100 points are permitted regardless of which calorie column applies to you.

4. Look up *everything* you eat under your calorie column, and add the numbers. Don't exceed 100 points per day.

Good Luck and Good Health!

Abbreviations Used in Food Descriptions

&	and
approx	approximately
c	cup(s)
ckd	cooked
cu in	cubic inch(es)
dia	diameter
drnd	drained
fl oz	fluid ounce(s)
fr	from
gr	gram(s)
hi-prot	high protein
inc	including
ind	individual
lg	large
liq	liquid
lo-cal	low calorie
lo-fat	low fat
med	medium
nt wt	net weight
pkt	packet, package
prep	prepared
pwd	powder
qt	quart
reg	regular
sl	slice(s)
sm	small
sq	square(s)
tbsp	tablespoon(s)
tsp	teaspoon(s)
w/	with
wo/	without

ALCOHOLIC BEVERAGES
Beer, Ale, and Wine

	1200	1500	1800	2100	2400	2800	3200	3600	TOT CAL	FAT CAL	S/FAT CAL	CHOL MG
Beer (including ale)—1 can or bottle (12 fl oz) (360 gm)	0	0	0	0	0	0	0	0	151	0	0	0
Beer, light—1 can or bottle (12 fl oz) (360 gm)	0	0	0	0	0	0	0	0	101	0	0	0
Beer, regular—1 can or bottle (12 fl oz) (356 gm)	0	0	0	0	0	0	0	0	146	0	0	0
Glug (inc glogg, gluhwein)—1 drink (4 fl oz) (116 gm)	0	0	0	0	0	0	0	0	113	0	0	0
Near beer—1 can or bottle (12 fl oz) (360 gm)	0	0	0	0	0	0	0	0	32	0	0	0
Sangria—1 drink (228 gm)	0	0	0	0	0	0	0	0	154	1	.1	0
Wine, Chinese—1 wine glass (3.5 fl oz) (100 gm)	0	0	0	0	0	0	0	0	70	0	0	0
Wine, cooking (cooked)—1 fl oz (29 gm)	0	0	0	0	0	0	0	0	2	0	0	0
Wine cooler—1 drink (7 fl oz) (210 gm)	0	0	0	0	0	0	0	0	101	1	.1	0
Wine, dessert, dry—1 glass (2 fl oz) (59 gm)	0	0	0	0	0	0	0	0	74	0	0	0
Wine, dessert, sweet (inc marsala, port, tokay, Madeira, muscatel, angelica, sherry, sweet vermouth)—1 wine glass (3.5 fl oz) (100 gm)	0	0	0	0	0	0	0	0	153	0	0	0
Wine, light—1 wine glass (3.5 fl oz) (102 gm)	0	0	0	0	0	0	0	0	51	0	0	0
Wine spritzer—1 drink (146 gm)	0	0	0	0	0	0	0	0	61	0	0	0

Item												
Wine, table, all—1 glass (3.5 fl oz) (103 gm)	0								72	0	0	0
Wine, table, dry (inc burgundy, claret, chianti, sauterne, Rhine, champagne, homemade wine, dry sherry)—1 glass (3.5 fl oz) (100 gm)	0								70	0	0	0

Cocktails, Liqueurs, and Liquors

Item												
Alexander—1 cocktail (74 gm)	5	5	4	4	3	3	3	3	179	16	10.0	6
Bacardi cocktail—1 cocktail (63 gm)	0	0	0	0	0	0	0	0	118	0	0	0
Black Russian—1 cocktail (90 gm)	0	0	0	0	0	0	0	0	255	1	.4	0
Bloody Mary—1 cocktail (148 gm)	0	0	0	0	0	0	0	0	123	1	.1	0
Bourbon and soda (inc scotch & soda, rum & soda)—1 cocktail (116 gm)	0	0	0	0	0	0	0	0	105	0	0	0
Brandy (inc applejack, cognac, tequila)—(1 fl oz) (28 gm)	0	0	0	0	0	0	0	0	65	0	0	0
Coquito, Puerto Rican (coconut rum)—1 small goblet (4 fl oz) (125 gm)	48	42	37	34	32	29	27	25	297	103	69.7	107
Cordial or liqueur (inc amaretto, anisette, Benedictine, Chartreuse, cointreau, crème de menthe, curaçao, drambuie, grenadine, kahlua, kirsch—1 cordial glass (20 gm)	0	0	0	0	0	0	0	0	74	1	.2	0
Crème de menthe, 72 proof—1 glass (1.5 fl oz) (50 gm)	0	0	0	0	0	0	0	0	186	1	.1	0
Daiquiri—1 cocktail (61 gm)	0	0	0	0	0	0	0	0	113	0	0	0
Fruit punch, alcoholic (inc champagne, fruit punch, champagne punch)—1 punch cup (4 fl oz) (116 gm)	0	0	0	0	0	0	0	0	101	0	0	0
Gibson—1 cocktail (71 gm)	0	0	0	0	0	0	0	0	159	0	0	0
Gimlet—1 cocktail (71 gm)	0	0	0	0	0	0	0	0	132	0	0	0
Gin—1 jigger (42 gm)	0	0	0	0	0	0	0	0	110	0	0	0
Gin & Tonic—1 cocktail (225 gm)	0	0	0	0	0	0	0	0	171	0	0	0

	1200	1500	1800	2100	2400	2800	3200	3600	TOT CAL	FAT CAL	S/FAT CAL	CHOL MG
Cocktails, Liqueurs, and Liquors (cont.)												
Gin Rickey—1 cocktail (205 gm)	0	0	0	0	0	0	0	0	114	0	0	0
Gold Cadillac—1 cocktail (125 gm)	10	9	8	7	6	6	5	5	394	32	19.6	11
Grasshopper, prep w/cream or ice cream—1 cocktail (64 gm)	11	9	8	7	7	6	5	5	164	32	19.9	11
Highball—1 cocktail (160 gm)	0	0	0	0	0	0	0	0	104	0	0	0
Irish Coffee (inc Coffee Royale)—(1 fl oz) (26 gm)	4	3	3	3	3	2	2	2	26	12	7.3	5
Liqueur, coffee, 53 proof—1 glass (1.5 fl oz) (52 gm)	0	0	0	0	0	0	0	0	174	1	.4	0
Liqueur, coffee with cream, 34 proof—1 jigger (1.5 fl oz) (47 gm)	19	16	14	13	11	10	9	8	154	67	40.9	7
Long Island Iced Tea—1 drink (5 fl oz) (150 gm)	0	0	0	0	0	0	0	0	143	0	.1	0
Mai Tai—1 cocktail (126 gm)	0	0	0	0	0	0	0	0	310	1	.4	0
Manhattan—1 cocktail (57 gm)	0	0	0	0	0	0	0	0	128	0	0	0
Margarita—1 cocktail (77 gm)	0	0	0	0	0	0	0	0	170	1	.2	0
Martini—1 cocktail (71 gm)	0	0	0	0	0	0	0	0	159	0	0	0
Mint Julep—1 cocktail (65 gm)	0	0	0	0	0	0	0	0	156	0	0	0
Old-fashioned—1 cocktail (60 gm)	0	0	0	0	0	0	0	0	155	0	0	0
Piña Colada—1 cocktail (133 gm)	9	7	6	6	5	4	4	4	231	23	19.7	0
Piña Colada, canned—1 cocktail (6.8 fl oz) (200 ml) (222 gm)	58	49	42	38	34	30	27	24	525	152	131.3	0
Piña Colada, prep from recipe—1 cocktail (4.5 oz) (141 gm)	5	4	4	3	3	3	2	2	262	23	11.1	0
Rum—1 jigger (42 gm)	0	0	0	0	0	0	0	0	97	0	0	0
Rum & Cola—1 cocktail (211 gm)	0	0	0	0	0	0	0	0	160	1	.1	0
Rum, hot buttered—1 drink (251 gm)	35	30	26	23	21	19	17	16	317	107	67.0	31

Screwdriver (inc Harvey Wallbanger, Slo-Screw)—1 cocktail (213 gm)	0	0	0	0	0	0	182	1	.1	0	
Singapore Sling—1 cocktail (225 gm)	0	0	0	0	0	0	228	1	.2	0	
Sloe Gin Fizz—1 cocktail (222 gm)	0	0	0	0	0	0	121	0	.1	0	
Stinger—1 cocktail (92 gm)	½	0	0	0	0	0	282	1	.6	0	
Tequila Sunrise—1 cocktail (172 gm)	0	0	0	0	0	0	189	2	.2	0	
Tom Collins (inc Vodka Collins)—1 cocktail (222 gm)	0	0	0	0	0	0	121	0	.1	0	
Vodka—1 jigger (42 gm)	0	0	0	0	0	0	97	0	0	0	
Whiskey (inc bourbon, Scotch, rye)—1 jigger (42 gm)	0	0	0	0	0	0	105	0	0	0	
Whiskey Sour (inc Scotch sour, vodka sour, apricot sour, brandy sour)—1 cocktail (90 gm)	0	0	0	0	0	0	122	1	.2	0	
White Russian—1 cocktail (100 gm)	4	3	3	2	2	2	268	11	6.9	4	

BABY FOODS

Cereal, barley, dry—.5 oz (14 gm)	—	—	—	—	—	—	52	5	—	—	
Cereal, barley, prep w/whole milk—1 oz (28 gm)	—	—	—	—	—	—	31	8	—	—	
Cereal, egg yolks & bacon, junior—1 jar (128 gm)	32	29	27	24	23	21	101	58	18.5	152	
Cereal, egg yolks & bacon, strained—1 jar (213 gm)	53	49	45	40	38	36	178	99	30.7	253	
Cereal, grits & egg yolks, strained—1 jar (128 gm)	—	—	—	—	—	—	73	26	—	—	
Cereal, hi-prot, dry—.5 oz (14 gm)	—	—	—	—	—	—	51	7	—	—	
Cereal, hi-prot, prep w/whole milk—1 oz (28 gm)	—	—	—	—	—	—	31	10	—	—	
Cereal, hi-prot, w/apple & orange, dry—.5 oz (14 gm)	—	—	—	—	—	—	53	8	—	—	
Cereal, hi-prot, w/apple & orange, prep w/whole milk—1 oz (28 gm)	—	—	—	—	—	—	32	10	—	—	

THE 2-IN-1 SYSTEM

Baby Foods (cont.)

	1200	1500	1800	2100	2400	2800	3200	3600	TOT CAL	FAT CAL	S/FAT CAL	CHOL MG
Cereal, mixed, dry—.5 oz (14 gm)	—	—	—	—	—	—	—	—	54	5	—	—
Cereal, mixed, prep w/whole milk—1 oz (28 gm)	—	—	—	—	—	—	—	—	32	9	—	—
Cereal, mixed, w/applesauce & bananas, junior—1 jar (220 gm)	—	—	—	—	—	—	—	—	183	8	—	—
Cereal, mixed, w/applesauce & bananas, strained—1 jar (135 gm)	—	—	—	—	—	—	—	—	111	6	—	—
Cereal, mixed, w/bananas, dry—0.5 oz (14 gm)	—	—	—	—	—	—	—	—	56	6	—	—
Cereal, mixed, w/bananas, prep w/ whole milk—1 oz (28 gm)	0	0	0	0	0	0	0	0	33	9	.2	0
Cereal, mixed, w/honey, dry—0.5 oz (14 gm)	—	—	—	—	—	—	—	—	55	6	—	—
Cereal, mixed, w/honey, prep w/whole milk—1 oz (28 gm)	—	—	—	—	—	—	—	—	33	9	—	—
Cereal, oatmeal, dry—0.5 oz (14 gm)	—	—	—	—	—	—	—	—	56	10	—	—
Cereal, oatmeal, prep w/whole milk—1 oz (28 gm)	—	—	—	—	—	—	—	—	33	11	—	—
Cereal, oatmeal, w/applesauce & bananas, junior—1 jar (220 gm)	—	—	—	—	—	—	—	—	165	14	—	—
Cereal, oatmeal, w/applesauce & bananas, strained—1 jar (135 gm)	1	1	1	½	½	½	½	½	99	8	1.6	0
Cereal, oatmeal, w/bananas, dry—0.5 oz (14 gm)	—	—	—	—	—	—	—	—	56	8	—	—
Cereal, oatmeal, w/bananas, prep w/whole milk—1 oz (28 gm)	—	—	—	—	—	—	—	—	33	10	—	—
Cereal, oatmeal, w/honey, dry—0.5 oz (14 gm)	—	—	—	—	—	—	—	—	55	9	—	—

Cereal, oatmeal, w/honey, prep w/whole milk—1 oz (28 gm)	—	—	—	—	—	—	—	—	33	10	—	—
Cereal, rice, dry—0.5 oz (14 gm)	—	—	—	—	—	—	—	—	56	6	—	—
Cereal, rice, prep w/whole milk—1 oz (28 gm)	—	—	—	—	—	—	—	—	33	9	—	—
Cereal, rice, w/applesauce & bananas, strained—1 jar (135 gm)	1	1	½	½	½	½	½	½	107	5	1.4	0
Cereal, rice, w/bananas, dry—0.5 oz (14 gm)	—	—	—	—	—	—	—	—	57	5	—	—
Cereal, rice, w/bananas, prep w/whole milk—1 oz (28 gm)	—	—	—	—	—	—	—	—	33	9	—	—
Cereal, rice, w/honey, dry—0.5 oz (14 gm)	—	—	—	—	—	—	—	—	56	4	—	—
Cereal, rice, w/honey, prep w/whole milk—1 oz (28 gm)	—	—	—	—	—	—	—	—	33	8	—	—
Cereal, rice, w/mixed fruit, junior—1 jar (220 gm)	—	—	—	—	—	—	—	—	186	5	—	—
Cereal, toasted oat rings—1 serving (14 gm)	1	1	½	½	½	½	½	½	56	8	1.5	0
Cereal, w/eggs, strained—1 jar (128 gm)	13	12	11	10	10	9	9	8	74	17	5.7	66
Cereal, w/egg yolks, junior—1 jar (213 gm)	14	14	12	11	11	9	9	8	110	34	11.8	—
Cereal, w/egg yolks, strained—1 jar (128 gm)	16	14	14	13	12	11	11	10	66	21	7.0	81
Cookies—1 cookie (7 gm)	—	—	—	—	—	—	—	—	28	8	2.2	—
Cookies, arrowroot—1 cookie (6 gm)	—	—	—	—	—	—	—	—	24	7	1.6	—
Dessert, apple betty, junior—1 jar (220 gm)	—	—	—	—	—	—	—	—	153	0	—	—
Dessert, apple betty, strained—1 jar (135 gm)	—	—	—	—	—	—	—	—	97	0	—	—
Dessert, banana pudding, junior—½ jar (7.5 oz-7.75 oz) (108 gm)	6	5	5	5	5	4	4	4	72	8	2.5	31

	1200	1500	1800	2100	2400	2800	3200	3600	TOT CAL	FAT CAL	S/FAT CAL	CHOL MG
Baby Foods (cont.)												
Dessert, banana pudding, strained—½ jar (4.5 oz–4.75 oz) (66 gm)	4	3	3	3	3	3	3	2	47	5	1.5	19
Dessert, caramel pudding, junior—1 jar (213 gm)	—	—	—	—	—	—	—	—	167	17	—	—
Dessert, caramel pudding, strained—1 jar (135 gm)	—	—	—	—	—	—	—	—	104	8	—	—
Dessert, cherry vanilla pudding, junior—1 jar (220 gm)	5	4	4	4	3	3	3	3	152	4	2.6	22
Dessert, cherry vanilla pudding, strained—1 jar (135 gm)	3	3	2	2	2	2	2	2	91	4	1.6	14
Dessert, ciruelas w/tapioca—½ jar (4.5 oz–4.75 oz) (66 gm)	0	0	0	0	0	0	0	0	46	1	.1	0
Dessert, cottage cheese with fruit, strained or junior—½ jar, strained (4.5 oz–4.75 oz) (67 gm)	2	2	2	2	2	1	1	1	49	5	2.9	6
Dessert, cottage cheese w/pineapple, junior—1 jar (220 gm)	—	—	—	—	—	—	—	—	172	14	—	—
Dessert, cottage cheese w/pineapple, strained—1 jar (135 gm)	—	—	—	—	—	—	—	—	94	10	—	—
Dessert, custard pudding, chocolate, junior—1 jar (220 gm)	17	15	14	13	12	11	10	10	195	32	17.8	59
Dessert, custard pudding, chocolate, strained—1 jar (128 gm)	11	9	9	8	7	7	6	6	107	19	11.5	35
Dessert, custard pudding, vanilla, junior—1 jar (220 gm)	13	11	10	9	8	7	7	6	196	45	23.4	18
Dessert, custard pudding, vanilla, strained—1 jar (128 gm)	7	6	5	5	4	4	4	3	109	23	11.6	10

Dessert, dutch apple, junior—1 jar (220 gm)	12	10	9	9	8	8	7	7	151	19	12.3	39
Dessert, dutch apple, strained—1 jar (135 gm)	7	6	6	5	5	5	4	4	92	11	7.0	25
Dessert, fruit dessert, wo/ascorbic acid, junior—1 jar (220 gm)	—	—	—	—	—	—	—	—	138	0	—	—
Dessert, fruit dessert, wo/ascorbic acid, strained—1 jar (135 gm)	—	—	—	—	—	—	—	—	79	0	—	—
Dessert, fruit pudding, orange, strained—1 jar (135 gm)	4	3	2	2	2	2	2	2	108	11	6.8	4
Dessert, fruit pudding, pineapple, junior 1 jar (220 gm)	—	—	—	—	—	—	—	—	192	8	—	—
Dessert, fruit pudding, pineapple, strained—1 jar (128 gm)	—	—	—	—	—	—	—	—	104	4	—	—
Dessert, mango dessert w/vitamin C—1 jar (4.75 oz) (131 gm)	0	0	0	0	0	0	0	0	105	2	.5	0
Dessert, peach cobbler, junior—1 jar (220 gm)	—	—	—	—	—	—	—	—	147	0	—	—
Dessert, peach cobbler, strained—1 jar (135 gm)	—	—	—	—	—	—	—	—	88	0	—	—
Dessert, peach melba, junior—1 jar (220 gm)	—	—	—	—	—	—	—	—	132	0	—	—
Dessert, peach melba, strained—1 jar (135 gm)	—	—	—	—	—	—	—	—	81	0	—	—
Dessert, pineapple dessert, junior—½ jar (7.5 oz-7.75 oz) (108 gm)	0	0	0	0	0	0	0	0	94	4	.3	0
Dessert, pineapple dessert, strained—½ jar (4.5 oz-4.75 oz) (66 gm)	½	½	½	½	½	½	0	0	53	2	1.1	0
Dessert, pineapple orange, strained—1 jar (128 gm)	—	—	—	—	—	—	—	—	89	0	—	—
Dessert, tropical fruit, junior—1 jar (220 gm)	—	—	—	—	—	—	—	—	131	0	—	—
Dessert, tutti-frutti pudding, junior—½ jar (7.75 oz) (108 gm)	3	3	3	3	2	2	2	2	76	4	1.2	16

	1200	1500	1800	2100	2400	2800	3200	3600	TOT CAL	FAT CAL	S/FAT CAL	CHOL MG
Baby Foods (cont.)												
Dessert, tutti-frutti pudding, strained —½ jar (4.5 oz) (66 gm)	2	2	2	1	1	1	1	1	47	2	.5	10
Dinner, beef lasagna, toddler—1 jar (177 gm)	—	—	—	—	—	—	—	—	137	34	—	—
Dinner, beef noodle, junior—1 jar (213 gm)	7	6	5	5	5	4	4	4	122	36	9.6	17
Dinner, beef noodle, strained—1 jar (128 gm)	5	4	4	4	3	3	3	3	68	20	8.1	9
Dinner, beef & rice, toddler—1 jar (177 gm)	—	—	—	—	—	—	—	—	146	46	—	—
Dinner, beef stew, toddler—1 jar (177 gm)	8	7	6	6	5	5	4	4	90	19	9.3	22
Dinner, beef w/vegetables, toddler—1 jar (6.25 oz) (177 gm)	10	9	8	7	7	6	6	5	120	33	16.1	21
Dinner, chicken & rice—½ jar (108 gm)	3	3	2	2	2	2	2	2	55	9	2.6	11
Dinner, chicken noodle, junior—1 jar (213 gm)	9	8	7	7	6	6	6	5	109	27	7.7	35
Dinner, chicken, noodles & vegetables, toddler—1 jar (6.25 oz) (177 gm)	11	10	10	9	8	8	7	7	113	27	8.1	50
Dinner, chicken noodle, strained—1 jar (128 gm)	6	5	5	4	4	4	4	3	67	17	4.6	23
Dinner, chicken soup—1 jar (4.5 oz–4.75 oz) (131 gm)	3	3	3	2	2	2	2	2	66	20	5.7	5
Dinner, chicken soup, cream of, strained —1 jar (128 gm)	—	—	—	—	—	—	—	—	74	18	—	—
Dinner, chicken soup, strained—1 jar (128 gm)	—	—	—	—	—	—	—	—	64	20	—	—

Food											
Dinner, chicken stew, toddler—1 jar (6.25 oz) (177 gm)	16	14	13	12	11	10	10	9	138	59	17.5	51
Dinner, high-meat, beef & all vegetables, strained—jar (128 gm)	14	12	11	10	9	8	8	7	96	48	24.2	23
Dinner, high-meat, beef & vegetables, junior—1 jar (128 gm)	15	13	12	11	10	9	8	7	108	53	26.5	23
Dinner, high-meat, chicken & vegetables, junior—1 jar (128 gm)	11	10	9	8	7	7	6	6	117	63	18.5	20
Dinner, high-meat, chicken & vegetables, strained—1 jar (128 gm)	8	7	7	6	6	5	5	4	100	41	11.5	20
Dinner, high-meat, cottage cheese w/pineapple, strained—1 jar (135 gm)	—	—	—	—	—	—	—	—	157	27	—	—
Dinner, high-meat, ham & vegetables, junior—1 jar (128 gm)	9	8	7	7	6	6	5	5	98	38	13.0	23
Dinner, high-meat, ham & vegetables, strained—1 jar (128 gm)	10	8	8	7	6	6	5	5	97	40	13.6	23
Dinner, high-meat, turkey & vegetables, junior—1 jar (128 gm)	11	10	8	8	7	6	6	5	115	58	18.5	18
Dinner, high-meat, turkey & vegetables, strained—1 jar (128 gm)	11	10	8	8	7	6	6	5	111	55	18.5	18
Dinner, high-meat, veal & vegetables, junior—1 jar (128 gm)	11	10	9	8	7	7	6	6	93	36	14.9	28
Dinner, high-meat, veal & vegetables, strained—1 jar (128 gm)	10	9	8	7	7	6	6	5	89	31	13.9	24
Dinner, lamb & noodles, junior—1 jar (213 gm)	—	—	—	—	—	—	—	—	138	42	—	—
Dinner, macaroni & bacon, junior—1 jar (213 gm)	—	—	—	—	—	—	—	—	160	64	—	—
Dinner, macaroni & cheese, junior—1 jar (213 gm)	13	12	10	9	8	7	7	6	130	39	25.5	14
Dinner, macaroni & cheese, strained—1 jar (128 gm)	8	7	6	5	5	4	4	4	76	24	15.3	8
Dinner, macaroni & ham, junior—1 jar (213 gm)	—	—	—	—	—	—	—	—	127	26	—	—

THE 2-IN-1 SYSTEM

	1200	1500	1800	2100	2400	2800	3200	3600	TOT CAL	FAT CAL	S/FAT CAL	CHOL MG
Baby Foods (cont.)												
Dinner, macaroni, tomato & beef, junior—1 jar (213 gm)	6	5	4	4	4	3	3	3	125	22	9.6	9
Dinner, macaroni, tomato & beef, strained—1 jar (128 gm)	3	3	3	2	2	2	2	2	71	13	5.8	5
Dinner, mixed vegetable, junior—1 jar (213 gm)	—	—	—	—	—	—	—	—	71	1	—	—
Dinner, mixed vegetable, strained—1 jar (128 gm)	—	—	—	—	—	—	—	—	52	1	—	—
Dinner, spaghetti & tomato & meat, junior—1 jar (213 gm)	6	5	5	4	4	4	3	3	135	24	10.4	11
Dinner, spaghetti & tomato & meat, toddler—1 jar (177 gm)	5	5	4	4	3	3	3	3	133	16	8.6	9
Dinner, split pea & ham, junior—1 jar (213 gm)	—	—	—	—	—	—	—	—	152	25	—	—
Dinner, split pea w/vegetables, ham or bacon, junior—½ jar (7.5 oz–7.75 oz) (108 gm)	3	2	2	2	2	2	2	1	77	13	4.6	5
Dinner, turkey, rice & vegetables, junior—½ jar (7.5 oz–7.75 oz) (108 gm)	4	3	3	3	2	2	2	2	53	14	4.2	11
Dinner, turkey, rice & vegetables, strained—½ jar (4.5 oz–4.75 oz) (66 gm)	2	2	2	2	2	1	1	1	32	8	2.5	7
Dinner, turkey & rice, junior—1 jar (213 gm)	7	6	6	5	5	4	4	4	104	26	8.3	21
Dinner, turkey & rice, strained—1 jar (128 gm)	4	4	3	3	3	3	2	2	63	15	4.8	13

Food												
Dinner, vegetables & bacon, junior—1 jar (213 gm)	13	11	10	9	8	7	6	6	150	74	26.8	9
Dinner, vegetables & bacon, strained—1 jar (128 gm)	7	6	5	4	4	4	3	3	88	38	13.7	4
Dinner, vegetables & beef, junior—1 jar (213 gm)	8	7	6	5	5	4	4	4	113	32	13.4	11
Dinner, vegetables & beef, strained—1 jar (128 gm)	5	4	4	3	3	3	3	2	67	23	9.2	6
Dinner, vegetables & chicken, junior—1 jar (213 gm)	6	5	5	4	4	4	4	3	106	21	5.8	21
Dinner, vegetables & chicken, strained—1 jar (128 gm)	4	4	3	3	3	3	3	3	55	13	3.4	17
Dinner, vegetables, dumplings & beef, junior—1 jar (213 gm)	—	—	—	—	—	—	—	—	103	15	—	—
Dinner, vegetables, dumplings & beef, strained—1 jar (128 gm)	—	—	—	—	—	—	—	—	61	11	—	—
Dinner, vegetables, noodles & chicken, junior—1 jar (213 gm)	—	—	—	—	—	—	—	—	137	43	—	—
Dinner, vegetables, noodles & chicken, strained—1 jar (128 gm)	—	—	—	—	—	—	—	—	81	30	—	—
Dinner, vegetables, noodles & turkey, junior—1 jar (213 gm)	—	—	—	—	—	—	—	—	110	29	—	—
Dinner, vegetables, noodles & turkey, strained—1 jar (128 gm)	—	—	—	—	—	—	—	—	56	14	—	—
Dinner, vegetables & ham, junior—1 jar (213 gm)	7	6	5	5	4	4	4	3	110	32	11.5	11
Dinner, vegetables & ham, strained—1 jar (128 gm)	4	3	3	3	3	2	2	2	62	20	6.9	6
Dinner, vegetables & ham, toddler—1 jar (177 gm)	10	8	7	7	6	5	5	5	128	47	16.6	14
Dinner, vegetables & lamb, junior—1 jar (213 gm)	9	7	6	6	5	5	4	4	108	33	15.3	11
Dinner, vegetables & lamb, strained—1 jar (128 gm)	6	5	4	4	4	3	3	3	67	23	10.4	8

	1200	1500	1800	2100	2400	2800	3200	3600	TOT CAL	FAT CAL	S/FAT CAL	CHOL MG
Baby Foods (cont.)												
Dinner, vegetables & liver, junior—1 jar (213 gm)	—	—	—	—	—	—	—	—	93	11	—	—
Dinner, vegetables & liver, strained—1 jar (128 gm)	9	9	8	8	7	7	7	7	50	5	1.2	56
Dinner, vegetables & turkey, junior—1 jar (213 gm)	7	6	5	5	5	4	4	4	101	24	7.7	21
Dinner, vegetables & turkey, strained—1 jar (128 gm)	4	4	3	3	3	3	2	2	54	14	4.6	13
Dinner, vegetables & turkey, toddler—1 jar (177 gm)	—	—	—	—	—	—	—	—	141	55	—	—
Egg yolks, strained—1 jar (94 gm)	135	125	117	111	106	100	95	91	191	147	43.9	739
Formula, Enfamil—1 fl oz (31 gm)	4	4	3	3	3	2	2	2	21	11	9.2	2
Formula, Enfamil with Iron—1 fl oz (31 gm)	4	4	3	3	3	2	2	2	21	11	9.2	2
Formula, Gerber—1 fl oz (30 gm)	2	2	2	2	1	1	1	1	20	9	2.3	7
Formula, Isomil—1 fl oz (31 gm)	2	1	1	1	1	1	1	1	20	10	3.7	0
Formula, Isomil SF—1 fl oz (31 gm)	2	1	1	1	1	1	1	1	20	10	3.7	0
Formula, I-Soyalac—1 fl oz (31 gm)	2	1	1	1	1	1	1	1	20	10	3.7	0
Formula, Lofenalac—1 fl oz (31 gm)	½	½	½	½	0	0	0	0	21	7	.9	0
Formula, Mulumil—1 fl oz (31 gm)	3	3	2	2	2	2	2	1	20	8	6.5	2
Formula, Nursoy—1 fl oz (31 gm)	2	2	1	1	1	1	1	1	20	10	3.7	0
Formula, Nutramigen—1 fl oz (31 gm)	½	½	½	½	0	0	0	0	21	7	.9	0
Formula, PM 60/40—1 fl oz (31 gm)	2	1	1	1	1	1	1	1	20	10	3.6	0
Formula, Portagen—1 fl oz (31 gm)	3	3	2	2	2	2	2	1	21	9	7.4	0
Formula, Pregestimil—1 fl oz (31 gm)	2	2	1	1	1	1	1	1	21	7	3.1	0
Formula, Prosobee—1 fl oz (31 gm)	2	2	2	2	1	1	1	1	21	10	5.3	0
Formula, RSF—1 fl oz (31 gm)	2	2	1	1	1	1	1	1	21	10	4.1	0

Food											
Formula, Similac—1 fl oz (31 gm)	2	2	2	1	1	1	1	20	10	4.1	2
Formula, Similac Advance—1 fl oz (31 gm)	1	1	1	1	1	½	½	17	8	1.2	2
Formula, Similac with Iron—1 fl oz (31 gm)	2	2	2	1	1	1	1	20	10	4.1	2
Formula, Similac with Whey—1 fl oz (31 gm)	2	2	1	1	1	1	1	20	10	4.1	0
Formula, Similac with Whey and Iron—1 fl oz (31 gm)	2	2	2	1	1	1	1	20	10	4.1	2
Formula, SMA with Iron—1 fl oz (31 gm)	2	2	2	1	1	1	1	20	10	4.1	2
Formula, Soyalac—1 fl oz (31 gm)	2	1	1	1	1	1	1	20	10	3.7	0
Fruit, apple & blueberry, junior—1 jar (220 gm)	—	—	—	—	—	—	—	137	4	—	—
Fruit, apple & blueberry, strained—1 jar (135 gm)	—	—	—	—	—	—	—	82	3	—	—
Fruit, apple & raspberry w/sugar, junior—1 jar (220 gm)	½	0	0	0	0	0	0	127	4	.6	0
Fruit, apple & raspberry w/sugar, strained—1 jar (135 gm)	0	0	0	0	0	0	0	79	2	.4	0
Fruit, apples & cranberries w/tapioca—1 tbsp (15 gm)	0	0	0	0	0	0	0	9	0	.1	0
Fruit, apples & pears, junior—½ jar (7.5 oz–7.75 oz) (108 gm)	0	0	0	0	0	0	0	43	0	0	0
Fruit, apples & pears, strained—½ jar (4.5 oz–4.75 oz) (66 gm)	0	½	½	0	0	0	0	27	1	.1	0
Fruit, applesauce & apricots, junior—1 jar (220 gm)	½	0	0	0	0	0	0	104	5	.8	0
Fruit, applesauce & apricots, strained—1 jar (135 gm)	0	0	0	0	0	0	0	60	3	.4	0
Fruit, applesauce & cherries, junior—1 jar (220 gm)	—	—	—	—	—	—	—	106	0	—	—
Fruit, applesauce & cherries, strained—1 jar (135 gm)	—	—	—	—	—	—	—	65	0	—	—

	1200	1500	1800	2100	2400	2800	3200	3600	TOT CAL	FAT CAL	S/FAT CAL	CHOL MG
Baby Foods *(cont.)*												
Fruit, applesauce, junior—1 jar (213 gm)	0	0	0	0	0	0	0	0	79	4	.5	0
Fruit, applesauce & pineapple, junior—1 jar (213 gm)	0	0	0	0	0	0	0	0	83	2	.4	0
Fruit, applesauce & pineapple, strained—1 jar (128 gm)	—	—	—	—	—	—	—	—	48	1	—	—
Fruit, applesauce, strained—1 jar (128 gm)	0	0	0	0	0	0	0	0	53	2	.4	0
Fruit, applesauce w/other fruits (inc cherries)—½ jar (4.5 oz—4.75 oz) (66 gm)	0	0	0	0	0	0	0	0	32	0	0	0
Fruit, apricot w/tapioca, junior—1 jar (220 gm)	—	—	—	—	—	—	—	—	139	0	—	—
Fruit, apricot w/tapioca, strained—1 jar (135 gm)	—	—	—	—	—	—	—	—	80	0	—	—
Fruit, bananas & pineapple w/tapioca, junior—1 jar (135 gm)	0	0	0	0	0	0	0	0	91	1	.4	0
Fruit, bananas & pineapple w/tapioca, strained—1 jar (220 gm)	½	½	½	½	½	0	0	0	143	2	.8	0
Fruit, bananas w/tapioca, junior—1 jar (220 gm)	1	1	1	½	½	½	½	½	147	4	1.6	0
Fruit, bananas w/tapioca, strained—1 jar (135 gm)	0	0	0	0	0	0	0	0	77	1	.4	0
Fruit, guava & papaya w/tapioca, strained—1 jar (128 gm)	—	—	—	—	—	—	—	—	80	1	—	—
Fruit, guava w/tapioca, strained—1 jar (128 gm)	—	—	—	—	—	—	—	—	86	0	—	—

Food										
Fruit, mango w/tapioca, strained—1 jar (135 gm)	—	—	—	—	—	—	109	3	—	—
Fruit, papaya & applesauce w/tapioca, strained—1 jar (128 gm)	—	—	—	—	—	—	89	1	—	—
Fruit, peaches w/sugar, junior—1 jar (220 gm)	0	0	0	0	0	0	157	4	.4	0
Fruit, peaches w/sugar, strained—1 jar (135 gm)	0	0	0	0	0	0	96	2	.3	0
Fruit, pear juice—1 can or bottle (4.2 fl oz) (130 gm)	0	0	0	0	0	0	59	0	0	0
Fruit, pears, junior—1 jar (213 gm)	0	0	0	0	0	0	93	2	.2	0
Fruit, pears & pineapple, junior—1 jar (213 gm)	½	0	0	0	0	0	93	4	.8	0
Fruit, pears & pineapple, strained—1 jar (128 gm)	½	0	0	0	0	½	52	1	.4	0
Fruit, pears, strained—1 jar (128 gm)	0	0	0	0	0	0	53	2	.1	0
Fruit, pineapple juice—1 can or bottle (4.2 fl oz) (130 gm)	0	0	0	0	0	0	73	1	.1	0
Fruit, plums w/tapioca, wo/ascorbic acid, junior—1 jar (220 gm)	0	0	0	0	0	0	163	0	—	—
Fruit, plums w/tapioca, wo/ascorbic acid, strained—1 jar (135 gm)	½	0	0	0	0	0	96	0	—	—
Fruit, prunes w/tapioca, wo/ascorbic acid, junior—1 jar (220 gm)	0	0	0	0	0	0	155	2	.2	0
Fruit, prunes w/tapioca, wo/ascorbic acid, strained—1 jar (135 gm)	0	0	0	0	0	0	94	1	.1	0
Juice, apple—1 jar (130 gm)	0	0	0	0	0	0	61	1	.3	0
Juice, apple & cherry—1 jar (130 gm)	0	0	0	0	0	0	53	3	.4	0
Juice, apple & grape—1 jar (130 gm)	0	0	0	0	0	0	60	2	.1	0
Juice, apple & peach—1 jar (130 gm)	0	0	0	0	0	0	55	1	.1	0
Juice, apple & plum—1 jar (130 gm)	0	0	0	0	0	0	63	0	.3	0
Juice, apple & prune—1 jar (130 gm)	0	0	0	0	0	0	94	2	.1	0
Juice, mixed fruit—1 jar (130 gm)	0	0	0	0	0	0	61	1	.1	0
Juice, orange—1 jar (130 gm)	0	0	0	0	0	0	58	3	.4	0

	1200	1500	1800	2100	2400	2800	3200	3600	TOT CAL	FAT CAL	S/FAT CAL	CHOL MG
Baby Foods (cont.)												
Juice, orange & apple—1 jar (130 gm)	0	0	0	0	0	0	0	0	56	3	.1	0
Juice, orange, apple & banana—1 jar (130 gm)	0	0	0	0	0	0	0	0	61	1	.1	0
Juice, orange & apricot—1 jar (130 gm)	0	0	0	0	0	0	0	0	60	1	.1	0
Juice, orange & banana—1 jar (130 gm)	—	0	0	0	0	0	0	0	65	1	—	—
Juice, orange & pineapple—1 jar (130 gm)	0	0	0	0	0	0	0	0	63	1	.1	0
Juice, prune & orange—1 jar (130 gm)	0	0	0	0	0	0	0	0	91	4	.2	0
Meat, beef & beef heart, strained—½ jar (3.5 oz) (50 gm)	14	12	11	11	10	9	9	8	47	20	9.4	60
Meat, beef, junior—1 jar (99 gm)	16	14	13	12	11	10	9	9	105	44	23.1	40
Meat, beef, strained—1 jar (99 gm)	16	14	13	12	11	10	9	9	106	48	23.0	40
Meat, beef w/beef heart, strained—1 jar (99 gm)	—	—	—	—	—	—	—	—	93	40	18.5	—
Meat, chicken, junior—1 jar (99 gm)	19	17	15	14	13	12	11	11	148	86	22.0	60
Meat, chicken sticks, junior—1 jar (71 gm)	20	18	16	15	14	12	11	11	134	92	29.9	46
Meat, chicken, strained—1 jar (99 gm)	17	15	14	13	12	11	10	10	128	70	18.1	56
Meat, ham, junior—1 jar (99 gm)	17	15	13	12	11	11	10	9	123	59	19.9	50
Meat, ham, junior—½ jar (3.5 oz) (50 gm)	8	7	7	6	6	5	5	5	63	30	10.1	25
Meat, ham, strained—1 jar (99 gm)	15	13	12	11	11	10	9	9	110	51	17.3	48
Meat, ham, strained—½ jar (3.5 oz) (50 gm)	8	7	6	6	5	5	5	4	56	26	8.7	24
Meat, lamb, junior—1 jar (99 gm)	18	16	14	13	12	11	10	10	111	47	22.8	50
Meat, lamb, strained—1 jar (99 gm)	17	15	14	13	12	11	10	9	102	42	20.6	50
Meat, liver beef, strained (inc liver & bacon)—½ jar (3.5 oz) (50 gm)	17	16	15	14	13	13	12	11	51	17	6.2	92

Food												
Meat, liver, strained—1 jar (99 gm)	182	12.2	33	100	23	24	25	26	28	29	31	34
Meat, meat sticks, junior—1 jar (71 gm)	413	37.2	94	130	53	56	59	62	66	70	74	81
Meat, meat sticks or frankfurters—2 sticks (20 gm)	14	10.5	26	37	4	4	4	4	5	5	6	7
Meat, pork, junior—½ jar (3.5 oz) (50 gm)	24	10.8	32	62	5	5	5	6	6	7	8	9
Meat, pork, strained—1 jar (99 gm)	48	21.3	64	123	9	10	11	12	12	14	15	17
Meat, turkey, junior—1 jar (99 gm)	49	20.6	63	128	9	10	11	12	13	13	15	17
Meat, turkey sticks, junior—1 jar (71 gm)	46	30.0	91	129	11	11	12	14	15	16	18	20
Meat, turkey, strained—1 jar (99 gm)	46	17.0	52	113	8	9	9	10	11	12	13	15
Meat, veal, junior—1 jar (99 gm)	40	21.3	44	109	8	9	10	11	11	12	14	16
Meat, veal, strained—1 jar (99 gm)	40	20.4	42	100	8	9	9	10	11	12	13	15
Pretzels—1 pretzel (6 gm)	—	—	1	24	—	—	—	—	—	—	—	—
Teething biscuits—1 biscuit (11 gm)	0	.2	5	43	0	0	0	0	0	0	0	0
Vegetables, beets, junior—½ jar (7.5 oz—7.75 oz) (108 gm)	0		1	37	0	0	0	0	0	0	0	0
Vegetables, beets, strained—1 jar (128 gm)	0	.3	1	43	0	0	0	0	0	0	0	0
Vegetables, carrots, buttered, junior—1 jar (213 gm)	—	—	11	70	—	—	—	—	—	—	—	—
Vegetables, carrots, buttered, strained—1 jar (128 gm)	—	—	7	46	—	—	—	—	—	—	—	—
Vegetables, carrots, junior—1 jar (213 gm)	0	.5	4	67	0	0	0	0	0	0	0	0
Vegetables, carrots, strained—1 jar (128 gm)	0	.3	2	34	0	0	0	0	0	0	0	0
Vegetables, corn, creamed, junior—1 jar (213 gm)	2	1.9	7	138	1	1	1	1	1	1	1	1
Vegetables, corn, creamed, strained—1 jar (128 gm)	1	1.2	5	73	½	½	½	½	½	½	½	½
Vegetables, garden vegetable, strained—1 jar (128 gm)	—	—	3	48	—	—	—	—	—	—	—	—

THE 2-IN-1 SYSTEM

Baby Foods (cont.)

	1200	1500	1800	2100	2400	2800	3200	3600	TOT CAL	FAT CAL	S/FAT CAL	CHOL MG
Vegetables, green beans, buttered, junior—1 jar (206 gm)	—	—	—	—	—	—	—	—	67	16	—	—
Vegetables, green beans, buttered, strained—1 jar (128 gm)	—	—	—	—	—	—	—	—	42	9	—	—
Vegetables, green beans, creamed, junior—1 jar (213 gm)	—	—	—	—	—	—	—	—	68	8	—	—
Vegetables, green beans, junior—1 jar (206 gm)	0	0	0	0	0	0	0	0	51	3	.4	0
Vegetables, green beans, junior—1 jar (128 gm)	0	0	0	0	0	0	0	0	32	1	.3	0
Vegetables, mixed vegetables, junior—1 jar (213 gm)	1	1	1	1	½	½	½	½	88	7	1.9	0
Vegetables, mixed vegetables, strained 1 jar (128 gm)	½	½	½	0	0	0	0	0	52	5	.8	0
Vegetables, peas, buttered, junior—1 jar (206 gm)	—	—	—	—	—	—	—	—	123	23	—	—
Vegetables, peas, buttered, strained—1 jar (128 gm)	—	—	—	—	—	—	—	—	72	13	—	—
Vegetables, peas, creamed, junior—½ jar (7.5 oz–7.75 oz) (108 gm)	5	4	4	3	3	3	3	2	57	18	10.2	4
Vegetables, peas, creamed, strained—1 jar (128 gm)	6	5	4	4	4	3	3	3	68	22	11.5	5
Vegetables, peas, junior—½ jar (7.5 oz–7.75 oz) (108 gm)	½	0	0	0	0	0	0	0	43	3	.6	0
Vegetables, peas, strained—1 jar (128 gm)	—	—	—	—	—	—	—	—	52	4	—	—
Vegetables, spinach, creamed, junior—1 jar (213 gm)	9	8	7	6	6	5	5	4	90	27	17.3	11

Food											
Vegetables, spinach, creamed, strained —1 jar (128 gm)	5	4	3	3	3	2	2	48	15	8.1	6
Vegetables, squash, buttered, junior—1 jar (213 gm)	—	—	—	—	—	—	—	63	12	—	—
Vegetables, squash, buttered, strained—1 jar (128 gm)	—	½	½	—	—	—	—	37	4	—	—
Vegetables, squash, junior—1 jar (213 gm)	½	0	0	0	0	0	0	51	4	.8	0
Vegetables, squash, strained—1 jar (128 gm)	0	0	0	0	0	0	0	30	2	.4	0
Vegetables, sweet potatoes, buttered, junior—1 jar (220 gm)	—	—	—	—	—	—	—	126	14	—	—
Vegetables, sweet potatoes, buttered, strained—1 jar (135 gm)	—	—	—	—	—	—	—	76	9	—	—
Vegetables, sweet potatoes, junior—1 jar (220 gm)	½	0	0	0	0	0	0	133	3	.6	0
Vegetables, sweet potatoes, strained—1 jar (135 gm)	0	0	0	0	0	0	0	77	2	.4	0

BEVERAGES (NONALCOHOLIC)
Coffee and Tea

Food											
Coffee, acid neutralized, fr powdered instant—1 coffee cup (6 fl oz) (180 gm)	0	0	0	0	0	0	0	1	0	0	0
Coffee & chicory, made fr ground—1 coffee cup (6 fl oz) (180 gm)	0	0	0	0	0	0	0	2	0	0	0
Coffee & chicory, made fr powdered instant (inc Luzianne)—1 tsp dry yields 6 fl oz (180 gm)	0	0	0	0	0	0	0	1	0	0	0
Coffee & cocoa (mocha), made fr powdered mix, w/whitener, presweetened—4 tsp dry yields 6 fl oz (180 gm)	6	5	4	4	3	3	2	46	15	13.0	0

Coffee and Tea (cont.)

	1200	1500	1800	2100	2400	2800	3200	3600	TOT CAL	FAT CAL	S/FAT CAL	CHOL MG
Coffee, decaffeinated & chicory, made fr powdered instant—1 tsp dry yields 6 fl oz (180 gm)	0	0	0	0	0	0	0	0	1	0	0	0
Coffee, decaffeinated, made fr ground —1 coffee cup (6 fl oz) (180 gm)	0	0	0	0	0	0	0	0	2	0	0	0
Coffee, decaffeinated, made fr powdered instant—1 coffee cup (6 fl oz) (180 gm)	0	0	0	0	0	0	0	0	1	0	0	0
Coffee, decaffeinated, w/cereal (inc w/barley)—1 coffee cup (6 fl oz) (180 gm)	0	0	0	0	0	0	0	0	1	0	0	0
Coffee, espresso, decaffeinated—1 espresso cup (2 fl oz) (60 gm)	0	0	0	0	0	0	0	0	1	0	0	0
Coffee, espresso (inc demitasse)—1 espresso cup (2 fl oz) (60 gm)	0	0	0	0	0	0	0	0	1	0	0	0
Coffee, made fr ground, equal parts reg & decaffeinated—1 coffee cup (6 fl oz) (180 gm)	0	0	0	0	0	0	0	0	2	0	0	0
Coffee, made fr ground, reg, flavored—1 coffee cup (6 fl oz) (180 gm)	0	0	0	0	0	0	0	0	2	0	0	0
Coffee, made fr ground, reg—1 coffee cup (6 fl oz) (180 gm)	0	0	0	0	0	0	0	0	2	0	0	0
Coffee, made fr liquid concentrate—1 coffee cup (6 fl oz) (180 gm)	0	0	0	0	0	0	0	0	0	0	0	0
Coffee, made fr powdered instant—1 coffee cup (6 fl oz) (180 gm)	0	0	0	0	0	0	0	0	1	0	0	0
Coffee, made fr powdered mix, presweetened, no whitener—2½ tsp dry yields 6 fl oz (180 gm)	0	0	0	0	0	0	0	0	27	0	0	0

Food												
Coffee, made fr powdered mix w/ whitener and lo-cal sweetener, instant (inc Sugar-Free Café au Lait, French Style Coffee and Orange Cappuccino)—1 coffee cup (6 fl oz) (180 gm)	6	5	5	4	4	3	3	3	29	18	14.1	0
Coffee, made fr powdered mix, w/ whitener & sugar, instant (inc Café au Lait, French Style Coffee, Orange Cappuccino, Amaretto, and Viennese)—4 tsp dry yields 6 fl oz (180 gm)	9	7	6	6	5	5	4	4	56	23	19.9	0
Coffee, prelightened, no sugar (inc fr vending machine)—1 vending machine cup (180 gm)	6	5	4	4	3	3	3	2	26	14	13.1	0
Coffee, presweetened with sugar, prelightened (inc coffee, light, w/sugar, fr vending machine)—1 vending machine cup (180 gm)	6	5	4	4	3	3	3	2	50	14	12.7	0
Coffee, reg, w/cereal (inc barley)—1 coffee cup (6 fl oz) (180 gm)	0	0	0	0	0	0	0	0	1	0	0	0
Coffee, Turkish (inc Mexican coffee)—1 Turkish cup (4 fl oz) (120 gm)	0	0	0	0	0	0	0	0	45	0	0	0
Instant, w/sugar, cappuccino flavor, prep w/water—6 fl oz water & 2 rounded tsp (192 gm)	7	6	5	5	4	4	3	3	62	19	16.5	0
Instant, w/sugar, French flavor, prep w/water—6 fl oz water & 2 rounded tsp (189 gm)	12	10	9	8	7	6	5	5	57	31	26.5	0
Instant, w/sugar, mocha flavor, prep w/water—6 fl oz water & 2 rounded tsp (188 gm)	6	5	5	4	4	3	3	3	51	17	14.5	0
Postum—1 coffee cup (6 fl oz) (180 gm)	0	0	0	0	0	0	0	0	11	0	0	0
Tea, chamomile—1 teacup (6 fl oz)	0	0	0	0	0	0	0	0	2	0	0	0
Tea, herb—1 teacup (6 fl oz) (180 gm)	0	0	0	0	0	0	0	0	2	0	0	0

	1200	1500	1800	2100	2400	2800	3200	3600	TOT CAL	FAT CAL	S/FAT CAL	CHOL MG
Coffee and Tea *(cont.)*												
Tea, leaf, decaffeinated (inc lemon-flavored, caffeine-reduced)—1 teacup (6 fl oz) (180 gm)	0	0	0	0	0	0	0	0	2	0	0	0
Tea, leaf, decaffeinated, presweetened w/sugar (inc fruit-flavored)—1 teacup (6 fl oz) (184 gm)	0	0	0	0	0	0	0	0	97	0	0	0
Tea, leaf (inc tea bags, Japanese green tea, Chinese tea bags, mint tea bags, green tea bags, Ceylon breakfast tea, spiced and flavored teas)—1 teacup (6 fl oz) (180 gm)	0	0	0	0	0	0	0	0	2	0	0	0
Tea, leaf, presweetened with lo-cal sweetener (inc lemon-flavored)—1 teacup (6 fl oz) (184 gm)	0	0	0	0	0	0	0	0	5	0	0	0
Tea, leaf, presweetened w/sugar (inc lemon-flavored)—6 fl oz (180 gm)	0	0	0	0	0	0	0	0	95	0	0	0
Tea, made fr caraway seeds—1 teacup (6 fl oz) (180 gm)	0	0	0	0	0	0	0	0	2	0	0	0
Tea, made fr frozen concentrate, decaffeinated, presweetened w/lo-cal sweetener (inc lemon-flavored)—1 teacup (6 fl oz) (184 gm)	0	0	0	0	0	0	0	0	5	0	0	0
Tea, made fr frozen concentrate (inc lemon-flavored)—1 teacup (6 fl oz) (180 gm)	0	0	0	0	0	0	0	0	2	0	0	0
Tea, made fr plain powdered instant—1 teacup (6 fl oz) (180 gm)	0	0	0	0	0	0	0	0	2	0	0	0

Food											
Tea, made fr powdered instant, decaffeinated (inc lemon-flavored)—1 teacup (6 fl oz) (180 gm)	0	0	0	4	0	0	0	0	0	0	0
Tea, made fr powdered instant, decaffeinated, presweetened w/lo-cal sweetener (inc lemon or other fruit-flavored, Lipton Fruit Tea)—1 teacup (6 fl oz) (184 gm)	0	0	0	4	0	0	0	0	0	0	0
Tea, made fr powdered instant, lemon-flavored—1 teacup (6 fl oz) (180 gm)	0	0	0	5	0	0	0	0	0	0	0
Tea, made fr powdered instant, lemon-flavored, presweetened w/sugar—1 teacup (6 fl oz) (184 gm)	0	.1	0	63	0	0	0	0	0	0	0
Tea, made fr powdered instant, presweetened w/lo-cal sweetener—1 teacup (6 fl oz) (184 gm)	0	0	0	4	0	0	0	0	0	0	0
Tea, made fr powdered instant, presweetened w/lo-cal sweetener, lemon-flavored—1 teacup (6 fl oz) (184 gm)	0	0	0	4	0	0	0	0	0	0	0
Tea, made fr powdered instant, presweetened w/sugar—1 teacup (6 fl oz) (184 gm)	0	0	0	19	0	0	0	0	0	0	0
Tea, made fr powdered instant, presweetened with sugar, decaffeinated—1 teacup (6 fl oz) (184 gm)	0	0	0	19	0	0	0	0	0	0	0
Tea, Russian—1 teacup (6 fl oz) (184 gm)	0	0	0	111	0	0	0	0	0	0	0
Tea, spiced, presweetened, made fr dry mix—1 teacup (6 fl oz) (184 gm)	0	.1	0	62	0	0	0	0	0	0	0

Fruit Drinks

	1200	1500	1800	2100	2400	2800	3200	3600	TOT CAL	FAT CAL	S/FAT CAL	CHOL MG
Apple-cherry drink (inc apple-grape, apple-raspberry, apple-pineapple)—1 c (8 fl oz) (250 gm)	0	0	0	0	0	0	0	0	116	0	0	0
Apple cider-flavored drink, made fr powdered mix, lo-cal, w/vitamin C added—1 c (8 fl oz) (240 gm)	0	0	0	0	0	0	0	0	2	0	0	0
Apple cider-flavored drink, made fr powdered mix, w/sugar & vitamin C added—1 cup (8 fl oz) (250 gm)	0	0	0	0	0	0	0	0	66	0	0	0
Apple drink—1 c (8 fl oz) (250 gm)	0	0	0	0	0	0	0	0	118	0	0	0
Apple drink w/vitamin C added—1 c (8 fl oz) (250 gm)	0	0	0	0	0	0	0	0	117	0	0	0
Apricot-pineapple juice drink—1 c (8 fl oz) (250 gm)	0	0	0	0	0	0	0	0	128	1	.1	0
Banana-orange drink—1 c (8 fl oz) (250 gm)	0	0	0	0	0	0	0	0	126	0	.1	0
Black cherry drink—1 c (8 fl oz) (250 gm)	0	0	0	0	0	0	0	0	117	0	0	0
Black cherry drink w/vitamin C added—1 c (8 fl oz) (250 gm)	0	0	0	0	0	0	0	0	117	0	0	0
Cherry drink w/vitamin C added—1 c (8 fl oz) (250 gm)	0	0	0	0	0	0	0	0	117	0	0	0
Citrus drink w/vitamin C added—1 c (8 fl oz) (250 gm)	0	0	0	0	0	0	0	0	126	0	0	0
Citrus fruit juice drink (60% fruit juice) (inc 5 Alive Citrus)—1 c (8 fl oz) (247 gm)	0	0	0	0	0	0	0	0	113	1	.1	0

Food						Cal						
Cranberry-apple juice drink, bottled—6 fl oz glass (184 gm)	0	0	0	0	0	123	0	0	0	0	0	0
Cranberry-apple juice drink, lo-cal, w/vitamin C added—1 c (8 fl oz) (240 gm)	0	0	0	0	0	46	0	0	0	0	0	0
Cranberry-apple juice drink w/vitamin C added (inc cocktail, raspberry-cranberry juice drink, cranberry-apricot juice drink)—1 c (8 fl oz) (253 gm)	0	0	0	0	0	170	0	0	0	0	0	0
Cranberry-apricot juice drink, bottled—6 fl oz glass (184 gm)	0	0	0	0	0	118	0	0	0	0	0	0
Cranberry-grape juice drink, bottled—6 fl oz glass (184 gm)	0	0	0	0	0	103	0	0	2	.5	0	0
Cranberry juice cocktail, bottled, lo-cal, w/saccharin & corn sweetener—6 fl oz glass (178 gm)	0	0	0	0	0	33	0	0	0	0	0	0
Cranberry juice cocktail, prep w/water fr frozen—6 fl oz glass (187 gm)	0	0	0	0	0	102	0	0	0	0	0	0
Cranberry juice drink, lo-cal w/vitamin C added (inc lo-cal cranberry juice cocktail)—1 c (8 fl oz) (240 gm)	0	0	0	0	0	46	0	0	0	0	0	0
Cranberry juice drink w/vitamin C added (inc cocktail)—1 c (8 fl oz) (253 gm)	0	0	0	0	0	147	0	0	0	0	0	0
Fluid replacement, electrolyte solution (inc Pedialyte)—1 c (8 fl oz) (240 gm)	0	0	0	0	0	24	0	0	0	0	0	0
Fruit drink (inc fruit punches & fruit ades, Hawaiian Punch made fr canned or frozen)—1 c (8 fl oz) (248 gm)	0	0	0	0	0	112	0	0	0	0	0	0
Fruit drink, lo-cal—1 c (8 fl oz) (240 gm)	0	0	0	0	0	43	0	0	0	0	0	0
Fruit drinks, fruit punches, & fruit ades, lo-cal, w/vitamin C added (inc lo-cal Hi-C, Kool-Aid, Wyler's)—1 c (8 fl oz) (240 gm)	0	0	0	0	0	43	0	0	0	0	0	0

Fruit Drinks (cont.)

	1200	1500	1800	2100	2400	2800	3200	3600	TOT CAL	FAT CAL	S/FAT CAL	CHOL MG
Fruit-flavored beverage, low sugar (inc Gatorade, Quick Kick)—1 c (8 fl oz) (240 gm)	0	0	0	0	0	0	0	0	55	0	0	0
Fruit-flavored drink, lo-cal, calcium fortified (inc Supri drink mix)—1 c (8 fl oz) (240 gm)	0	0	0	0	0	0	0	0	3	0	0	0
Fruit-flavored drink, made fr powdered mix, lo-cal, w/vitamin C added (inc Kool-Aid, Wyler's)—1 c (8 fl oz) (240 gm)	0	0	0	0	0	0	0	0	3	0	0	0
Fruit-flavored drink, made fr powdered mix w/high vitamin C added, lo-cal (inc Sugar-Free Tang)—1 c (8 fl oz) (240 gm)	0	0	0	0	0	0	0	0	6	0	0	0
Fruit-flavored drink, made fr powdered mix, w/sugar & vitamin C added (inc Kool-Aid, Wyler's)—1 c (8 fl oz) (250 gm)	0	0	0	0	0	0	0	0	89	0	.2	0
Fruit-flavored drink, made fr powdered mix w/vitamin C added, no sugar or lo-cal sweetener (inc Kool-Aid, Wyler's)—1 c (8 fl oz) (240 gm)	0	0	0	0	0	0	0	0	85	0	.2	0
Fruit-flavored drink, noncarbonated, made from lo-cal powdered mix (inc Sugar-Free Crystal Light)—1 c (8 fl oz) (240 gm)	0	0	0	0	0	0	0	0	3	0	0	0
Fruit-flavored drink, noncarbonated, made fr powdered mix, w/sugar (inc Flavor-Aid)—1 c (8 fl oz) (240 gm)	0	0	0	0	0	0	0	0	85	0	.2	0

Food										
Fruit-flavored drink, made fr powdered mix, mainly sugar, w/high vitamin C added (inc Keen, Tang Instant Breakfast Juice Drink)—1 tbsp dry yields 8 fl oz (240 gm)	0	0	0	110	0	0	0	0	0	0
Fruit punch drink, canned—6 fl oz glass (186 gm)	0	0	0	87	0	0	0	0	0	0
Fruit punch drink, prep w/water fr frozen—3/4 c (185 gm)	0	0	0	85	0	0	0	0	0	0
Fruit punch flavor drink, powder, w/added sodium, prep w/water—1 c water & 2 rounded tsp (262 gm)	0	.2	1	97	0	0	0	0	0	0
Fruit punch flavor drink, powder, wo/added sodium, prep w/water—1 c water & 2 rounded tsp (262 gm)	0	.2	1	97	0	0	0	0	0	0
Fruit punch, fruit drinks, or fruit ades, w/vitamin C added (inc Hi-C) 1 c (8 fl oz) (247 gm)	0	0	0	111	0	0	0	0	0	0
Fruit punch juice drink, prep w/water fr frozen—3/4 c (186 gm)	0	.4	3	92	0	0	0	0	0	0
Fruit punch, made w/fruit juice & soda—1 c (8 fl oz) (245 gm)	0	0	1	107	0	0	0	0	0	0
Fruit punch, made w/soda, fruit juice & sherbet or ice cream—1 c (8 fl oz) (256 gm)	4	5.9	10	163	2	2	2	2	3	3
Grapeade & grape drink—1 c (8 fl oz) (250 gm)	0	0	0	113	0	0	0	0	0	0
Grape drink, lo-cal—1 c (8 fl oz) (240 gm)	0	0	0	43	0	0	0	0	0	0
Grape drink w/vitamin C added—1 c (8 fl oz) (250 gm)	0	0	0	112	0	0	0	0	0	0
Grapefruit juice drink—1 c (8 fl oz) (250 gm)	0	.1	1	128	0	0	0	0	0	0
Grapefruit juice drink w/vitamin C added—1 c (8 fl oz) (250 gm)	0	.1	1	128	0	0	0	0	0	0

111

	1200	1500	1800	2100	2400	2800	3200	3600	TOT CAL	FAT CAL	S/FAT CAL	CHOL MG
Fruit Drinks (cont.)												
Grape juice drink—1 c (8 fl oz) (250 gm)	0	0	0	0	0	0	0	0	135	0	0	0
Grape juice drink, canned—6 fl oz glass (188 gm)	0	0	0	0	0	0	0	0	94	0	0	0
Guava drink—1 c (8 fl oz) (250 gm)	½	0	0	0	0	0	0	0	130	2	.6	0
Guava juice drink w/vitamin C added (inc Ocean Spray Mauna La'i)—1 c (8 fl oz) (253 gm)	½	0	0	0	0	0	0	0	132	2	.6	0
Lemonade flavor drink, powder, prep w/water—1 c water & 2 tbsp pwd (266 gm)	0	0	0	0	0	0	0	0	113	1	.4	0
Lemonade-flavored drink, made fr powdered mix, lo-cal, w/vitamin C added (inc Sugar-Free Country Time)—1 c (8 fl oz) (240 gm)	0	0	0	0	0	0	0	0	5	0	0	0
Lemonade-flavored drink, made fr powdered mix, w/sugar & vitamin C added (inc Country Time)—1 c (8 fl oz) (250 gm)	0	0	0	0	0	0	0	0	92	0	.1	0
Lemon-limeade—1 c (8 fl oz) (248 gm)	0	0	0	0	0	0	0	0	105	0	0	0
Limeade—1 c (8 fl oz) (248 gm)	0	0	0	0	0	0	0	0	103	0	0	0
Orange-apricot juice drink—1 c (8 fl oz) (249 gm)	0	0	0	0	0	0	0	0	124	2	.1	0
Orange breakfast drink, made fr frozen concentrate (inc Orange Plus, Awake)—1 c (8 fl oz) (250 gm)	0	0	0	0	0	0	0	0	116	2	.4	0
Orange drink & orangeade w/vitamin C added—1 c (8 fl oz) (249 gm)	0	0	0	0	0	0	0	0	124	0	0	0

Orange drink, breakfast type, w/ juice & pulp, prep w/water from frozen—6 fl oz glass (188 gm)	0	0	0	84	0	0	0	0	0	0
Orange drink, canned—6 fl oz glass (186 gm)	0	0	0	94	0	0	0	0	0	0
Orange drink (inc orangeade, Yabba Dabba Dew, Sunny Delight)—1 c (8 fl oz) (249 gm)	0	0	0	124	0	0	0	0	0	0
Orange flavor drink, breakfast type, prep w/water—6 fl oz (186 gm)	0	.2	1	86	0	0	0	0	0	0
Orange flavor drink, breakfast type, w/pulp, prep with water fr frozen—6 fl oz glass (186 gm)	0	.4	3	91	0	0	0	0	0	0
Orange-lemon drink—1 c (8 fl oz) (249 gm)	0	0	0	124	0	0	0	0	0	0
Papaya juice drink—1 c (8 fl oz) (248 gm)	0	1.0	3	141	0	0	0	½	½	½
Pineapple & grapefruit juice drink, canned—3/4 c (188 gm)	0	.1	1	88	0	0	0	0	0	0
Pineapple-grapefruit juice drink—1 c (8 fl oz) (250 gm)	0	.2	2	118	0	0	0	0	0	0
Pineapple-grapefruit juice drink w/vitamin C added—1 c (8 fl oz) (250 gm)	0	.2	2	118	0	0	0	0	0	0
Pineapple-orange juice drink—1 c (8 fl oz) (250 gm)	0	0	0	125	0	0	0	0	0	0
Pineapple-orange juice drink w/vitamin C added—1 c (8 fl oz) (250 gm)	0	0	0	125	0	0	0	0	0	0
Strawberry-flavored drink—1 c (8 fl oz) (250 gm)	0	0	0	125	0	0	0	0	0	0
Strawberry-flavored drink w/vitamin C added—1 c (8 fl oz) (250 gm)	0	.2	0	89	0	0	0	0	0	0

Miscellaneous Drinks

	1200	1500	1800	2100	2400	2800	3200	3600	TOT CAL	FAT CAL	S/FAT CAL	CHOL MG
Cereal beverage (inc Pero, Break Away)—1 coffee cup (6 fl oz) (180 gm)	0	0	0	0	0	0	0	0	11	1	.2	0
Cereal beverage w/beet roots, fr powdered instant (inc Kafix)—1 coffee cup (6 fl oz) (180 gm)	0	0	0	0	0	0	0	0	11	0	0	0
Chicory—1 coffee cup (6 fl oz) (180 gm)	0	0	0	0	0	0	0	0	1	0	0	0
Clam & tomato juice, canned—5.5 fl oz can (166 gm)	0	0	0	0	0	0	0	0	77	1	.1	0
Cocoa, dry powder, high-medium fat, processed w/alkali—1 tbsp (5 gm)	2	2	2	1	1	1	1	1	14	9	5.1	0
Cocoa, dry powder, low-medium fat, processed w/alkali—1 tbsp (5 gm)	2	1	1	1	1	1	1	1	12	6	3.4	0
Cocoa mix, aspartame sweetened— 6 fl oz water & 0.53 oz pkt (192 gm)	1	1	1	1	1	1	1	1	48	4	2.4	1
Cocoa mix, w/added nutrients, prep w/water—6 fl oz water & 1 pkt (209 gm)	7	6	5	5	4	4	3	3	120	27	16.2	0
Eggnog flavor mix, powder—2 heaping tsp (28 gm)	1	1	1	1	1	1	½	½	111	3	.7	3
Eggnog flavor mix, prep w/milk fr powder—1 c milk & 2 heaping tsp (272 gm)	26	22	20	18	16	15	13	12	260	76	46.3	33
Thirst-quencher drink, bottled—1 c (241 gm)	0	0	0	0	0	0	0	0	60	1	0	0
Wine, light, nonalcoholic—1 fl oz (31 gm)	0	0	0	0	0	0	0	0	2	0	0	0

Food									
Wine, nonalcoholic—1 fl oz (29 gm)	0	0	0	0	0	2	0	0	0
Soft Drinks									
Ale type (inc Ale-8)—1 can (12 fl oz) (360 gm)	0	0	0	0	0	32	0	0	0
Carbonated, malt (inc Malta India, Martita)—1 can (12 fl oz) (360 gm)	0	0	0	0	0	32	0	0	0
Carbonated water, sweetened (inc tonic, quinine water)—1 can (12 fl oz) (366 gm)	0	0	0	0	0	124	0	0	0
Carbonated water, unsweetened (inc flavored, club soda, Perrier, seltzer water)—1 can (12 fl oz) (355 gm)	0	0	0	0	0	0	0	0	0
Chocolate-flavored soda—1 can (12 fl oz) (369 gm)	0	0	0	0	0	155	0	0	0
Chocolate-flavored soda, sugar-free (inc Canfield's Diet Chocolate Fudge Soda)—1 can (12 fl oz) (355 gm)	1	1	1	½	½	7	6	2.2	0
Cola type—1 can (12 fl oz) (369 gm)	0	0	0	0	0	151	0	0	0
Cola type, decaffeinated—1 can (12 fl oz) (369 gm)	0	0	0	0	0	151	0	0	0
Cola type, sugar-free—1 can (12 fl oz) (355 gm)	1	1	1	½	½	7	6	2.2	0
Cola type, sugar free, decaffeinated—1 can (12 fl oz) (355 gm)	1	1	1	½	½	7	6	2.2	0
Cola w/chocolate flavor—1 can (12 fl oz) (369 gm)	0	0	0	0	0	151	0	0	0
Cola w/chocolate flavor, sugar-free (inc caffeine-free)—1 can (12 fl oz) (355 gm)	1	1	1	½	½	7	6	2.2	0
Cola w/fruit or vanilla flavor—1 can (12 fl oz) (369 gm)	0	0	0	0	0	151	0	0	0

THE 2-IN-1 SYSTEM

	1200	1500	1800	2100	2400	2800	3200	3600	TOT CAL	FAT CAL	S/FAT CAL	CHOL MG
Soft Drinks (cont.)												
Cola w/fruit or vanilla flavor, sweetened w/lo-cal sweetener—1 can (12 fl oz) (355 gm)	1	1	1	1	1	½	½	½	7	6	2.2	0
Cream soda (inc Almond Smash)—1 can (12 fl oz) (371 gm)	0	0	0	0	0	0	0	0	189	0	0	0
Cream soda, sugar-free—1 can (12 fl oz) (355 gm)	1	1	1	1	1	½	½	½	7	6	2.2	0
Fruit-flavored, caffeine-containing (inc Mellow Yellow, Mountain Dew, Big Red)—1 can (12 fl oz) (372 gm)	0	0	0	0	0	0	0	0	156	0	0	0
Fruit-flavored, no caffeine added (inc orange, lemon, lime, cherry, grape, strawberry, Tom Collins mixer, 7-Up, Sprite)—1 can (12 fl oz) (372 gm)	0	0	0	0	0	0	0	0	156	0	0	0
Fruit-flavored, sugar-free—1 can (12 fl oz) (355 gm)	1	1	1	1	1	½	½	½	7	6	2.2	0
Fruit-flavored, w/fruit juice (inc Slice, Mandarin Orange Slice, Apple Slice, Cherry Cola Slice)—1 can (12 fl oz) (372 gm)	0	0	0	0	0	0	0	0	156	0	0	0
Fruit-flavored, w/fruit juice, sweetened w/lo-cal sweetener (inc Diet Slice, Diet Mandarin Orange Slice, Diet Apple Slice)—1 can (12 fl oz) (355 gm)	1	1	1	1	1	½	½	½	7	6	2.2	0
Ginger ale—1 can (12 fl oz) (366 gm)	0	0	0	0	0	0	0	0	113	0	0	0

								Cal			
Ginger ale, sugar-free—1 can (12 fl oz) (355 gm)	1	1	1	1	1	½	1	7	6	2.2	0
Grape soda—1 can (12 fl oz) (372 gm)	0	0	0	0	0	0	0	161	0	0	0
Lemon-lime soda—1 can (12 fl oz) (368 gm)	0	0	0	0	0	0	0	149	0	0	0
Mavi drink—1 can (12 fl oz) (369 gm)	0	0	0	0	0	0	0	140	0	0	0
Orange—1 can (12 fl oz) (372 gm)	0	0	0	0	0	0	0	177	0	0	0
Pepper type, decaffeinated—1 can (12 fl oz) (369 gm)	0	0	0	0	0	0	0	151	0	0	0
Pepper type (inc Dr. Pepper)—1 can (12 fl oz) (369 gm)	0	0	0	0	0	0	0	151	0	0	0
Pepper type, sugar-free—1 can (12 fl oz) (355 gm)	1	1	1	1	1	½	1	7	6	2.2	0
Pepper type, sugar-free, decaffeinated—1 can (12 fl oz) (355 gm)	1	1	1	1	1	½	1	7	6	2.2	0
Root beer—1 can (12 fl oz) (369 gm)	0	0	0	0	0	0	0	151	0	0	0
Root beer, sugar-free—1 can (12 fl oz) (355 gm)	1	1	1	1	1	½	1	7	6	2.2	0
Tonic water, sugar-free—1 can (12 fl oz) (355 gm)	1	1	1	1	1	½	1	7	6	2.2	0

BREAD, ROLLS, etc.
Bread

								Cal			
Barley—1 sl (26 gm)	1	1	1	1	1	1	1	69	8	2.5	1
Batter—1 reg sl (33 gm)	6	5	4	4	4	3	3	94	21	6.7	17
Black—1 reg sl (26 gm)	0	0	0	0	0	0	0	64	3	.4	0
Black, toasted—1 reg sl (24 gm)	0	0	0	0	0	0	0	65	3	.4	0
Boston brown—1 sl (45 gm)	1	1	1	½	½	½	½	90	4	1.3	1
Boston brown bread, w/white corn meal—1 sl, 3¼" dia x ½" (45 gm)	0	0	0	0	0	0	0	95	5	0	0
Boston brown bread, w/yellow corn meal—1 sl, 3¼" dia x ½" (45 gm)	0	0	0	0	0	0	0	95	5	0	0

Bread (cont.)

	1200	1500	1800	2100	2400	2800	3200	3600	TOT CAL	FAT CAL	S/FAT CAL	CHOL MG
Bran (inc Granola, Branola, Honey Bran)—1 sl (36 gm)	6	5	5	4	4	4	3	3	110	26	7.3	17
Bran, toasted (inc Granola, Branola, Honey Bran)—1 sl (33 gm)	6	5	5	4	4	4	3	3	112	27	7.4	17
Bran, w/raisins—1 slice (36 gm)	5	5	4	4	4	3	3	3	110	24	6.5	15
Bran, w/raisins, toasted—1 sl (33 gm)	5	5	4	4	4	3	3	3	111	24	6.5	15
Bread crumbs, dry, grated—1 c (100 gm)	5	4	4	3	3	3	2	2	392	41	9.4	5
Bread sticks (inc sesame sticks)—1 med stick (10 gm)	1	1	1	1	1	1	½	½	43	12	2.3	0
Bread stuffing (inc homemade stuffing)—1 c moist (200 gm)	21	18	15	14	12	11	10	9	599	234	46.9	1
Bread-stuffing mix, cooked, w/water & table fat, dry & crumbly—1 c (140 gm)	78	67	59	53	48	44	40	37	501	275	143.6	90
Bread-stuffing mix, cooked, w/water, table fat & egg, moist—1 c (200 gm)	73	63	57	51	47	43	39	37	416	230	117.9	132
Bread-stuffing mix, dry—1 c coarse crumbs (70 gm)	3	2	2	2	2	2	1	1	260	24	5.5	3
Buckwheat—1 med sl (27 gm)	1	1	1	1	1	1	1	1	72	9	2.6	1
Buckwheat, toasted—1 med sl (25 gm)	1	1	1	1	1	1	1	1	74	9	2.7	1
Cheese (inc onion cheese)—1 sl (26 gm)	2	2	2	1	1	1	1	1	72	11	4.2	2
Cheese, toasted (inc onion cheese)—1 sl (24 gm)	2	2	1	1	1	1	1	1	76	11	4.1	1

Food												
Cinnamon—1 sl (26 gm)	1	1	1	1	½	½	½	½	70	7	1.8	0
Cinnamon, toasted—1 sl (24 gm)	1	1	1	1	½	½	½	½	75	8	1.9	0
Corn & molasses (inc Anadama bread)—1 sl (32 gm)	3	2	2	2	2	2	1	1	85	17	5.6	2
Corn & molasses, toasted (inc Anadama bread)—1 sl (29 gm)	3	2	2	2	2	2	1	1	86	17	5.7	2
Corn bread, home recipe, degermed yellow meal, w/lard—1 cu in (8 gm)	2	1	1	1	1	1	1	1	18	5	1.5	6
Corn bread, home recipe, degermed yellow meal, w/vegetable shortening—1 cu in (8 gm)	1	1	1	1	1	1	1	1	18	5	1.2	6
Corn bread, home recipe, whole ground yellow meal, w/lard—1 cu in (8 gm)	2	2	2	2	1	1	1	1	17	5	1.7	7
Corn bread, home recipe, whole ground yellow meal, w/vegetable shortening—1 cu in (8 gm)	2	1	1	1	1	1	1	1	17	5	1.4	6
Corn bread (inc jalapeño corn bread)—1 piece (2½" x 2½" x 1½") (65 gm)	16	14	12	11	11	10	9	9	161	51	19.2	45
Corn bread, made from home recipe—1 piece (2½" x 2½" x 1½") (65 gm)	14	12	11	10	9	9	8	8	197	63	18.9	36
Corn bread, spoon bread, home recipe, white ground cornmeal & al, w/lard—1 c (240 gm)	90	80	73	68	63	59	55	52	468	247	94.4	307
Corn bread, spoon bread, home recipe, white ground cornmeal, w/vegetable shortening—1 c (240 gm)	80	72	66	61	57	53	50	47	468	247	77.9	293
Corn bread stuffing—1 c, dry type (141 gm)	28	25	23	22	20	19	18	17	317	85	24.3	110
—1 c, moist type (203 gm)	40	36	33	31	29	27	26	24	456	123	35.0	158
Corn flour patties or tarts, fried—2 patties (20 gm)	½	½	0	0	0	0	0	0	46	5	.7	0
Corn pone, fried—1 piece (61 gm)	6	5	4	4	3	3	3	2	157	59	13.4	0
Corn pone, home recipe, white ground meal, w/lard—1 sector (60 gm)	4	3	3	3	2	2	2	2	122	29	8.2	2

	THE 2-IN-1 SYSTEM											
	1200	1500	1800	2100	2400	2800	3200	3600	TOT CAL	FAT CAL	S/FAT CAL	CHOL MG
Bread *(cont.)*												
Corn pone, home recipe, white ground meal, w/vegetable shortening—1 sector (60 gm)	3	2	2	2	1	1	1	1	122	29	5.8	0
Corn pone (inc hoecake)—1 piece (⅛ of 9″ dia x 3/4″ pone) (61 gm)	3	2	2	2	1	1	1	1	133	29	5.8	0
Cottonseed, toasted—1 sl (41 gm)	4	3	3	3	2	2	2	2	130	23	7.7	3
Cracked wheat (inc Honey Wheat, Roman Meal, Wheatberry, crushed wheat)—1 thin sl (20 gm)	½	½	½	½	0	0	0	0	53	4	.9	0
———1 reg sl (26 gm)	½	½	½	½	½	0	0	0	68	5	1.1	0
Cracked wheat, toasted (inc Honey Wheat, Roman Meal, Wheatberry)—1 thin slice (18 gm)	½	½	½	0	0	0	0	0	56	4	.8	0
———1 reg sl (24 gm)	½	½	½	½	½	0	0	0	75	6	1.1	0
Croutons (inc seasoned bread crumbs)—1 c (30 gm)	1	1	1	1	1	1	½	½	94	10	2.4	0
Cuban (inc Spanish, Portuguese)—1 med sl (20 gm)	1	½	½	½	½	½	0	0	58	5	1.2	0
Cuban, toasted (inc Spanish, Portuguese)—1 med sl (18 gm)	1	½	½	½	½	½	0	0	61	6	1.2	0
Dough, fried (inc Indian fried)—1 sl or roll (26 gm)	5	5	4	4	3	3	3	2	97	44	12.0	1
Egg chalah—1 sl (23 gm)	6	6	5	5	5	4	4	4	75	23	4.3	28
Egg chalah, toasted—1 sl (21 gm)	6	6	5	5	5	5	4	4	76	24	4.3	29
Flat bread, toasted (inc Greek, Syrian flat bread, pita, Sahara, Arab, pocket)—1 small pita (6½″ dia) (55 gm)	½	½	½	0	0	0	0	0	172	5	.8	0

———1 large pita (9″ dia) (103 gm)	1	1	½	½	½	½	½	½	322	9	1.4	0
Flat (inc Greek, Syrian flat bread, pita, Sahara, Arab, pocket)—1 small pita (6½″ dia) (60 gm)	½	½	0	0	0	0	0	0	166	5	.7	0
———1 large pita (112 gm)	1	1	½	½	½	½	½	½	309	8	1.4	0
Flat, whole wheat (inc roti)—1 small pita (6½″ dia) (60 gm)	4	3	3	2	2	2	2	1	190	55	8.1	0
———1 large pita (9″ dia) (112 gm)	7	6	5	4	4	3	3	3	355	103	15.2	0
Flat, whole wheat, toasted (incl pita, roti)—1 small pita (6½″ dia) (55 gm)	4	3	3	3	2	2	2	2	215	62	9.2	0
———1 large pita (9″ dia) (102 gm)	8	6	6	5	4	4	3	3	399	116	17.1	0
French or Vienna (inc Hawaiian sandwich bread)—1 med slice (4¾″ x 4″ x ½″) (25 gm)	1	1	½	½	½	½	½	½	73	7	1.5	0
French or Vienna, toasted (inc Hawaiian sandwich bread)—1 med sl (23 gm)	1	1	1	½	½	½	½	½	78	7	1.6	0
Fruit and nut (inc date nut, banana nut)—1 sl (56 gm)	14	13	11	11	10	9	8	8	217	91	19.5	37
Fruit, wo/nuts (inc lemon bread, banana bread)—1 sl (41 gm)	11	9	8	8	7	7	6	6	150	54	13.7	29
Garlic—1 med sl (4¾″ x 4″ x ½″) (29 gm)	3	3	2	2	2	2	1	1	100	34	6.8	0
Garlic, toasted (inc Texas toast)—1 med sl (26 gm)	3	3	2	2	2	2	1	1	99	33	6.7	0
Gluten—1 sl (26 gm)	2	2	1	1	1	1	1	1	69	9	2.2	5
Gluten, toasted—1 sl (24 gm)	2	2	2	1	1	1	1	1	72	9	2.3	6
High protein—1 sl (26 gm)	1	1	1	1	½	½	½	½	72	8	1.8	0
High protein, toasted—1 sl (24 gm)	1	1	1	1	½	½	½	½	73	8	1.8	0
Hush puppies (inc fried corn bread)—1 hush puppy (22 gm)	7	6	6	5	5	5	4	4	73	34	9.1	21
Irish soda—1 sl (30 gm)	2	2	2	2	2	2	1	1	85	13	3.1	7

	THE 2-IN-1 SYSTEM											
	1200	1500	1800	2100	2400	2800	3200	3600	TOT CAL	FAT CAL	S/FAT CAL	CHOL MG
Bread *(cont.)*												
Italian, Grecian, Armenian (inc sesame bread)—1 med sl (20 gm)	0	0	0	0	0	0	0	0	55	1	.2	0
Italian, Grecian, Armenian, toasted (inc sesame bread)—1 med slice (18 gm)	0	0	0	0	0	0	0	0	55	1	.2	0
Low-fat, 98% fat-free (inc Bonnie) 1—1 sl (24 gm)	0	0	0	0	0	0	0	0	66	2	.3	0
Low-fat, 98% fat-free, toasted (inc Bonnie)—1 sl (22 gm)	0	0	0	0	0	0	0	0	67	2	.3	0
Low-sodium—1 reg sl (26 gm)	1	1	1	1	½	½	½	½	70	7	1.8	0
Low-sodium, toasted—1 reg sl (24 gm)	1	1	1	1	½	½	½	½	71	8	1.8	0
Milk & honey (inc Arnold's)—1 sl (28 gm)	1	1	1	1	1	½	½	½	74	6	1.6	1
Milk & honey, toasted (inc Arnold's)—1 sl (25 gm)	1	1	1	1	1	½	½	½	72	5	1.6	1
Multigrain—1 reg sl (26 gm)	1	½	½	½	½	½	0	0	63	7	1.2	0
Multigrain, reduced-calorie, high-fiber—1 reg sl (26 gm)	1	1	1	1	1	½	½	½	53	5	.5	3
Multigrain, reduced-calorie, high-fiber, toasted—1 reg sl (24 gm)	1	1	1	1	1	½	½	½	53	5	.5	3
Multigrain, toasted—1 reg sl (24 gm)	1	½	½	½	½	½	½	0	68	8	1.3	0
Multigrain, w/raisins—1 reg sl (26 gm)	1	½	½	½	½	½	½	0	65	6	1.3	0
Multigrain, w/raisins, toasted—1 reg sl (24 gm)	1	½	½	½	½	½	½	0	66	6	1.3	0

Nut—1 sl (49 gm)	12	11	10	9	9	8	7	7	169	71	15.0	37
Oatmeal—1 sl (25 gm)	1	1	1	1	1	1	1	1	66	10	2.8	1
Oatmeal, toasted—1 sl (23 gm)	1	1	1	1	1	1	1	1	67	10	2.9	1
Onion—1 sl (26 gm)	1	1	1	½	½	½	½	½	66	7	1.7	0
Potato—1 reg sl (26 gm)	1	1	1	1	½	½	½	½	70	7	1.8	0
Potato, toasted—1 reg sl (24 gm)	1	1	1	1	½	½	½	½	71	8	1.8	0
Pumpernickel (inc w/raisins)—1 reg sl (26 gm)	0	0	0	0	0	0	0	0	64	3	.4	0
Pumpernickel, toasted (inc w/raisins) —1 reg sl (24 gm)	0	0	0	0	0	0	0	0	65	3	.4	0
Pumpkin (inc w/raisins and/or nuts)—1 sl (60 gm)	8	8	7	7	6	6	5	5	179	39	7.0	34
Raisin (inc w/ or wo/nuts, cinnamon-raisin, Weight Watchers) —1 reg sl (26 gm)	1	1	½	½	½	½	½	½	68	7	1.5	0
Raisin, toasted (inc w/ or wo/ nuts, cinnamon-raisin)—1 reg sl (24 gm)	1	1	1	½	½	½	½	½	76	7	1.7	0
Reduced-calorie, high-fiber, white or other kind (inc Fresh Horizons, New World, Less, Roman Light, Lite Loaf, Sunbeam Lite, Lite 'N Up)—1 thin sl (23 gm)	0	0	0	0	0	0	0	0	63	2	.3	0
Reduced-calorie, high-fiber, white or other kind toasted (inc Fresh Horizons, New World, Less, Roman Light, Lite Loaf, Sunbeam Lite, Lite 'N Up—1 thin slice (21 gm)	0	0	0	0	0	0	0	0	63	2	.3	0
Rusk (hard-crisp bread or toast)—1 rusk (9 gm)	1	1	1	1	1	1	1	½	38	7	2.1	1
Rye, American, 1/3 rye & 2/3 clear flour—1 reg sl/lb loaf (25 gm)	0	0	0	0	0	0	0	0	61	3	.4	0

	1200	1500	1800	2100	2400	2800	3200	3600	TOT CAL	FAT CAL	S/FAT CAL	CHOL MG
Bread *(cont.)*												
Rye, American, 1/3 rye & 2/3 clear flour, toasted—1 reg sl/lb loaf (22 gm)	0	0	0	0	0	0	0	0	62	3	.4	0
Rye (inc corn rye)—1 thin sl (20 gm)	0	0	0	0	0	0	0	0	49	2	.3	0
—1 reg sl (26 gm)	0	0	0	0	0	0	0	0	63	3	.3	0
Rye, pumpernickel—1 reg sl (32 gm)	0	0	0	0	0	0	0	0	79	4	.5	0
Rye, reduced-calorie, high-fiber (inc **Less**)—1 thin sl (20 gm)	0	0	0	0	0	0	0	0	35	3	.4	0
—1 reg sl (26 gm)	0	0	0	0	0	0	0	0	45	4	.5	0
Rye, reduced-calorie, high-fiber, toasted (inc **Less**)—1 thin slice (18 gm)	0	0	0	0	0	0	0	0	34	3	.4	0
—1 reg sl (24 gm)	0	0	0	0	0	0	0	0	46	4	.5	0
Rye, toasted—1 thin sl (18 gm)	0	0	0	0	0	0	0	0	51	2	.2	0
—1 reg sl (24 gm)	0	0	0	0	0	0	0	0	68	3	.3	0
Salt-rising—1 sl (24 gm)	1	½	½	½	½	½	0	0	64	5	1.2	0
Salt-rising, toasted—1 sl (22 gm)	1	½	½	½	½	½	0	0	65	5	1.2	0
Sourdough—1 med sl (4¾" x 4" x ½") (25 gm)	1	1	½	½	½	½	½	½	73	7	1.5	0
Sourdough, toasted—1 med sl (23 gm)	1	1	1	½	½	½	½	½	78	7	1.6	0
Soy—1 sl (26 gm)	2	1	1	1	1	1	1	1	69	9	3.4	1
Soy, toasted—1 sl (24 gm)	2	1	1	1	1	1	1	1	71	10	3.4	1
Spanish coffee (inc Majorca)—1 piece (85 gm)	11	10	9	9	8	7	7	7	293	68	10.5	42
Spoon bread—½ c (94 gm)	29	26	24	22	21	19	18	17	166	70	26.6	110

Sprouted wheat—1 sl (26 gm)	1	½	½	½	½	½	½	0	63	6	1.3	0
Sprouted wheat, toasted—1 sl (24 gm)	1	½	½	½	½	½	½	0	64	6	1.3	0
Sunflower meal—1 med sl (27 gm)	2	2	1	1	1	1	1	1	75	12	4.0	1
Sunflower meal, toasted—1 med sl (25 gm)	2	2	1	1	1	1	1	1	77	13	4.1	1
Sweet potato—1 sl (25 gm)	4	4	3	3	3	3	3	2	73	14	3.1	17
Triticale—1 sl (25 gm)	½	½	0	0	0	0	0	0	61	5	.7	0
Triticale, toasted—1 slice (23 gm)	½	½	½	0	0	0	0	0	61	5	.8	0
Wheat germ—1 sl (28 gm)	1	1	1	1	1	1	1	1	78	13	3.2	0
Wheat germ, toasted—1 sl (25 gm)	1	1	1	1	1	1	1	1	77	12	3.1	0
White and whole wheat (inc half and half, Health Nut, Golden Meal, Country Grain)—1 thin sl (20 gm)	1	½	½	½	½	½	0	0	51	5	1.2	0
———1 reg sl (26 gm)	1	1	½	½	½	½	½	½	66	7	1.5	0
White & whole wheat, toasted (inc half and half, Health Nut, Golden Meal, Country Grain)—1 reg sl (24 gm)	1	1	1	½	½	½	½	½	72	7	1.7	0
White, diet (inc lo-cal diet bread, Arnold Diet, Hollywood Light, Profile, Weight Watchers, Contour)—1 sl (24 gm)	0	0	0	0	0	0	0	0	66	2	.3	0
White (inc diet sliced or very thin sliced, Weight Watchers, Pepperidge Farm, Hollywood Light, Arnold Diet Sliced, Dutch Hearth)—1 very thin sl (15 gm)	½	½	½	½	½	0	0	0	41	4	1.0	0
———1 thin sl (20 gm)	1	1	½	½	½	½	½	½	54	6	1.4	0
———1 reg sl (26 gm)	1	1	1	1	½	½	½	½	70	7	1.8	0
White, made fr home recipe or purchased at a bakery—1 reg sl (32 gm)	3	2	2	2	2	1	1	1	93	16	5.3	2
White, made fr home recipe or purchased at a bakery, toasted—1 reg sl (29 gm)	3	2	2	2	2	1	1	1	94	16	5.4	2

	THE 2-IN-1 SYSTEM											
	1200	1500	1800	2100	2400	2800	3200	3600	TOT CAL	FAT CAL	S/FAT CAL	CHOL MG
Bread *(cont.)*												
White, toasted (inc diet sliced or very thin sliced, Weight Watchers, Pepperidge Farm, Hollywood Light, Arnold Diet Sliced, Dutch Hearth—1 very thin sl (13 gm)	½	½	½	½	½	0	0	0	41	4	1.0	0
——1 thin sl (17 gm)	1	½	½	½	½	½	½	0	53	6	1.3	0
——1 reg sl (24 gm)	1	1	1	1	½	½	½	½	75	8	1.9	0
Whole wheat, high-fiber—1 sl (28 gm)	1	½	½	½	½	½	½	0	68	8	1.3	0
Whole wheat, high-fiber, toasted (inc Less)—1 sl (25 gm)	1	1	½	½	½	½	½	½	72	8	1.4	0
Whole wheat (inc dark bread, brown bread, Hollywood Dark, Honey Graham)—1 thin sl (23 gm)	½	½	½	½	½	0	0	0	56	6	1.1	0
——1 reg sl (26 gm)	1	½	½	½	½	½	0	0	63	7	1.2	0
Whole wheat, made fr home recipe or purchased at a bakery—1 thin sl (27 gm)	2	1	1	1	1	1	1	1	74	15	3.9	0
——1 reg sl (30 gm)	2	2	2	1	1	1	1	1	82	17	4.3	1
Whole wheat, made fr home recipe or purchased at a bakery, toasted—1 thin sl (25 gm)	2	1	1	1	1	1	1	1	76	15	4.0	0
——1 reg sl (27 gm)	2	2	2	1	1	1	1	1	82	17	4.3	1
Whole wheat, made w/water, soft crumb—1 reg sl (28 gm)	1	1	1	1	½	½	½	½	67	6	1.3	1
Whole wheat, reduced-calorie, high-fiber (inc Roman Light, Less)—1 reg sl (28 gm)	0	0	0	0	0	0	0	0	53	3	.5	0

Whole wheat, reduced-calorie, high-fiber, toasted (inc Roman Light, Less)—1 reg sl (25 gm)	0	0	0	0	0	0	0	0	53	3	.4	0
Whole wheat, toasted (inc dark bread, brown bread, Hollywood Dark, Honey Graham)—1 thin sl (21 gm)	1	½	½	½	½	½	0	0	61	7	1.2	0
———1 reg sl (24 gm)	1	½	½	½	½	½	½	0	69	8	1.3	0
Whole wheat, toasted, made w/water, soft crumb—1 reg sl (24 gm)	1	1	1	1	½	½	½	½	69	6	1.4	1
Whole wheat, w/nuts—1 sl (49 gm)	12	11	10	9	9	8	7	7	160	65	14.7	38
Whole wheat, w/raisins—1 sl (26 gm)	½	½	½	½	½	0	0	0	65	6	1.1	0
Whole wheat, w/raisins, toasted—1 sl (24 gm)	1	½	½	½	½	½	0	0	70	7	1.2	0
Zucchini (inc squash bread, carrot bread, w/nuts)—1 sl (40 gm)	10	9	8	7	7	6	6	6	145	66	10.8	33

Rolls

Baked from mix, made w/water—1 roll, 2½″ dia (35 gm)	2	1	1	1	1	1	1	1	105	14	3.3	1
Cheese—1 roll (41 gm)	4	4	3	3	2	2	2	2	124	26	8.7	2
Cracked wheat—1 med roll (36 gm)	1	1	1	½	½	½	½	½	95	7	1.6	0
Diet—1 roll (28 gm)	1	1	1	½	½	½	½	½	81	8	1.6	0
Egg bread—1 roll (28 gm)	6	5	5	4	4	4	3	3	98	30	7.3	17
French or Vienna—1 med roll (45 gm)	1	1	1	1	1	1	1	½	131	12	2.6	0
French or Vienna, toasted—1 med roll (41 gm)	1	1	1	1	1	1	1	½	131	12	2.6	0
Garlic—1 med roll (35 gm)	2	2	1	1	1	1	1	1	104	18	4.3	0
Hoagie, submarine, or sandwich (inc steak roll, torpedo roll)—1 small (62 gm)	2	1	1	1	1	1	1	1	180	17	3.6	0
———1 med (100 gm)	3	2	2	2	1	1	1	1	290	27	5.8	0
———1 lg (inc hoagie, submarine) (135 gm)	4	3	3	2	2	2	2	1	392	36	7.9	0

Rolls (cont.)

	1200	1500	1800	2100	2400	2800	3200	3600	TOT CAL	FAT CAL	S/FAT CAL	CHOL MG
Hoagie, submarine, or sandwich, toasted—1 small (56 gm)	2	1	1	1	1	1	1	1	178	17	3.6	0
—1 med (91 gm)	3	2	2	2	2	1	1	1	290	27	5.8	0
—1 lg (inc hoagie, submarine) (123 gm)	4	3	3	2	2	2	2	1	392	36	7.9	0
Multigrain—1 pan or dinner (28 gm)	4	3	3	3	3	2	2	2	79	14	4.4	12
—1 hamburger or sandwich (43 gm)	6	5	5	4	4	4	4	3	122	21	6.8	19
Multigrain, toasted—1 pan or dinner (25 gm)	4	3	3	3	3	2	2	2	78	14	4.3	12
—1 hamburger or sandwich (39 gm)	6	5	5	4	4	4	4	3	121	21	6.8	19
Roll dough, frozen, unraised,—1 roll, 2⅜" x 2" (28 gm)	1	1	1	1	1	1	1	1	75	13	3.0	1
Rolls and buns, partially baked, brown & serve, unbrown—1 roll, 2½" dia (26 gm)	2	2	2	1	1	1	1	1	78	16	3.8	2
Rye—1 pan or dinner (28 gm)	0	0	0	0	0	0	0	0	68	3	.4	0
—1 med (36 gm)	0	0	0	0	0	0	0	0	87	4	.5	0
Sour dough—1 med (45 gm)	1	1	1	1	1	1	1	½	131	12	2.6	0
Sweet, cinnamon bun, frosted—1 med (approx 2"–2½" sq or 2"–2½" dia) (55 gm)	8	7	6	5	5	4	4	4	185	47	12.1	14
Sweet, cinnamon bun, no frosting—1 med (approx 2"–2½" sq or 2"–2½" dia) (55 gm)	8	7	7	6	5	5	5	4	174	45	12.4	18
Sweet (inc butterhorn, Portuguese sweet bread)—1 med (55 gm)	8	7	7	6	5	5	5	4	174	45	12.4	18

Food											
Sweet, toasted (inc butterhorn, Portuguese sweet bread)—1 med (50 gm)	18	12.4	45	174	4	5	5	5	6	7	8
Sweet, w/fruit and nuts, frosted—1 med (approx 2"–2½" sq or 2½" dia) (55 gm)	0	6.0	34	174	1	1	1	2	2	2	3
Sweet, w/fruit and nuts, no frosting—1 medium (approx 2"–2½" sq or 2"–2½" dia) (55 gm)	0	4.5	28	161	1	1	1	1	1	2	2
Sweet, with fruit, frosted (inc hot cross buns)—1 med (approx 2"–2½" sq or 2"–2½" dia) (55 gm)	0	5.1	23	168	1	1	1	1	1	2	2
Sweet, w/fruit, no frosting (inc w/jelly, sweet bread, Mexican sweet bread, apple rolls, kolaches, lakvar, kuchen)—1 med (approx 2"–2½" sq or 2"–2½" dia) (55 gm)	0	3.4	14	151	1	1	1	1	1	1	2
Sweet, w/nuts, frosted (inc caramel roll)—1 med (approx 2"–2½" sq or 2"–2½" dia) (55 gm)	14	12.1	53	185	4	4	4	5	5	6	8
Sweet, w/nuts, no frosting—1 med (approx 2"–2½" sq or 2"–2½" dia) (55 gm)	17	13.5	66	188	4	5	5	6	6	7	9
White, hard (inc kaiser)—1 small (25 gm)	0	1.6	7	78	½	½	½	½	½	1	1
—1 med (50 gm)		3.2	14	156	1	1	1	1	1	1	1
White, hard, toasted (inc kaiser)—1 med (46 gm)	0	3.2	15	158	1	1	1	1	1	1	1
White, soft (inc potato roll, onion roll, roll made fr mix, soft, seeded roll, brown 'n' serve, hamburger, frankfurter, white roll)—1 pan or dinner roll (28 gm)	0	3.5	14	83	1	1	1	1	1	1	2
—1 hamburger, frankfurter, onion roll (43 gm)	0	5.3	22	128	1	1	2	2	2	2	2

	1200	1500	1800	2100	2400	2800	3200	3600	TOT CAL	FAT CAL	S/FAT CAL	CHOL MG
Rolls *(cont.)*												
White, soft, made fr home recipe or purchased at a bakery—1 pan or dinner (25 gm)	4	3	3	3	2	2	2	2	84	20	5.6	8
—1 hamburger roll (38 gm)	6	5	4	4	4	3	3	3	128	30	8.6	12
White, soft, toasted—1 crescent (28 gm)	2	1	1	1	1	1	1	1	92	16	3.8	0
—1-foot long (86 gm)	5	4	4	3	3	3	2	2	282	48	11.6	0
Whole wheat—1 med (36 gm)	1	1	1	1	1	½	½	½	93	9	2.0	0
Whole wheat, toasted—1 med (33 gm)	1	1	1	1	1	½	½	½	93	9	2.0	0
Bagels												
Multigrain, w/raisins, toasted—1 med (3" dia) (50 gm)	½	½	0	0	0	0	0	0	138	5	.7	0
Pumpernickel (inc rye)—1 med (3" dia) (55 gm)	½	0	0	0	0	0	0	0	152	4	.6	0
Pumpernickel, toasted—1 med (3" dia) (50 gm)	½	0	0	0	0	0	0	0	157	4	.6	0
Raisin—1 med (3" dia) (55 gm)	3	3	3	3	2	2	2	2	156	7	1.5	16
Raisin, toasted—1 med (3" dia) (50 gm)	3	3	3	2	2	2	2	2	149	6	1.5	15
Water bagels, (inc flavored bialy)—1 med (3" dia) (55 gm)	4	3	3	3	3	3	3	2	164	8	1.7	19
Whole wheat—1 med (3" dia) (55 gm)	½	½	½	0	0	0	0	0	152	5	.8	0
Whole wheat, toasted—1 med (3" dia) (50 gm)	½	½	½	0	0	0	0	0	157	5	.8	0
Whole wheat, w/raisins—1 med (3" dia) (55 gm)	½	½	½	0	0	0	0	0	153	5	.8	0

Biscuits and Scones

Food											
Biscuit dough, fried—1 piece (43 gm)	7	6	5	4	4	4	3	3	153	62	15.6	0
Biscuit dough, raw—1 c (170 gm)	11	9	8	7	6	6	5	5	471	98	25.1	0
Biscuits, baking powder or buttermilk type, made fr home recipe—1 med (2″ dia) (30 gm)	5	4	4	3	3	3	2	2	112	47	11.8	0
Biscuits, baking powder or buttermilk type, made fr mix—1 med (2″ dia) (30 gm)	3	2	2	2	2	1	1	1	102	26	6.2	0
Biscuits, baking powder or buttermilk type, made fr refrigerated dough—1 reg biscuit (19 gm)	1	1	1	1	1	1	1	1	55	12	3.0	0
—1 crescent (28 gm)	2	2	1	1	1	1	1	1	82	17	4.3	0
Biscuits, cheese—1 biscuit (2″ dia) (30 gm)	10	8	7	7	6	5	5	4	115	56	20.2	6
Biscuits, cinnamon-raisin—1 biscuit (3″ dia) (64 gm)	8	7	6	5	5	4	4	3	224	73	18.1	0
Biscuits, fr mix, made w/milk—1 biscuit, (2″ dia (28 gm)	3	2	2	2	1	1	1	1	91	23	5.7	0
Biscuits, whole wheat—1 med (2″ dia) (30 gm)	4	4	3	3	3	2	2	2	94	33	9.4	2
Scones—1 scone (42 gm)	19	17	15	14	13	12	11	11	151	61	19.3	64
Scones, whole wheat—1 scone (42 gm)	18	16	15	14	13	12	11	11	145	61	19.2	63

Croissants

Food											
Brioche (inc Pan de Huevo)—1 piece (77 gm)	25	22	21	19	18	17	16	15	269	96	22.0	97
Cheese—1 Sara Lee (42 gm)	13	11	10	9	8	8	7	7	154	74	17.9	30
Chocolate—1 Vie de France (56 gm)	21	18	16	15	14	12	11	10	213	114	35.2	35
Fruit (inc Sara Lee Apple, Strawberry & Cinnamon-Nut-Raisin)—1 Sara Lee Strawberry (80 gm)	20	18	16	15	14	12	11	11	276	129	27.9	50

	1200	1500	1800	2100	2400	2800	3200	3600	TOT CAL	FAT CAL	S/FAT CAL	CHOL MG
Croissants (cont.)												
——1 Sara Lee Apple (92 gm)	23	20	18	17	16	14	13	12	317	148	32.0	58
——1 Sara Lee Cinnamon-Nut-Raisin (92 gm)	23	20	18	17	16	14	13	12	317	148	32.0	58
Nut—1 Vie de France (56 gm)	15	14	12	11	10	9	9	8	209	102	21.4	38
Sara Lee Wheat 'n' Honey—1 fresh (56 gm)	16	14	12	11	10	10	9	8	205	97	21.3	39
Muffins and Popovers												
Matzo, fritters—2 fritters (30 gm)	16	14	13	12	11	10	10	9	110	64	16.3	55
Muffins, bran (inc w/raisins & nuts)—1 muffin (2⅝" dia) (50 gm)	13	12	11	10	10	9	8	8	126	45	13.5	48
Muffins, corn, baked fr home recipe, degermed cornmeal & vegetable shortening—1 muffin (2⅜" dia) (40 gm)	8	7	6	6	6	5	5	4	126	36	11.0	21
Muffins, corn, baked fr home recipe, whole-grain cornmeal & lard—1 muffin (2⅜" dia) (40 gm)	10	9	8	7	7	6	6	5	115	37	13.7	24
Muffins, corn, baked fr home recipe, whole-grain cornmeal & vegetable-shortening—1 muffin (2⅜" dia) (40 gm)	8	7	7	6	6	5	5	4	115	37	10.9	22
Muffins, corn, baked fr home recipe, degermed cornmeal & lard—1 muffin (2⅜" dia) (40 gm)	10	9	8	7	7	6	6	5	126	36	14.2	23
Muffins, corn bread, sticks, rounds—1 muffin (52 gm)	14	13	12	11	10	9	9	8	165	47	15.2	49

Food												
Muffins, corn bread, sticks, rounds, made fr home recipe—1 muffin (52 gm)	11	10	9	8	8	7	6	6	158	50	15.1	29
Muffins, corn bread, sticks, rounds, toasted—1 muffin (47 gm)	14	13	12	11	10	9	9	8	166	47	15.3	49
Muffins, corn, made fr mix, w/egg & milk—1 muffin (40 gm)	8	7	7	6	6	5	5	5	130	38	11.0	23
Muffins, English, bran (inc Branola)— 1 muffin (58 gm)	1	1	1	1	1	1	1	½	127	12	2.7	0
Muffins, English, bran, toasted (inc Branola)—1 muffin (53 gm)	1	1	1	1	1	1	1	1	133	13	2.8	0
Muffins, English, bran, w/raisins—1 muffin (58 gm)	1	1	1	1	1	1	1	½	134	11	2.5	0
Muffins, English, bran, w/raisins, toasted—1 muffin (53 gm)	1	1	1	1	1	1	1	½	143	12	2.6	0
Muffins, English, cracked wheat (inc Roman Meal)—1 muffin (58 gm)	1	1	1	1	1	1	1	½	158	12	2.6	0
Muffins, English, cracked wheat, toasted (inc Roman Meal)—1 muffin (50 gm)	1	1	1	1	1	1	1	½	150	11	2.5	0
Muffins, English (inc sourdough)— 1 muffin (58 gm)	1	1	1	1	1	½	½	½	132	10	2.3	0
Muffins, English, toasted (inc sourdough)— 1 muffin (50 gm)	1	1	1	1	1	½	½	½	133	11	2.3	0
Muffins, English, whole wheat—1 muffin (58 gm)	1	1	1	1	1	1	1	1	130	14	2.8	0
Muffins, English, whole wheat, toasted— 1 muffin (50 gm)	1	1	1	1	1	1	1	½	126	14	2.7	0
Muffins, English, w/raisins—1 muffin (61 gm)	1	1	1	1	1	½	½	½	146	10	2.3	0
Muffins, English, w/raisins, toasted—1 muffin (53 gm)	1	1	1	1	1	½	½	½	148	10	2.3	0
Muffins, fruit and/or nuts (inc blueberry, cranberry-nut)—1 muffin (2⅝" dia) (58 gm)	14	12	11	11	10	9	9	8	165	46	14.1	49

	1200	1500	1800	2100	2400	2800	3200	3600	TOT CAL	FAT CAL	S/FAT CAL	CHOL MG
Muffins and Popovers (cont.)												
Muffins, multigrain, w/nuts (inc Orowheat Health Nut Muffins)—1 muffin (66 gm)	17	15	14	13	12	11	10	10	211	94	18.2	57
Muffins, oatmeal (inc Granola)—1 muffin (2⅝" dia) (47 gm)	8	7	6	6	5	5	5	4	112	29	9.2	23
Muffins, plain (inc matzo meal)—1 muffin (2⅝" dia) (47 gm)	8	7	7	6	6	5	5	5	139	41	10.5	24
Muffins, pumpkin, w/raisins—1 muffin (2⅝" dia) (58 gm)	8	8	7	7	6	6	5	5	181	39	7.1	34
Muffins, whole wheat (inc graham)—1 muffin (2⅝" dia) (47 gm)	11	9	9	8	7	7	6	6	141	53	15.2	26
Popovers, baked fr home recipe—1 popover (2¾" dia) (40 gm)	14	13	12	11	11	10	9	9	90	33	11.6	59
Popovers, baked fr home recipe—1 popover (2¾" dia) (40 gm)	14	13	12	11	11	10	9	9	90	33	11.6	59
Popovers (inc Dutch Baby, Yorkshire pudding)—1 popover (31 gm)	11	10	9	9	8	8	7	7	70	24	8.5	47
Flour, Baking Powder, and Yeast												
Baking powder, w/straight phosphate—1 tbsp (13 gm)	0	0	0	0	0	0	0	0	15	0	0	0
Baking powder, cream of tartar, w/tartaric acid—1 tbsp (10 gm)	0	0	0	0	0	0	0	0	7	0	0	0
Biscuit mix, dry—1 c (120 gm)	14	12	10	9	8	7	6	6	509	136	31.4	1
Buckwheat flour, dark—1 c sifted (98 gm)	0	0	0	0	0	0	0	0	326	22	0	0

134

Food										
Buckwheat flour, light—1 c sifted (98 gm)	0	0	0	0	0	0	340	11	0	0
Corn flour, white variety—1 c (117 gm)	1	1	1	1	1	1	431	27	3.1	0
Corn flour, yellow variety—1 c (117 gm)	1	1	1	1	1	1	431	27	3.1	0
Cornmeal, bolted, white, whole grain—1 cup (122 gm)	1	1	1	1	2	2	442	37	4.1	0
Cornmeal, bolted, yellow, whole grain—1 c (122 gm)	1	1	1	1	2	2	442	37	4.1	0
Cornmeal, self-rising, white, degermed, wo/wheat flour—1 c (141 gm)	0	0	0	0	0	0	491	14	0	0
Cornmeal, self-rising, white, whole grain, wo/wheat flour—1 c (134 gm)	1	1	1	1	2	2	465	39	4.2	0
Cornmeal, self-rising, yellow, degermed, wo/wheat flour—1 c (141 gm)	0	0	0	0	0	0	491	14	0	0
Cornmeal, self-rising, yellow, whole grain, wo/wheat flour—1 c (134 gm)	1	1	1	1	2	2	465	39	4.2	0
Cornmeal, white, cooked, degermed, wo/salt—1 c (240 gm)	0	0	0	0	0	0	120	5	0	0
Cornmeal, white, dry, degermed—1 c (138 gm)	0	0	0	0	0	0	502	15	0	0
Cornmeal, yellow, dry, degermed—1 c (138 gm)	0	0	0	0	0	0	502	15	0	0
Flour, white—1 c (125 gm)	½	½	½	1	1	1	455	11	1.8	0
Flour, whole wheat—1 c (120 gm)	1	1	1	1	1	1	400	22	3.0	0
Peanut flour, defatted—1 c (60 gm)	0	0	4	0	5	5	196	3	.4	0
Peanut flour, lo-fat—1 c (60 gm)	3	3	4	0	5	6	257	118	16.4	0
Rye, whole-grain flour, dark—1 c unsifted, spooned (128 gm)	0	0	0	0	0	0	419	30	0	0
Rye, whole-grain flour, light—1 c unsifted, spooned (102 gm)	0	0	0	0	0	0	364	9	0	0
Wheat flours, cake, patent, or pastry—1 c sifted, spooned (96 gm)	0	0	0	0	0	0	349	7	0	0

	1200	1500	1800	2100	2400	2800	3200	3600	TOT CAL	FAT CAL	S/FAT CAL	CHOL MG
Flour, Baking Powder, and Yeast (cont.)												
Wheat flours, patent, self-rising—1 c sifted, spooned (115 gm)	0	0	0	0	0	0	0	0	405	10	0	0
Yeast, baker's, compressed—0.6 oz pkg (18 gm)	0	0	0	0	0	0	0	0	15	1	0	0
Yeast, baker's, dry, active—¼ oz pkg (7 gm)	0	0	0	0	0	0	0	0	20	1	0	0
Yeast, brewer's, debittered—1 tbsp (8 gm)	0	0	0	0	0	0	0	0	23	1	0	0
Yeast extract spread (inc Vegemite, Marmite, Promite)—1 tsp (6 gm)	0	0	0	0	0	0	0	0	41	0	0	0
Yeast (inc brewer's yeast) dry—1 tbsp, dry (8 gm)	0	0	0	0	0	0	0	0	23	1	.2	0
CAKES and PIES **Cakes**												
Angel food, chocolate wo/icing—1 piece (½ of 10" dia) (57 gm)	1	½	½	½	½	½	½	0	146	3	1.3	0
Angel food, w/icing—1 piece (½ of 10" dia) (77 gm)	0	0	0	0	0	0	0	0	210	1	.1	0
Angel food, wo/icing—1 piece (½ of 10" dia) (57 gm)	0	0	0	0	0	0	0	0	147	1	.1	0
Applesauce, lo-cal, wo/icing (inc Weight Watchers Apple Raisin Spice Cake)—1 individual cake (74 gm)	9	8	8	7	7	6	6	6	174	36	8.0	36
Applesauce, w/icing—1 piece (½ of 10" dia) (108 gm)	17	15	13	12	11	10	9	8	399	120	29.0	26

Food												
Applesauce, wo/icing (inc w/ or wo/ nuts, rhubarb, blueberry, apricot, blackberry, apple crunch cake)—1 piece (1/12 of 10" dia) (87 gm)	16	14	12	11	10	9	9	8	314	103	25.9	29
Banana, w/icing—1 piece (1/12 of 10" dia) (108 gm)	13	12	11	10	9	8	8	7	310	71	15.5	40
Banana, wo/icing—1 piece (1/12 of 10" dia) (87 gm)	12	11	10	9	9	8	7	7	247	68	14.8	38
Batter, raw, chocolate—1 tbsp (14 gm)	4	3	3	3	2	2	2	2	43	16	5.5	8
Batter, raw, not chocolate—1 tbsp (15 gm)	3	3	2	2	2	2	2	2	44	13	3.2	9
Boston cream pie—1 piece (1/12 of 8" dia) (69 gm)	17	15	14	13	12	11	10	10	207	55	14.9	64
Butter, w/icing—1 piece (1/12 of 2-layer, 8" or 9" dia) (139 gm)	60	52	47	43	39	36	33	31	523	185	90.2	126
Butter, wo/icing—1 piece (1/10 of 8" or 9" dia) (53 gm)	26	23	21	19	18	16	15	14	194	81	35.4	67
Caramel, iced, butter, (icing made w/butter)—1 cu in (6 gm)	3	2	2	2	2	2	2	1	23	8	4.1	6
Caramel, uniced, butter—1 cu in (5 gm)	2	2	2	2	2	2	1	1	17	7	3.5	6
Carrot, lo-cal (inc Weight Watchers)—1 individual cake (85 gm)	11	9	8	8	7	6	6	5	179	59	19.0	16
Carrot, w/icing (inc carrot cupcakes w/icing)—1 piece (1/12 of 10" dia) (134 gm)	37	33	29	27	25	23	21	20	548	254	47.0	102
Carrot, wo/icing (inc zucchini cake, carrot pudding)—1 piece (1/12 of 10" dia) (111 gm)	30	26	24	22	21	19	18	17	478	253	34.3	92
Cheesecake, chocolate—1 piece (1/12 of 9" dia) (128 gm)	83	72	64	58	53	48	44	41	500	290	139.5	136
Cheesecake (inc cream cheese pie)—1 piece (1/12 of 9" dia) (128 gm)	56	49	43	39	36	33	30	28	407	224	90.9	101

	1200	1500	1800	2100	2400	2800	3200	3600	TOT CAL	FAT CAL	S/FAT CAL	CHOL MG
Cakes (cont.)												
Cheesecake, lo-cal (inc Weight Watchers)—1 individual cake (approx 3½" x 3½" x 1") (113 gm)	24	21	18	16	15	13	12	11	241	91	46.9	23
Cheesecake, lo-cal, with fruit (inc Weight Watchers)—1 individual cake (approx 3½" x 3½" x 1") (113 gm)	22	19	17	15	14	12	11	10	232	85	42.7	20
Cheesecake w/fruit—1 piece (1/12 of 9" dia) (142 gm)	43	38	34	31	28	25	23	22	385	174	70.6	78
Chiffon, chocolate, w/icing—1 piece (1/12 of 10" dia) (92 gm)	35	31	28	25	24	22	20	19	325	138	44.9	95
Chiffon, chocolate, wo/icing—1 piece (1/12 of 10" dia) (66 gm)	27	24	22	20	19	18	17	16	226	97	28.5	92
Chiffon, w/icing—1 piece (1/12 of 10" dia) (92 gm)	23	21	19	18	17	16	15	14	314	111	20.4	91
Chiffon, wo/icing—1 piece (1/12 of 10" dia) (66 gm)	24	22	20	19	18	17	16	15	254	97	17.4	106
Chocolate, devil's food or fudge, pudding type mix (oil, eggs, & water added to dry mix), wo/icing or filling (inc Pillsbury Plus)—1 piece (1/10 of 8" or 9" dia) (53 gm)	18	16	15	13	12	11	11	10	194	88	21.2	55
Chocolate, devil's food or fudge, pudding type mix (oil, eggs & water added to dry mix), w/icing, coating, or filling (inc Pillsbury Plus)—1 piece (1/12 of 2-layer, 8" or 9" dia) (139 gm)	35	31	28	26	24	22	20	19	524	196	46.7	92

Food											
Chocolate, devil's food or fudge, standard type mix (eggs & water added to dry mix), wo/icing or filling—1 piece (¹⁄₁₀ of 8″ or 9″ dia) (42 gm)	10	9	8	7	7	6	6	5	131	37	11.9	29
Chocolate, devil's food or fudge, standard type mix (eggs & water added to dry mix), w/icing, coating, or filling (inc Jiffy)—1 piece (¹⁄₁₂ of 2-layer, 8″ or 9″ dia) (109 gm)	19	17	15	14	13	12	11	10	378	101	27.9	43
Chocolate, devil's food or fudge, w/icing, coating, or filling, made fr home recipe or purchased ready-to-eat—1 piece (¹⁄₁₂ of 2-layer, 8″ or 9″ dia) (109 gm)	27	24	21	19	17	16	14	13	408	144	47.0	41
Chocolate, devil's food or fudge, wo/icing or filling, made fr home recipe or purchased ready-to-eat—1 piece (¹⁄₁₀ of 8″ or 9″ dia) (42 gm)	14	12	11	10	9	8	8	7	152	65	23.6	25
Chocolate, made w/mayonnaise or salad dressing, w/icing, coating, or filling—1 piece (¹⁄₁₂ of 2-layer, 8″ or 9″ dia) (139 gm)	12	10	9	8	7	6	6	5	483	130	25.2	6
Chocolate, made w/mayonnaise or salad dressing, wo/icing or filling—1 piece (¹⁄₁₀ of 8″ or 9″ dia) (53 gm)	4	3	3	3	3	2	2	2	160	46	7.8	4
Chocolate, w/icing, lo-cal (inc Weight Watchers Chocolate Cake)—1 individual cake (71 gm)	12	11	10	9	9	8	7	7	228	46	11.3	44
Coffee cake, fr mix, w/eggs & milk—1 cu in (8 gm)	2	1	1	1	1	1	1	1	25	6	2.0	5
Coffee cake, crumb or quick-bread type (inc cinnamon cake)—1 piece (¹⁄₁₂ of 9″ square) (58 gm)	12	11	10	9	9	8	8	8	137	38	9.9	52

Cakes *(cont.)*

	1200	1500	1800	2100	2400	2800	3200	3600	TOT CAL	FAT CAL	S/FAT CAL	CHOL MG
Coffee cake, crumb or quick-bread type, w/fruit (inc apple pastry cake)—1 piece (½ of 9" square) (65 gm)	13	11	10	10	9	9	8	8	159	38	10.0	52
Coffee cake, yeast type (inc w/ or wo/ nuts, coffee bread w/icing)—1 piece (½ of 9" square) (47 gm)	7	6	6	5	5	4	4	4	149	38	10.6	16
Coffee cake, yeast type, made fr home recipe or purchased at a bakery (inc w/ or wo/nuts, coffee bread w/icing)—1 piece (½ of 9" square) (47 gm)	7	6	6	5	5	4	4	4	160	41	10.3	16
Cottage pudding, butter, no sauce—1 piece, (12 cu in) (54 gm)	18	16	14	13	12	11	10	10	178	50	25.6	44
Cottage pudding, butter, w/chocolate sauce—1 piece, (12 cu in w/sauce) (74 gm)	19	17	15	14	13	12	11	10	227	55	28.0	44
Cottage pudding, butter, w/strawberry sauce—1 piece, (12 cu in w/sauce) (70 gm)	18	16	15	13	12	11	10	10	197	51	26.1	44
Cream (inc Italian rum-cream)—1 piece (1/10 of 8" dia) (51 gm)	29	25	23	21	19	18	16	15	191	71	41.0	68
Cupcakes, chocolate, w/icing or filling (inc Ho Ho, Ding Dong, Yodels, Ring Ding, Big Wheels, Funny Bones, Suzy Q, Swiss Cake Roll)—1 cupcake (2¾" dia) (46 gm)	8	7	6	6	5	5	4	4	157	42	11.2	18
Cupcakes, chocolate, wo/icing or filling (inc chocolate crunch bar, w/ or wo/nuts)—1 cupcake (2¾" dia) (33 gm)	7	7	6	6	5	5	4	4	101	28	8.6	23

Food												
Cupcakes, not chocolate, w/icing or filling (inc Tasty-Kake Kandy Cakes, Twinkies, Crazy Bones, Zingers, Raisin Cakes, Snackin' Cakes)—1 cupcake (2¾" dia) (48 gm)	8	7	6	6	5	5	4	4	169	46	11.0	19
Cupcakes, not chocolate, wo/icing or filling—1 cupcake (2¾" dia) (33 gm)	7	6	6	5	5	4	4	4	107	31	7.8	22
Dobos Torte (inc seven-layer cake)—1 piece (1/12 of 8½" dia) (123 gm)	74	66	60	55	52	48	45	42	501	255	80.7	243
Fruitcake, light or dark—1 piece (1/12 of 7" dia) (113 gm)	18	16	14	13	12	11	10	9	433	141	26.7	38
Funnel cake—1 cake (6" dia) (90 gm)	29	25	23	21	20	18	17	16	286	139	35.3	83
German chocolate—1 piece (1/12 of 2-layer, 8" or 9" dia) (109 gm)	38	34	30	28	26	23	22	20	379	194	55.0	90
Gingerbread—1 piece (1/10 of 8" dia) (69 gm)	13	11	10	9	9	8	7	7	214	67	16.8	35
Gingerbread, baked fr mix, w/water,—1 cu in (6 gm)	½	½	½	½	0	0	0	0	17	4	.9	0
Gingerbread, butter—1 cu in (7 gm)	2	2	2	1	1	1	1	0	19	5	3.0	4
Graham cracker—1 piece (1/10 of 9" dia) (45 gm)	13	12	11	10	9	9	8	8	156	63	15.1	43
Honey spice, baked fr mix, w/eggs & water—1 cu in (7 gm)	1	1	1	1	1	1	1	1	23	6	1.9	4
Icebox w/fruit and whipped cream—1 piece (1/10 of cake) (83 gm)	28	26	24	22	21	20	19	18	173	38	15.7	134
Ice cream roll (or cake), chocolate—1 piece (1/10 of roll) (34 gm)	12	10	9	8	7	7	6	6	102	45	19.6	18
Ice cream roll (or cake), not chocolate—1 piece (1/10 of roll) (34 gm)	9	8	7	7	6	6	5	5	100	36	14.6	18
Jelly roll—1 piece (1/10 of roll) (51 gm)	22	20	19	18	17	16	15	15	149	23	6.7	120
Johnnycake—1 piece (49 gm)	11	10	9	9	8	8	7	7	134	32	9.1	45

	1200	1500	1800	2100	2400	2800	3200	3600	TOT CAL	FAT CAL	S/FAT CAL	CHOL MG
Cakes *(cont.)*												
Lemon, w/icing—1 piece (1/12 of 2-layer, 8" or 9" dia) (139 gm)	22	19	17	16	15	13	12	12	494	130	29.9	55
Lemon, wo/icing—1 piece (1/10 of 8" or 9" dia) (53 gm)	17	15	14	13	12	11	10	10	203	91	19.7	55
Marble, baked fr mix, w/eggs & water, w/boiled white icing—1 cu in (5 gm)	1	1	1	1	1	1	1	1	17	4	1.1	3
Marble, w/icing—1 piece (1/12 of 2-layer, 8" or 9" dia) (87 gm)	22	20	18	16	15	13	12	11	301	95	34.8	45
Marble, wo/icing—1 piece (1/10 of 8" or 9" dia) (40 gm)	10	9	8	7	7	6	6	6	134	41	12.3	29
Oatmeal—1 piece (1/10 of cake) (78 gm)	15	13	11	10	10	9	8	7	280	96	24.1	26
Oatmeal, w/icing—1 piece (1/10 of cake) (110 gm)	9	8	7	6	6	5	5	4	382	65	15.7	13
Plum pudding (inc date pudding)—1 piece (42 gm)	13	11	10	9	8	8	7	6	129	52	21.3	22
Poor Man's (spice-type)—1 piece (1/10 of 8" square) (48 gm)	7	6	5	4	4	4	3	3	153	36	13.8	4
Pound 'wo/icing (inc toasted cake, yogurt honey pound cake, butter rum, whiskey cake, lemon pound cake with glaze)—1 piece (1/10 of loaf) (91 gm)	37	33	30	28	27	25	24	22	351	132	29.7	150
Pound, chocolate, w/icing—1 piece (1/10 of loaf) (91 gm)	36	32	29	26	24	22	21	19	381	181	51.4	87
Pumpkin—1 piece (1/10 of 8" square) (51 gm)	10	9	8	7	7	6	6	6	151	43	11.0	32
Raisin-nut, w/icing—1 piece (1/12 of 2-layer, 8" dia) (133 gm)	30	26	23	21	20	18	17	16	494	156	42.8	69

Food												
Raisin-nut, wo/icing (inc prune cake, date-nut, raisin-nut cake)—1 piece (1/10 of 8" dia) (59 gm)	15	13	12	11	10	9	9	8	221	78	21.3	37
Ravani (made w/farina)—1 piece (1/10 of cake) (56 gm)	8	7	6	6	5	5	5	4	176	41	8.5	24
Rice flour (inc coconut mochiko, Filipino cake)—1 piece (45 gm)	24	20	17	15	14	12	11	10	147	60	53.0	0
Rum-flavored (Sopa Borracha)—1 slice (4" x 3" x 1¾") (185 gm)	38	36	33	32	30	29	27	26	504	40	11.9	213
Shortcake, biscuit type, w/fruit—1 biscuit w/fruit (162 gm)	13	11	10	9	8	7	6	6	364	119	29.5	1
Shortcake, biscuit type, w/whipped cream & fruit—1 biscuit w/fruit & whipped cream (169 gm)	27	23	20	18	16	14	13	12	391	154	54.7	17
Shortcake, sponge type, w/fruit—1 cake w/fruit (156 gm)	34	32	30	28	27	25	24	23	296	37	10.7	189
Shortcake, sponge type, w/whipped cream & fruit (inc strawberry cake, shortcake—1 cake with fruit & whipped cream (163 gm)	42	38	36	33	31	29	28	26	314	66	29.9	186
Soybean—1 piece (1/10 of 8" dia) (68 gm)	9	8	7	6	5	5	4	4	236	91	19.9	2
Spice, w/icing (inc walnut cake w/whipped cream, Little Debbie Apple Spice, Stir-n-Frost Spice Cake)—1 piece (1/12 of 2-layer, 8" or 9" dia) (109 gm)	25	22	20	18	17	15	14	13	374	102	34.7	61
Spice, wo/icing—1 piece (1/10 of 8" or 9" dia) (42 gm)	12	10	9	8	8	7	7	6	146	45	15.6	30
Sponge, chocolate, w/icing—1 piece (1/12 of 10" dia) (92 gm)	43	40	37	34	33	31	29	28	293	69	27.6	200
Sponge, chocolate, wo/icing—1 piece (1/12 of 10" dia) (66 gm)	34	32	30	28	27	25	24	23	198	40	13.7	181
Sponge, w/icing—1 piece (1/12 of 10" dia) (92 gm)	31	29	27	25	24	22	21	20	281	47	14.4	158

	1200	1500	1800	2100	2400	2800	3200	3600	TOT CAL	FAT CAL	S/FAT CAL	CHOL MG
Cakes (cont.)												
Sponge, wo/icing (inc shortcake, sponge type, plain)—1 piece (¼₂ of 10″ dia) (66 gm)	31	29	27	26	24	23	22	21	193	33	9.6	172
Torte (inc fruit tortes)—1 piece (¼₂ of torte) (76 gm)	30	26	24	22	20	19	17	16	224	102	38.6	81
Upside-down (all fruits)—1 piece (¼₂ of 9″ dia) (121 gm)	20	18	16	14	13	12	11	10	374	133	29.4	45
White, coconut icing, butter—1 cu in (6 gm)	2	2	2	1	1	1	1	1	20	6	3.9	2
White, pudding type mix (oil, egg white & water added to dry mix), w/icing (inc Pillsbury Plus, Duncan Hines, Betty Crocker)—1 piece (¼₂ of 2-layer, 8″ or 9″ dia) (146 gm)	20	17	15	13	12	10	9	8	565	196	45.3	1
White, pudding type mix (oil, egg white & water added to dry mix), wo/icing (inc Pillsbury Plus, Duncan Hines, Betty Crocker)—1 piece (¼₀ of 8″ or 9″ dia) (57 gm)	13	11	9	8	7	7	6	5	253	137	28.8	0
White, standard type mix (egg whites & water added to mix), w/icing (inc Jiffy, Washington)—1 piece (¼₂ of 2-layer, 8″ or 9″ dia) (107 gm)	13	11	9	8	7	7	6	5	429	106	28.6	1
White, standard type mix (egg whites & water added to mix), wo/icing (inc Jiffy, Washington)—1 piece (¼₀ of 8″ or 9″ dia) (42 gm)	5	4	3	3	3	2	2	2	141	34	10.7	0

Food												
White, w/icing, made fr home recipe or purchased ready-to-eat (inc wedding cake)—1 piece (1/12 of 2-layer, 8" or 9" dia) (107 gm)	14	12	10	9	8	7	7	6	408	129	31.3	2
White, wo/icing, made fr home recipe or purchased ready-to-eat—1 piece (1/10 of 8" or 9" dia) (42 gm)	7	6	5	5	4	4	3	3	154	60	15.8	1
Whole wheat, w/fruit & nuts (inc apple nut loaf, whole wheat banana carob cake)—1 piece (1/10 of loaf) (63 gm)	12	11	10	9	8	7	7	6	242	83	17.1	29
Yellow, pudding type mix (oil, eggs & water added to dry mix), w/icing (inc Pillsbury Plus, Duncan Hines, Betty Crocker Super Moist)—1 piece (139 gm)	32	28	26	24	22	20	19	18	538	201	39.9	92
Yellow, pudding type mix (oil, eggs & water added to dry mix), wo/icing (inc Pillsbury Plus, Duncan Hines, Betty Crocker Super Moist)—1 piece (1/10 of 8" or 9" dia) (53 gm)	16	14	13	12	11	11	10	9	203	91	17.1	55
Yellow, standard type mix (eggs & water added to dry mix), w/icing (inc Jiffy, Washington)—1 piece (1/12 of 2-layer, 8" or 9" dia) (99 gm)	16	14	13	12	11	10	9	9	349	94	22.8	39
Yellow, standard type mix (eggs & water added to dry mix), wo/icing (inc Jiffy, Washington)—1 piece (1/10 of 8" or 9" dia) (39 gm)	8	7	7	6	6	5	5	5	127	36	9.2	26
Yellow, w/icing, made fr home recipe or purchased ready-to-eat (inc Jiffy, Washington)—1 piece (1/12 of 2-layer, 8" or 9" dia) (99 gm)	18	16	14	13	12	11	10	9	364	105	27.4	37
Yellow, wo/icing, made home recipe or purchased ready-to-eat—1 piece (1/10 of 8" or 9" dia) (39 gm)	9	8	7	7	6	6	5	5	139	44	12.3	23

Pies

	1200	1500	1800	2100	2400	2800	3200	3600	TOT CAL	FAT CAL	S/FAT CAL	CHOL MG
Apple, fried pie (inc McDonald's)—1 pie (86 gm)	18	15	13	12	11	9	8	8	311	167	41.4	0
Apple (inc apple-peach, apple-berry)—1 piece (⅛ of 9″ dia) (151 gm)	22	18	16	14	13	11	10	9	457	198	49.0	0
Apple, individual size or tart—1 tart (117 gm)	19	16	14	12	11	10	9	8	378	174	43.1	0
Apple, lo-cal (inc Weight Watchers Apple Pie)—1 ind serving (85 gm)	3	3	2	2	2	2	2	1	177	38	7.4	0
Apple, one crust (inc apple pie w/crumb topping)—1 piece (⅛ of 9″ dia) (151 gm)	14	12	10	9	8	7	6	6	373	128	31.1	0
Apple-sour cream—1 piece (⅛ of 9″ dia) (160 gm)	22	18	16	14	13	11	10	9	357	133	45.9	8
Apricot—1 piece (⅛ of 9″ dia) (151 gm)	20	17	15	13	12	10	9	8	428	185	45.6	0
Apricot, fried pie—1 pie (86 gm)	17	15	13	11	10	9	8	7	301	160	39.4	0
Apricot, individual size or tart—1 tart (117 gm)	18	15	13	11	10	9	8	7	354	161	39.6	0
Banana cream—1 piece (⅛ of 9″ dia) (145 gm)	30	27	24	22	21	19	18	17	311	124	38.5	85
Banana cream, individual size or tart—1 tart (117 gm)	26	23	20	19	17	16	15	14	282	123	36.1	62
Berry, not blackberry or blueberry (inc cranberry, Juneberry, gooseberry)—1 piece (⅛ of 9″ dia) (151 gm)	21	17	15	13	12	11	10	9	433	196	46.8	0
Berry, not blackberry or blueberry, individual size or tart—1 tart (117 gm)	17	14	12	11	10	9	8	7	343	157	38.0	0

Food												
Blackberry (inc boysenberry)—1 piece (⅛ of 9" dia) (150 gm)	19	16	14	12	11	10	9	8	410	181	43.5	0
Blackberry, individual size or tart—1 tart (117 gm)	17	14	12	11	10	9	8	7	341	157	37.9	0
Black bottom—1 piece (⅛ of 9" dia) (99 gm)	48	42	38	35	32	29	27	25	278	150	69.1	113
Blueberry (inc huckleberry)—1 piece (⅛ of 9" dia) (151 gm)	31	26	23	20	18	16	15	13	406	201	66.0	11
Blueberry, individual size or tart—1 tart (117 gm)	17	14	12	11	10	9	8	7	343	157	38.0	0
Blueberry, one crust—1 piece (⅛ of 9" dia) (137 gm)	12	10	9	8	7	6	5	5	292	108	26.3	0
Buttermilk—1 piece (⅛ of 9" dia) (145 gm)	47	41	37	34	31	29	27	25	560	250	64.4	117
Butterscotch, baked, w/vegetable shortening, wo/salt in filling—1 piece (⅛ of pie) (114 gm)	27	24	21	19	18	16	15	14	304	113	40.4	59
Cherry—1 piece (⅛ of 9" dia) (151 gm)	20	17	14	13	11	10	9	8	407	183	44.3	0
Cherry, fried pie (inc McDonald's)—1 pie (86 gm)	16	14	12	10	9	8	7	7	286	147	36.5	0
Cherry, individual size or tart—1 tart (117 gm)	17	15	13	11	10	9	8	7	339	160	39.0	0
Cherry, made w/cream cheese and sour cream—1 piece (⅛ of 9" dia) (160 gm)	55	48	43	39	35	32	29	27	457	193	92.3	91
Cherry, one crust—1 piece (⅛ of 9" dia) (137 gm)	11	10	8	7	7	6	5	5	311	105	25.7	0
Chiffon, chocolate—1 piece (⅛ of 9" dia) (99 gm)	42	37	34	31	29	27	25	24	321	145	45.6	138
Chiffon, not chocolate—1 piece (⅛ of 9" dia) (99 gm)	41	37	35	32	30	28	27	25	297	116	30.5	177
Chiffon, w/liqueur (inc grasshopper pie)—1 piece (⅛ of 9" dia) (99 gm)	48	42	38	35	33	30	28	26	350	178	64.4	125

	THE 2-IN-1 SYSTEM								TOT CAL	FAT CAL	S/FAT CAL	CHOL MG
	1200	1500	1800	2100	2400	2800	3200	3600				
Pies *(cont.)*												
Chocolate cream (inc chocolate meringue, chocolate icebox dessert, chocolate pudding pie)—1 piece (⅛ of 9″ dia) (145 gm)	47	41	36	33	31	28	26	24	406	185	70.4	98
Chocolate cream, individual size or tart —1 tart (117 gm)	37	32	29	26	24	22	20	19	356	169	59.5	70
Chocolate marshmallow—1 piece (⅛ of 9″ dia) (103 gm)	43	37	32	28	26	23	20	19	385	198	94.1	11
Chocolate meringue, baked, vegetable shortening—1 piece, (⅛ of pie) (114 gm)	30	27	24	22	20	18	17	16	287	123	46.1	64
Coconut cream (inc coconut custard)—1 piece (⅛ of 9″ dia) (145 gm)	103	90	80	73	67	61	56	52	570	321	160.0	205
Coconut cream, individual size or tart—1 tart (117 gm)	49	42	38	34	32	29	26	24	303	170	79.5	87
Custard (inc custard cream, egg pie)—1 piece (⅛ of 9″ dia) (136 gm)	39	35	31	29	27	25	23	22	283	134	44.3	123
Custard, individual size or tart (inc custard cream)—1 tart (117 gm)	28	25	23	21	20	19	17	16	223	76	27.7	101
Lemon cream (inc lemon icebox pie)—1 piece (⅛ of 9″ dia) (145 gm)	40	35	32	30	28	25	24	22	410	163	46.9	121
Lemon cream, individual size or tart—1 tart (117 gm)	32	28	26	23	22	20	19	17	348	147	41.2	88
Lemon, fried pie—1 pie (85 gm)	27	23	21	19	18	16	15	14	286	150	37.8	63
Lemon meringue (inc key lime pie)—1 piece (⅛ of 9″ dia) (137 gm)	34	30	28	26	24	22	21	20	341	134	33.6	122
Lemon meringue, individual size or tart —1 tart (117 gm)	31	28	25	23	22	20	19	17	329	142	36.3	96

Lemon (not cream or meringue)—1 piece (⅛ of 9″ dia) (99 gm)	31	28	25	23	22	20	19	17	386	145	35.9	97
Lemon pie filling—1 c (266 gm)	89	82	76	72	68	64	61	58	931	169	41.7	450
Mince—1 piece (⅛ of 9″ dia) (151 gm)	19	16	14	12	11	10	9	8	432	173	43.1	0
Mince, individual size or tart—1 tart (117 gm)	17	14	12	11	10	9	8	7	357	152	37.9	0
Oatmeal—1 piece (⅛ of 9″dia) (114 gm)	34	31	28	26	24	22	21	20	451	163	38.2	112
Peach—1 piece (⅛ of 9″ dia) (151 gm)	19	16	14	12	11	10	9	8	410	176	43.6	0
Peach, fried pie—1 pie (86 gm)	17	14	12	11	10	9	8	7	291	155	38.5	0
Peach, individual size or tart—1 tart (117 gm)	18	15	13	11	10	9	8	7	346	161	39.9	0
Peach, one crust—1 piece (⅛ of 9″ dia) (151 gm)	12	10	8	7	7	6	5	5	335	105	26.0	0
Peanut butter cream—1 piece (⅛ of 9″ dia) (145 gm)	36	31	28	26	24	22	20	19	444	194	51.3	83
Pear—1 piece (⅛ of 9″ dia) (151 gm)	19	16	14	12	11	9	9	8	407	171	41.9	0
Pear, individual size or tart—1 tart (117 gm)	17	14	12	11	10	9	8	7	339	152	37.5	0
Pecan (inc coconut-pecan pie, walnut pie)—1 piece (⅛ of 9″ dia) (115 gm)	28	25	22	21	19	17	16	15	488	237	36.9	75
Pecan, individual size or tart—1 small tart (57 gm)	14	13	11	10	10	9	8	8	248	123	20.9	33
Pie shell—1 pie shell (9″ dia) (172 gm)	63	53	46	41	37	32	29	26	958	574	142.5	0
Pie shell, graham cracker—1 pie shell (9″ dia) (210 gm)	60	50	44	39	35	31	27	25	1076	545	134.8	0
Pineapple—1 piece (⅛ of 9″ dia) (151 gm)	20	16	14	13	11	10	9	8	407	178	44.1	0
Pineapple cream (inc millionaire's pie, Hawaiian pie, sunshine pie)—1 piece (⅛ of 9″ dia) (145 gm)	28	24	22	20	19	17	16	15	304	114	33.9	80
Pineapple, individual size or tart—1 tart (117 gm)	18	15	13	12	10	9	8	7	344	162	40.2	0
Plum—1 piece (⅛ of 9″ dia) (151 gm)	20	17	15	13	12	10	9	8	434	181	45.0	0

	THE 2-IN-1 SYSTEM											
	1200	1500	1800	2100	2400	2800	3200	3600	TOT CAL	FAT CAL	S/FAT CAL	CHOL MG
Pies *(cont.)*												
Prune—1 piece (⅛ of 9″ dia) (151 gm)	15	12	11	10	9	8	7	6	460	134	33.2	0
Pudding, flavors other than chocolate—1 piece (⅛ of 9″ dia) (145 gm)	23	19	17	15	14	12	11	10	319	144	46.4	13
Pudding, flavors other than chocolate, w/chocolate coating, individual size—1 ind pie (142 gm)	36	31	27	24	21	19	17	15	520	270	78.8	8
Pumpkin—1 piece ⅛ of 9″ dia) (155 gm)	40	36	32	29	27	25	23	21	333	180	58.2	94
Pumpkin, individual size or tart—1 tart (117 gm)	31	27	24	22	21	19	17	16	282	155	47.7	65
Pumpkin pie mix, canned—½ c (135 gm)	½	½	½	0	0	0	0	0	141	2	.8	0
Raisin—1 piece (⅛ of 9″ dia) (151 gm)	19	16	14	12	11	10	9	8	420	171	42.6	0
Raisin, individual size or tart—1 tart (117 gm)	18	15	13	12	11	9	8	8	373	166	41.3	0
Raspberry—1 piece (⅛ of 9″ dia) (150 gm)	21	17	15	13	12	11	9	9	430	194	46.5	0
Raspberry cream—1 piece (⅛ of 9″ dia) (145 gm)	37	32	28	25	23	21	19	17	294	155	71.3	36
Raspberry, one crust—1 piece (⅛ of 9″ dia) (137 gm)	13	11	9	8	8	7	6	5	333	126	29.3	0
Rhubarb—1 piece (⅛ of 9″ dia) (151 gm)	23	19	17	15	13	12	11	10	428	209	52.0	0
Rhubarb, individual size or tart—1 tart (117 gm)	20	17	14	13	11	10	9	8	357	180	44.7	0
Rhubarb, one crust—1 piece (⅛ of 9″ dia) (137 gm)	14	12	10	9	8	7	7	6	313	130	32.2	0
Shoofly—1 piece (⅛ of 9″ dia) (114 gm)	20	18	16	15	13	12	11	11	393	115	29.0	47

Sour cream, raisin—1 piece (⅛ of 9″ dia) (145 gm)	66	57	50	46	42	38	34	32	522	298	115.5	94
Squash—1 piece (⅛ of 9″ dia) (155 gm)	30	27	24	22	21	19	18	17	299	119	37.1	87
Strawberry cream—1 piece (⅛ of 9″ dia) (145 gm)	34	29	25	23	21	19	17	16	292	140	64.2	33
Strawberry cream, individual size or tart—1 tart (117 gm)	28	24	21	19	17	15	14	13	277	140	55.5	23
Strawberry, individual size or tart—1 individual pie (120 gm)	15	13	11	10	9	8	7	6	319	142	34.9	0
Strawberry, one crust—1 piece (⅛ of 9″ dia) (168 gm)	17	14	12	11	10	9	8	7	387	155	38.0	0
Strawberry-rhubarb—1 piece (⅛ of 9″ dia) (151 gm)	23	19	17	15	13	12	10	10	431	209	51.6	0
Sweet potato—1 piece (⅛ of 9″ dia) (155 gm)	34	30	27	25	23	21	19	18	323	157	47.1	83
Tofu w/fruit—1 piece (⅛ of 9″ dia) (145 gm)	21	18	16	15	13	12	11	10	304	160	32.3	40
Tollhouse—1 piece (⅛ of 9″ dia) (115 gm)	54	46	41	37	33	30	27	25	615	400	99.0	62
Vanilla cream (inc butterscotch pie)—1 piece (⅛ of 9″ dia) (145 gm)	40	36	32	30	27	25	23	22	383	164	51.5	111
Vanilla wafer dessert base—1 c (129 gm)	50	44	40	37	34	32	30	28	560	270	58.2	152
Yogurt, frozen (inc all flavors)—1 piece (⅛ of 9″ dia) (145 gm)	42	36	31	27	25	22	19	18	355	182	94.0	3

CEREALS
Cold (Ready-to-Eat)

100% Bran—⅔ c (44 gm)	2	1	1	1	1	1	1	1	118	20	3.5	0
100% Natural Cereal, plain—½ c (52 gm)	36	30	26	23	21	19	17	15	244	101	81.7	0
100% Natural Cereal, w/apples & cinnamon—½ c (52 gm)	31	26	23	20	18	16	14	13	239	88	69.6	0

	THE 2-IN-1 SYSTEM											
	1200	1500	1800	2100	2400	2800	3200	3600	TOT CAL	FAT CAL	S/FAT CAL	CHOL MG
Cold (Ready-to-Eat) *(cont.)*												
100% Natural Cereal, w/raisins & dates—½ c (55 gm)	27	23	20	18	16	14	13	11	248	92	61.6	0
40% Bran Flakes, Kellogg—1 single serving box (1 oz) (28 gm)	½	½	½	0	0	0	0	0	91	5	.8	0
———1 c (39 gm)	½	½	½	½	½	0	0	0	127	7	1.1	0
40% Bran Flakes, Post—1 c (47 gm)	½	½	½	½	½	0	0	0	152	7	1.1	0
40% Bran Flakes, Ralston Purina—1 c (49 gm)	½	½	½	½	½	0	0	0	159	6	1.0	0
All-Bran—1 single serving box (1.25 oz) (35 gm)	½	½	½	½	0	0	0	0	87	6	.9	0
———⅔ cup (41 gm)	½	½	½	½	½	0	0	0	102	7	1.1	0
Almond Delight—1 c (38 gm)	1	1	1	½	½	½	½	½	147	19	1.6	0
Alpen—½ c (57 gm)	6	5	5	4	4	3	3	3	201	17	14.6	0
Alpha-Bits—1 c (34 gm)	1	½	½	½	½	½	½	0	133	7	1.3	0
Apple Jacks—1 single serving box (.75 oz) (21 gm)	0	0	0	0	0	0	0	0	81	1	.1	0
———1 c (28 gm)	0	0	0	0	0	0	0	0	108	1	.1	0
Apple Raisin Crisp—⅔ c (37 gm)	½	½	½	½	½	0	0	0	141	10	1.1	0
Banana-Flavored Frosted Flakes—1 c (38 gm)	1	½	½	½	½	½	½	0	147	12	1.3	0
Body Buddies, brown sugar & honey—1 c (33 gm)	1	1	1	1	½	½	½	½	128	10	1.9	0
Body Buddies, natural fruit flavor—1 c (33 gm)	1	1	1	1	½	½	½	½	128	10	1.9	0
Booberry—1 c (33 gm)	4	3	3	3	2	2	2	2	128	10	9.0	0
Bran Buds—⅔ c (56 gm)	1	1	1	1	1	½	½	½	144	12	2.0	0
Bran Chex—1 c (49 gm)	1	1	1	1	1	½	½	½	156	12	2.0	0

Food								Calories			
Bran Muffin Crisp—²⁄₃ c (40 gm)	1	1	½	½	½	½	½	131	9	1.4	0
Branola—½ c (55 gm)	20	17	14	13	11	10	9	233	50	44.4	0
Buc Wheats—1 c (39 gm)	1	1	1	1	1	½	½	151	12	2.1	0
C-3PO's—1 c (33 gm)	0	0	0	0	0	0	0	128	0	0	0
Cabbage Patch—1 c (25 gm)	0	0	0	0	0	0	0	97	0	0	0
Cap'n Crunch—1 c (37 gm)	9	8	7	6	5	5	4	156	31	20.1	0
Cap'n Crunch's Choco Crunch—1 c (37 gm)	9	8	7	6	5	5	4	144	24	20.4	0
Cap'n Crunch's Crunch Berries—1 c (35 gm)	8	6	6	5	4	4	3	146	26	17.4	0
Cap'n Crunch's Peanut Butter Crunch—1 c (35 gm)	8	6	6	5	4	4	3	154	41	17.1	0
Cheerios—1 single serving box (.75 oz) (21 gm)	1	1	1	1	1	1	½	82	12	2.2	0
Chex cereal—1 c (39 gm)	2	2	1	1	1	1	1	109	16	3.0	0
Cinnamon Toast Crunch—1 c (38 gm)	13	11	10	9	8	7	6	143	9	4.1	0
Circus Fun—1 c (28 gm)	3	3	2	2	2	2	2	161	35	29.9	0
Cocoa Krispies—1 single serving box (1 oz) (28 gm)	½	½	½	½	2	2	1	109	9	7.6	0
—1 c (36 gm)	½	½	½	½	½	0	0	108	4	.9	0
Cocoa Pebbles—1 c (32 gm)	4	3	3	2	2	2	2	139	5	1.1	0
Cocoa Puffs—1 c (30 gm)	½	½	½	½	2	2	2	131	16	8.5	0
Cookie-Crisp (inc all flavors)—1 c (30 gm)	½	½	½	½	½	0	0	117	5	.9	0
Corn Bran—1 c (36 gm)	1	½	½	½	½	½	½	120	10	1.1	0
Corn Chex—1 c (29 gm)	0	0	0	0	0	½	0	125	11	1.3	0
Corn Flakes, Kellogg—1 single serving box (.75 oz) (21 gm)	0	0	0	0	0	0	0	114	1	.1	0
—1 c (25 gm)	0	0	0	0	0	0	0	82	1	.1	0
Corn flakes, low-sodium—1 c (25 gm)	0	0	0	0	0	0	0	100	1	.1	0
Corn Flakes, Ralston Purina—1 c (25 gm)	0	0	0	0	0	0	0	98	1	.1	0

	THE 2-IN-1 SYSTEM											
	1200	1500	1800	2100	2400	2800	3200	3600	TOT CAL	FAT CAL	S/FAT CAL	CHOL MG
Cold (Ready-to-Eat) *(cont.)*												
Corn Pops—1 single serving box (.75 oz) (21 gm)	0	0	0	0	0	0	0	0	80	1	.1	0
——1 c (31 gm)	0	0	0	0	0	0	0	0	118	1	.1	0
Corn Puffs—1 c (19 gm)	½	½	½	½	½	0	0	0	74	4	1.1	0
Corn Total—1 c (33 gm)	½	½	½	0	0	0	0	0	126	7	.8	0
Count Chocula—1 c (33 gm)	4	3	3	3	2	2	2	2	128	10	9.0	0
Cracker Jack Cereal—1 c (28 gm)	4	4	3	3	3	2	2	2	109	14	9.8	0
Cracklin' Bran—⅔ c (40 gm)	15	13	11	10	9	8	7	6	153	53	33.8	0
Cracklin' Oat Bran—⅔ c (40 gm)	19	16	14	13	11	10	9	8	169	51	43.9	0
Crispix—1 c (31 gm)	0	0	0	0	0	0	0	0	120	0	0	0
Crispy Oatmeal and Raisin Chex—1 c (39 gm)	1	1	1	½	½	½	½	½	144	9	1.7	0
Crispy Rice—1 c (28 gm)	0	0	0	0	0	0	0	0	111	1	.3	0
Crispy Wheats 'n Raisins—1 c (43 gm)	½	½	½	½	0	0	0	0	150	6	.9	0
C W Post, plain—⅔ c (65 gm)	30	25	22	20	18	15	14	13	289	91	68.2	0
C W Post, w/raisins—⅔ c (69 gm)	29	25	21	19	17	15	13	12	299	89	66.1	0
Dairy Crisp—½ c (60 gm)	29	25	21	19	17	15	13	12	254	76	66.3	0
Dairy Crisp with Strawberry Bits—½ c (60 gm)	29	25	21	19	17	15	13	12	254	76	66.3	0
Donkey Kong—1 c (32 gm)	1	½	½	½	½	½	0	0	124	10	1.2	0
Donkey Kong Junior—1 c (33 gm)	1	½	½	½	½	½	0	0	128	10	1.2	0
Donuts Cereal, chocolate-flavored—1 c (33 gm)	12	10	9	8	7	6	6	5	140	31	27.2	0
Donuts Cereal, powdered—1 c (33 gm)	12	10	9	8	7	6	6	5	140	31	27.2	0
E.T.—1 c (38 gm)	18	16	13	12	11	9	8	8	174	60	41.6	0
Familia—½ c (61 gm)	13	11	10	9	8	7	6	6	237	35	29.9	0
Fiber One—⅔ c (38 gm)	1	1	1	1	1	1	1	1	148	18	3.0	0

Fortified oat flakes, (oat w/other grains) —1 c (48 gm)	1	1	1	1	1	1	½	½	177	6	2.3	0
Frankenberry—1 c (34 gm)	4	3	3	3	2	2	2	2	132	11	9.3	0
Froot Loops—1 single serving box (.75 oz) (21 gm)	1	1	½	½	½	½	½	½	82	4	1.5	0
———1 c (28 gm)	1	1	1	1	1	½	½	½	110	5	2.1	0
Froot loops, (corn w/other grains)—1 oz (28 gm)	1	1	1	1	½	½	½	½	111	5	1.9	0
Frosted Flakes, Kellogg—1 single serving box (1 oz) (28 gm)	0	0	0	0	0	0	0	0	107	1	.1	0
———1 c (35 gm)	0	0	0	0	0	0	0	0	133	1	.1	0
Frosted Flakes, Ralston Purina—1 c (38 gm)	½	½	½	0	0	0	0	0	149	5	.8	0
Frosted Mini-Wheats (inc all flavors)—⅔ c (40 gm)	0	0	0	0	0	0	0	0	144	4	.5	0
Frosted Mini-Wheats, sugar-frosted, brown sugar & cinnamon (wheat)—4 biscuits (31 gm)	0	0	0	0	0	0	0	0	111	3	.4	0
Frosted Rice Krinkles—1 c (45 gm)	0	0	0	0	0	0	0	0	173	1	.2	0
Frosted Rice Krispies—1 single serving box (⅞ oz) (25 gm)	0	0	0	0	0	0	0	0	96	1	.2	0
———1 c (45 gm)	0	0	0	0	0	0	0	0	172	1	.3	0
Frosty O's—1 c (34 gm)	1	1	1	1	1	1	1	1	133	15	2.8	0
Fruitful Bran—½ c (55 gm)	0	0	0	0	0	0	0	0	178	0	0	0
Fruit 'N' Fiber Harvest Medley—⅔ c (38 gm)	1	1	½	½	½	½	½	½	123	12	1.4	0
Fruit 'N' Fiber Mountain Trail—⅔ c (38 gm)	5	4	4	3	3	3	2	2	124	15	11.9	0
Fruit 'N' Fiber, w/apples & cinnamon—⅔ c (38 gm)	½	½	½	½	½	0	0	0	120	5	1.0	0
Fruit 'N' Fiber, w/dates, raisins & walnuts—⅔ c (37 gm)	1	1	1	1	½	½	½	½	117	10	1.8	0
Fruit 'N' Fiber, w/tropical fruit (pineapple, bananas & coconut)—⅔ c (40 gm)	5	4	4	3	3	2	2	2	127	13	10.9	0

Cold (Ready-to-Eat) (cont.)

	1200	1500	1800	2100	2400	2800	3200	3600	TOT CAL	FAT CAL	S/FAT CAL	CHOL MG
Fruity Pebbles—1 c (32 gm)	5	4	4	3	3	3	2	2	130	15	11.5	0
Ghost Busters—1 c (28 gm)	2	2	1	1	1	1	1	1	109	9	4.3	0
G-I Joe Action Stars—1 c (28 gm)	5	5	4	3	3	3	2	2	109	19	12.2	0
Golden Grahams—1 c (39 gm)	4	3	3	3	2	2	2	2	150	13	9.2	0
Golden Harvest Proteinola—⅔ c (56 gm)	36	31	27	23	21	19	17	15	257	95	82.0	0
Graham Crackos—1 c (30 gm)	0	0	0	0	0	0	0	0	108	2	.3	0
Granola, homemade—½ c (61 gm)	12	10	9	8	7	6	5	5	297	149	26.3	0
Grape Nut Flakes—1 c (39 gm)	½	0	0	0	0	0	0	0	140	4	.6	0
Grape Nuts—½ c (55 gm)	0	0	0	0	0	0	0	0	196	2	.3	0
Gremlins—1 c (28 gm)	7	6	5	4	4	4	3	3	109	18	15.5	0
Halfsies—1 c (33 gm)	1	1	1	1	1	½	½	½	132	12	2.2	0
Heartland Natural Cereal, plain—½ c (58 gm)	19	16	14	12	11	10	9	8	252	80	43.3	0
Heartland Natural Cereal, w/coconut—½ c (53 gm)	30	25	22	19	17	15	14	12	234	78	67.3	0
Heartland Natural Cereal, w/raisins—½ c (55 gm)	27	23	20	17	16	14	12	11	234	70	60.8	0
Honey and Nut Corn Flakes—1 c (38 gm)	1	1	1	1	1	½	½	½	151	19	2.2	0
Honey Bran—1 c (35 gm)	½	½	½	½	½	0	0	0	119	7	1.1	0
Honeycomb, plain—1 c (22 gm)	0	0	0	0	0	0	0	0	86	4	.4	0
Honeycomb, strawberry—1 c (22 gm)	0	0	0	0	0	0	0	0	86	4	.4	0
Honey Nut Cheerios—1 single serving box (21 gm)	½	½	½	0	0	0	0	0	79	4	.8	0
—1 c (33 gm)	1	½	½	½	½	½	0	0	125	7	1.2	0
Honey Smacks—1 single serving box (⅞ oz) (25 gm)	½	0	0	0	0	0	0	0	93	4	.6	0

———1 c (38 gm)	½	½	½	½	0	0	0	0	142	6	.9	0
Horizon—⅔ c (57 gm)	14	12	10	9	8	7	6	6	221	36	31.5	0
Just Right—1 c (43 gm)	1	½	½	½	½	½	0	0	152	8	1.2	0
Just Right, w/raisins, dates & nuts—⅔ c (37 gm)	½	½	½	½	½	0	0	0	141	9	1.0	0
Kaboom—1 c (34 gm)	2	1	1	1	1	1	1	1	132	7	3.5	0
King Vitamin—1 c (21 gm)	3	2	2	2	2	1	1	1	85	10	6.6	0
Kix—1 c (19 gm)	½	½	½	½	½	0	0	0	74	4	1.1	0
Life (plain & cinnamon)—1 c (44 gm)	1	1	½	½	½	½	½	½	162	8	1.4	0
Lucky Charms—1 c (32 gm)	1	1	1	1	1	½	½	½	125	11	2.0	0
Malt-O-Meal Crisp Rice—1 c (28 gm)	0	0	0	0	0	0	0	0	111	2	.5	0
Malt-O-Meal Puffed Rice—1 c (14 gm)	0	0	0	0	0	0	0	0	56	1	.2	0
Malt-O-Meal Puffed Wheat—1 c (12 gm)	0	0	0	0	0	0	0	0	44	1	.2	0
Malt-O-Meal Sugar Puffs—1 c (32 gm)	½	½	½	0	0	0	0	0	119	5	.8	0
Malt-O-Meal Toasted Oat Cereal—1 c (22 gm)	1	1	1	1	1	1	½	½	86	13	2.4	0
Marshmallow Krispies—1 c (33 gm)	0	0	0	0	0	0	0	0	125	0	0	0
Most—⅔ c (35 gm)	0	0	0	0	0	0	0	0	118	3	.5	0
Mr. T—1 c (28 gm)	7	6	5	4	4	4	3	3	118	18	15.5	0
Nature Valley Granola, toasted oat mixture—½ c (56 gm)	26	22	19	17	15	13	12	11	249	88	58.1	0
Nature Valley Granola, w/cinnamon & raisins—½ c (56 gm)	31	26	23	20	18	16	14	13	255	81	69.8	0
Nature Valley Granola, w/fruit & nuts—½ c (56 gm)	35	29	25	22	20	18	16	14	261	91	78.5	0
Nutri-Grain Almond Raisin—⅔ cup (34 gm)	1	½	½	½	½	½	½	0	129	15	1.3	0
Nutri-Grain Barley—1 c (41 gm)	0	0	0	0	0	0	0	0	153	3	.5	0
Nutri-Grain Corn—1 c (42 gm)	½	½	½	½	½	0	0	0	160	9	1.0	0
Nutri-Grain Rye—1 c (40 gm)	0	0	0	0	0	0	0	0	144	3	.3	0
Nutri-Grain Wheat and Raisins—1 c (44 gm)	0	0	0	0	0	0	0	0	155	0	0	0

	THE 2-IN-1 SYSTEM											
	1200	1500	1800	2100	2400	2800	3200	3600	TOT CAL	FAT CAL	S/FAT CAL	CHOL MG
Cold (Ready-to-Eat) *(cont.)*												
Nutri-Grain Wheat (inc Nutri-Grain)—1 c (44 gm)	½	0	0	0	0	0	0	0	158	4	.6	0
Oat flakes, fortified—1 c (48 gm)	1	½	½	½	½	½	0	0	177	6	1.2	0
O J's—1 c (28 gm)	7	6	5	4	4	4	3	3	118	18	15.5	0
Pac Man—1 c (33 gm)	0	0	0	0	0	0	0	0	124	3	.4	0
Popeye—1 c (14 gm)	0	0	0	0	0	0	0	0	56	1	.2	0
Product 19—1 single serving box (.75 oz) (21 gm)	0	0	0	0	0	0	0	0	80	1	.2	0
———1 c (33 gm)	0	0	0	0	0	0	0	0	126	2	.3	0
Quisp—1 c (30 gm)	6	5	4	4	3	3	3	2	124	19	13.3	0
Rainbow Brite—1 c (28 gm)	2	2	1	1	1	1	1	1	109	9	4.3	0
Raisin Bran, Kellogg—1 single serving box (1.25 oz) (35 gm)	½	½	½	½	0	0	0	0	109	6	.9	0
———⅔ c (38 gm)	½	½	½	½	½	0	0	0	119	7	1.0	0
Raisin bran (inc store brands)—⅔ c (38 gm)	½	½	½	½	0	0	0	0	118	6	.9	0
Raisin Bran, Post—⅔ c (38 gm)	½	½	½	½	½	0	0	0	117	6	1.0	0
Raisin Bran, Ralston Purina—⅔ c (38 gm)	0	0	0	0	0	0	0	0	121	2	.3	0
Raisin Grape Nuts—½ c (59 gm)	½	½	½	½	½	0	0	0	207	5	1.0	0
Raisin Life—⅔ c (38 gm)	1	½	½	½	½	½	0	0	140	6	1.2	0
Raisin Nut Bran—⅔ c (38 gm)	0	0	0	0	0	0	0	0	147	36	.3	0
Raisin, Rice, and Rye—1 c (46 gm)	0	0	0	0	0	0	0	0	155	1	.2	0
Raisin Squares—1 c (28 gm)	0	0	0	0	0	0	0	0	99	0	0	0
Rice Chex—1 c (33 gm)	0	0	0	0	0	0	0	0	130	1	.3	0
Rice Flakes (inc store brands) (see also Crispy Rice)—1 c (27 gm)	0	0	0	0	0	0	0	0	107	1	.2	0

Rice Krispies—1 single serving box (⅝ oz) (18 gm)	0	0	0	0	0	0	0	0	71	1	.3	0
———1 c (27 gm)	0	0	0	0	0	0	0	0	107	2	.4	0
Rice, puffed—1 c (14 gm)	0	0	0	0	0	0	0	0	56	1	.2	0
Rocky Road—1 c (43 gm)	16	13	11	10	9	8	7	7	182	41	35.5	0
Shredded Wheat, 100%—1 rectangular biscuit (24 gm)	½	0	0	0	0	0	0	0	85	4	.6	0
Shredded Wheat 'N Bran—1 c (43 gm)	1	1	1	1	1	½	½	½	167	14	2.2	0
Shredded Wheat Toasted Wheat and Raisins—⅔ c (49 gm)	1	1	1	1	1	1	1	1	173	15	2.9	0
S'mores Crunch—1 c (38 gm)	3	3	2	2	2	2	2	1	161	18	7.7	0
Smurfberry Crunch—1 c (33 gm)	1	½	½	½	½	½	0	0	132	10	1.2	0
Special K—1 single serving box (⅝ oz) (18 gm)	0	0	0	0	0	0	0	0	70	1	.1	0
———1 c (33 gm)	0	0	0	0	0	0	0	0	129	1	.3	0
Strawberry Krispies—1 c (45 gm)	0	0	0	0	0	0	0	0	175	0	0	0
Strawberry Shortcake—1 c (28 gm)	½	0	0	0	0	0	0	0	118	4	.6	0
Sugar Smacks (wheat)—1 oz (28 gm)	—	—	—	—	—	—	—	—	106	5	—	0
Sugar-Sparkled Flakes—1 c (38 gm)	0	0	0	0	0	0	0	0	146	1	.1	0
Sugar-Sparkled Rice Krinkles—1 c (32 gm)	0	0	0	0	0	0	0	0	123	1	.1	0
Sun Flakes—1 c (28 gm)	3	3	2	2	2	2	2	1	109	9	7.6	0
Super Golden Crisp—1 c (33 gm)	0	0	0	0	0	0	0	0	123	3	.4	0
Super Sugar Crisp, (wheat)—1 c (33 gm)	—	—	—	—	—	—	—	—	123	3	—	0
Tasteeos—1 c (24 gm)	½	½	½	½	½	0	0	0	94	6	1.1	0
Team—1 c (42 gm)	½	½	½	0	0	0	0	0	164	7	.8	0
Toasted Wheat 'n Raisins—⅔ c (38 gm)	1	1	1	1	1	½	½	½	134	12	2.2	0
Toasties—1 c (23 gm)	0	0	0	0	0	0	0	0	89	0	.0	0
Toasty O's—1 c (28 gm)	1	1	1	1	1	1	1	1	109	16	3.0	0
Total—1 c (33 gm)	½	½	½	½	0	0	0	0	116	6	.9	0
Trix—1 c (28 gm)	1	1	½	½	½	½	½	½	108	4	1.5	0

	1200	1500	1800	2100	2400	2800	3200	3600	TOT CAL	FAT CAL	S/FAT CAL	CHOL MG
Cold (Ready-to-Eat) (cont.)												
Uncle Sam's Hi Fiber Cereal—½ c (55 gm)	36	30	26	23	21	18	16	15	252	93	80.5	0
Waffelos, (wheat w/other grains)—1 c (30 gm)	—	—	—	—	—	—	—	—	121	12	—	0
Wheat Chex—1 c (46 gm)	2	2	2	1	1	1	1	1	169	10	4.9	0
Wheat germ, plain—½ c (56 gm)	4	3	3	3	2	2	2	2	214	54	9.3	0
Wheat germ, w/sugar & honey—½ c (56 gm)	3	3	2	2	2	2	1	1	211	40	7.0	0
Wheaties—1 c (29 gm)	½	0	0	0	0	0	0	0	101	4	.6	0
Wheat 'n Raisin Chex—⅔ c (36 gm)	0	0	0	0	0	0	0	0	123	3	.4	0
Wheat, puffed, plain—1 c (12 gm)	0	0	0	0	0	0	0	0	44	1	.2	0
Whole wheat, cracked (inc home ground)—1 c (120 gm)	1	1	1	1	1	1	1	½	396	19	2.7	0
Cooked												
Barley, cooked (no fat added in cooking) (inc egg barley)—1 c, ckd (162 gm)	½	½	½	0	0	0	0	0	167	5	.8	0
Barley, pearled, light—1 c (200 gm)	0	0	0	0	0	0	0	0	698	18	0	0
Barley, pearled, pot or scotch—1 c (200 gm)	0	0	0	0	0	0	0	0	696	20	0	0
Buckwheat groats, cooked (no fat added in cooking) (inc kasha)—1 c, ckd (168 gm)	1	½	½	½	½	½	0	0	95	6	1.2	0
Bulgur, canned, fr hard red winter wheat, seasoned, w/onion, salt & MSG—1 c (135 gm)	5	5	4	3	3	3	2	2	246	41	12.1	0

Food												
Bulgur, cooked or canned (no fat added in cooking) (inc wheat pilaf)—1 c, ckd (135 gm)	½	½	½	½	½	½	0	0	227	9	1.1	0
Bulgur, parboiled wheat, dry, commercial, fr white wheat—1 c (155 gm)	0	0	0	0	0	0	0	0	553	17	0	0
Bulgur, parboiled wheat, dry, commercial, made fr club wheat—1 c (175 gm)	0	0	0	0	0	0	0	0	628	22	0	0
Corn grits, instant, plain, prep w/water (corn)—1 pkt prep (137 gm)	0	0	0	0	0	0	0	0	82	2	.1	0
Corn grits, instant, w/artificial cheese flavor, prep w/water (corn)—1 pkt prep (142 gm)	—	—	—	—	—	—	—	—	107	8	—	1
Corn grits, instant, w/imitation bacon bits, prep w/water (corn)—1 pkt prep (141 gm)	—	—	—	—	—	—	—	—	104	5	—	0
Corn grits, instant, w/imitation ham bits, prep w/water (corn, soy)—1 pkt prep (141 gm)	—	—	—	—	—	—	—	—	103	4	—	0
Corn grits, white, reg & quick, enriched, cooked w/water, wo/salt (corn)—1 c (242 gm)	0	0	0	0	0	0	0	0	146	5	.4	0
Corn grits, white, reg & quick, enriched, dry (corn)—1 c (156 gm)	—	—	—	—	—	—	—	—	579	16	—	0
Corn grits, yellow, reg & quick, enriched, cooked w/water, wo/salt (corn)—1 c (242 gm)	0	0	0	0	0	0	0	0	146	5	.4	0
Corn grits, yellow, reg & quick, enriched (dry, corn)—1 c (156 gm)	—	—	—	—	—	—	—	—	579	16	—	0
Cornmeal dumpling (inc boiled cornbread)—2 dumplings (160 gm)	35	32	30	28	26	25	23	22	275	74	21.7	163
Cornmeal, made w/milk & sugar, Puerto Rican style (harina de maiz con leche)—1 c, ckd (227 gm)	20	17	15	13	12	11	10	9	306	61	35.4	25

	1200	1500	1800	2100	2400	2800	3200	3600	TOT CAL	FAT CAL	S/FAT CAL	CHOL MG
Cooked (cont.)												
Cornmeal mush, fried (inc made w/water, made w/milk)—1 strip or sl (51 gm)	2	1	1	1	1	1	1	1	84	17	3.8	0
Cornmeal mush, made w/milk—1 c, ckd (240 gm)	33	28	25	22	20	18	17	16	359	98	58.8	42
Cornmeal mush, made w/water (inc cornmeal mush, polenta)—1 c, ckd (240 gm)	½	½	½	½	0	0	0	0	243	9	.9	0
Cornmeal sticks, boiled (inc quanimes)—2 sticks (140 gm)	28	24	21	18	16	14	13	12	247	74	63.4	0
Cornstarch, dry—1 c (128 gm)	0	0	0	0	0	0	0	0	463	0	0	0
Cornstarch w/milk, eaten as a cereal (2 tbsp cornstarch in 2½ cups milk)—1 c (250 gm)	27	23	20	18	17	15	14	13	182	77	48.0	35
Cream of rice, cooked w/water, wo/salt—1 c (244 gm)	0	0	0	0	0	0	0	0	126	1	.2	0
Cream of rice, dry—1 c (173 gm)	—	—	—	—	—	—	—	—	641	7	—	0
Cream of wheat, instant, dry—1 c (178 gm)	—	—	—	—	—	—	—	—	651	23	—	0
Cream of wheat, instant, prep w/water, wo/salt—1 c (241 gm)	½	0	0	0	0	0	0	0	153	5	.6	0
Cream of wheat, mix 'n eat, apple, banana & maple flavor (wheat, corn)—1 pkt prep (150 gm)	—	—	—	—	—	—	—	—	132	4	—	0
Cream of wheat, mix 'n eat, plain, prep w/water (wheat, corn)—1 pkt prep (142 gm)	0	0	0	0	0	0	0	0	102	3	.4	0

Food												
Cream of wheat, quick, cooked w/water, wo/salt, (wheat)—1 c (239 gm)	½	0	0	0	0	0	0	0	129	5	.6	0
Cream of wheat, quick, dry, (wheat)—1 c (175 gm)	—	—	—	—	—	—	—	—	632	21	—	0
Cream of wheat, regular, cooked w/water, wo/salt, (wheat)—¾ c (188 gm)	0	0	0	0	0	0	0	0	100	4	.5	0
Cream of wheat, regular, dry (wheat)—1 c (173 gm)	—	—	—	—	—	—	—	—	640	23	—	0
Farina, enriched, cooked w/water, wo/salt (wheat)—1 c (233 gm)	0	0	0	0	0	0	0	0	116	2	.5	0
Farina, enriched, dry (wheat) (176 gm)	—	—	—	—	—	—	—	—	649	8	—	0
Grits, cooked, corn or hominy, made w/milk (inc reg, quick, or instant)—1 c, ckd (242 gm)	21	18	16	14	13	12	11	10	259	62	37.7	27
Maltex, cooked w/water, wo/salt (wheat)—1 c (249 gm)	—	—	—	—	—	—	—	—	180	10	—	0
Maltex, dry (wheat)—1 c (151 gm)	—	—	—	—	—	—	—	—	531	29	—	0
Malt-O-Meal, chocolate, dry, (wheat, barley)—1 c (165 gm)	—	—	—	—	—	—	—	—	607	13	—	0
Malt-O-Meal, plain & chocolate, cooked w/water, wo/salt, (wheat, barley)—1 c (240 gm)	—	—	—	—	—	—	—	—	122	3	—	0
Malt-O-Meal, plain, dry, (wheat, barley)—1 c (165 gm)	—	—	—	—	—	—	—	—	607	13	—	0
Maypo, cooked w/water, wo/salt, (oats w/other grains)—1 c (240 gm)	—	—	—	—	—	—	—	—	170	22	—	0
Maypo, dry, (oats w/other grains)—1 c (94 gm)	—	—	—	—	—	—	—	—	362	45	—	0
Millet, cooked (no fat added in cooking)—1 c, ckd (131 gm)	1	½	½	½	½	½	0	0	71	6	1.2	0
Multigrain cereal, (no fat added in cooking) (inc seven-grain cereals)—1 c, ckd (246 gm)	2	1	1	1	1	1	1	1	202	20	3.8	0

THE 2-IN-1 SYSTEM

Cooked (cont.)

	1200	1500	1800	2100	2400	2800	3200	3600	TOT CAL	FAT CAL	S/FAT CAL	CHOL MG
Oatmeal, made w/evaporated milk & sugar, Puerto Rican style (avena con leche)—1 c, ckd (210 gm)	28	24	21	19	17	16	14	13	303	90	50.6	35
Oatmeal, made w/milk (inc reg, instant, or quick)—1 c, ckd (234 gm)	23	20	17	16	14	13	12	11	148	67	40.9	29
Oats, instant, fortified, plain, prep w/water—1 pkt (177 gm)	1	1	1	½	½	½	½	½	104	15	1.6	0
Oats, instant, fortified, w/apples & cinnamon, prep w/water—1 pkt (149 gm)	—	—	—	—	—	—	—	—	135	14	—	0
Oats, instant, fortified, w/bran & raisins, dry (oats, wheat bran)—1 pkt (43 gm)	—	—	—	—	—	—	—	—	158	17	—	0
Oats, instant, fortified, w/cinnamon & spice, dry—1 pkt (46 gm)	—	—	—	—	—	—	—	—	177	17	—	0
Oats, instant, fortified, w/maple & brown sugar flavor, prep w/water—1 pkt (155 gm)	—	—	—	—	—	—	—	—	163	17	—	0
Oats, instant, fortified, w/raisins & spice, prep w/water—1 pkt (158 gm)	—	—	—	—	—	—	—	—	161	16	—	0
Oats, regular, quick & instant, unfortified, cooked w/water, wo/salt—1 c (234 gm)	2	1	1	1	1	1	1	1	145	22	4.0	0
Oats, regular, quick & instant, unfortified, dry—1 c (81 gm)	4	3	3	2	2	2	2	2	311	46	8.5	0
Ralston, cooked w/water, wo/salt (oats)—1 c (253 gm)	—	—	—	—	—	—	—	—	134	7	—	0
Ralston, dry (oats)—1 c (118 gm)	—	—	—	—	—	—	—	—	402	23	—	0

Food												
Rice, brown & wild (no fat added in cooking)—1 c, ckd (151 gm)	1	1	1	1	½	½	½	½	225	8	1.8	0
Rice, brown, (no fat added in cooking)—1 c, ckd (195 gm)	1	1	1	1	1	1	1	1	232	11	2.8	0
Rice cake (mochi)—1 cake (9 gm)	0	0	0	0	0	0	0	0	21	0	0	0
Rice, cooked, w/milk—1 c, ckd (200 gm)	21	18	16	14	13	12	11	10	306	62	37.9	27
Rice, cream of, made w/milk & sugar, Puerto Rican style—1 c, ckd (245 gm)	25	22	19	17	16	14	13	12	296	75	45.3	32
Rice, long (no fat added in cooking) (inc transparent noodles)—1 c, ckd (153 gm)	0	0	0	0	0	0	0	0	121	1	.3	0
Rice, sweet—1 c, ckd (175 gm)	0	0	0	0	0	0	0	0	212	1	.4	0
Rice, white & wild (no fat added in cooking)—1 c, ckd (151 gm)	0	0	0	0	0	0	0	0	160	2	.4	0
Rice, white, converted (no fat added in cooking) (inc Uncle Ben's)—1 c, ckd (175 gm)	0	0	0	0	0	0	0	0	186	2	.5	0
Rice, white, instant (no fat added in cooking) (inc Minute Rice, yellow rice)—1 c, ckd (165 gm)	0	0	0	0	0	0	0	0	180	0	0	0
Rice, white, reg (no fat added in cooking)—1 c, ckd (205 gm)	½	0	0	0	0	0	0	0	223	2	.6	0
Rice, wild, cooked (no fat added in cooking)—1 c, ckd (130 gm)	0	0	0	0	0	0	0	0	117	2	.3	0
Roman meal, plain, cooked w/water, wo/salt (wheat w/other grains)—1 c (241 gm)	—	—	—	—	—	—	—	—	147	9	—	0
Roman meal, plain, dry (wheat w/other grains)—1 c (94 gm)	—	—	—	—	—	—	—	—	302	18	—	0
Roman meal w/oats, cooked w/water, wo/salt (wheat w/other grains)—1 c (240 gm)	—	—	—	—	—	—	—	—	169	18	—	0

	1200	1500	1800	2100	2400	2800	3200	3600	TOT CAL	FAT CAL	S/FAT CAL	CHOL MG
Cooked (cont.)												
Roman meal w/oats, dry (wheat w/other grains)—1 c (100 gm)	—	—	—	—	—	—	—	—	340	36	—	0
Rye, cream of—1 c, ckd (251 gm)	0	0	0	0	0	0	0	0	105	3	.3	0
Wheat, cream of, made w/milk—1 c, ckd (243 gm)	24	21	18	16	15	14	12	11	243	72	43.1	31
Wheat, cream of, made w/milk & sugar, Puerto Rican style—1 c ckd (245 gm)	26	22	20	18	16	15	13	12	308	79	46.4	33
Wheatena, cooked w/water, w/salt (wheat)—1 c (243 gm)	—	—	—	—	—	—	—	—	135	10	—	0
Wheatena, dry (wheat)—1 c (141 gm)	—	—	½	½	½	½	½	—	503	36	—	0
Wheat hearts (no fat added in cooking)—1 c, ckd (245 gm)	1	½	½	½	½	½	0	0	151	9	1.3	0
Wheat, rolled (no fat added in cooking)—1 c, ckd (242 gm)	1	½	½	½	½	½	½	0	149	9	1.3	0
Whole wheat cereal, chocolate-flavored (no fat added in cooking) (inc cocoa wheat cereal, Chocolate Malt-O-Meal)—1 c, ckd (246 gm)	2	2	2	1	1	1	1	1	160	15	4.8	0
Whole wheat hot natural cereal, cooked w/water, wo/salt—1 c (242 gm)	1	1	1	1	1	½	½	½	151	8	2.2	0
Whole wheat hot natural cereal, dry —1 c (94 gm)	—	—	—	—	—	—	—	—	321	17	—	0
Wild rice, raw—1 c (160 gm)	0	0	0	0	0	0	0	0	565	10	0	0
CHEESE Natural												
Blue or Roquefort (inc Gorgonzola)—1 oz (28 gm)	24	21	18	16	15	13	12	11	99	72	47.0	21

Food												
Brick (inc beer, Elbinger, Zweizeitige, Wilstermarsch, Bondost,) —1 sl (1 oz) (28 gm)	25	21	19	17	15	14	13	12	104	75	47.3	26
Brie—1 oz (28 gm)	24	20	18	16	15	13	12	11	93	70	43.9	28
Camembert—1 oz (28 gm)	20	17	15	14	12	11	10	9	84	61	38.5	20
Caraway—1 oz (28 gm)	25	22	19	17	16	14	13	12	107	75	47.4	26
Cheddar or American type (inc coon, longhorn, Wisconsin, New York, pioneer, hoop, Tillamook, sharp cheese, chevres)—1 sl (1 oz) (28 gm)	28	24	21	19	17	16	14	13	113	84	53.2	29
Cheshire—1 oz (28 gm)	27	23	20	18	16	15	13	12	110	78	49.7	29
Colby—1 sl (1 oz) (28 gm)	27	23	20	18	17	15	13	12	110	81	50.9	27
Cottage, creamed, large or small curd (inc farmer's)—½ c (105 gm)	14	12	11	10	9	8	7	7	109	43	27.0	16
Cottage, dry curd (inc baker's & pressed cottage cheese, Dutch)—½ c (73 gm)	2	1	1	1	1	1	1	1	62	3	1.8	5
Cottage, lo-fat, 1% fat—4 oz (113 gm)	4	3	3	3	2	2	2	2	82	11	6.6	5
Cottage, lo-fat, 2% fat—4 oz (113 gm)	7	6	5	5	4	4	4	3	101	20	12.4	9
Cottage, lo-fat, w/fruit—½ c (113 gm)	3	3	2	2	2	2	2	2	76	10	6.4	3
Cottage, lo-fat, w/vegetables—½ c (113 gm)	3	3	2	2	2	2	2	2	76	10	6.4	3
Cottage, w/fruit (inc creamed or uncreamed, large or small curd)—½ c (113 gm)	12	10	9	8	7	7	6	5	140	35	21.9	13
Cottage, w/vegetables—½ c (113 gm)	14	12	11	10	9	8	7	7	107	43	27.0	16
Cream—1 tbsp (14 gm)	15	13	11	10	9	8	7	7	49	44	27.7	15
Cream, lo-fat—1 tbsp (15 gm)	8	7	6	5	5	4	4	4	35	24	15.0	8
Feta (inc goat cheese)—1 oz (28 gm)	21	18	16	14	13	12	11	10	74	54	37.7	25
Fontina—1 sl (1 oz) (28 gm)	26	23	20	18	17	15	14	13	109	78	48.4	32
Gjetost—1 oz (28 gm)	26	22	20	18	16	14	13	12	132	76	48.9	27
Gouda or Edam (inc Caciocavallo, Delft, ball cheese)—1 sl (1 oz) (28 gm)	24	20	18	16	15	13	12	11	100	70	44.3	27

Natural (cont.)

	1200	1500	1800	2100	2400	2800	3200	3600	TOT CAL	FAT CAL	S/FAT CAL	CHOL MG
Gruyère—1 sl (1 oz) (28 gm)	26	22	20	18	16	15	13	12	116	81	47.7	31
Havarti—1 sl (1 oz) (28 gm)	25	21	19	17	15	14	13	12	104	75	47.3	26
Limburger—1 oz (28 gm)	23	19	17	15	14	13	11	11	92	69	42.2	25
Monterey Jack—1 sl (1 oz) (28 gm)	25	22	19	17	16	14	13	12	105	76	48.0	25
Mozzarella (inc pizza cheese)—1 sl (1 oz) (28 gm)	14	12	11	10	9	8	7	7	78	43	27.4	15
Mozzarella, part skim milk—1 oz (28 gm)	14	12	11	10	9	8	7	7	72	41	25.8	16
Mozzarella, whole milk—1 oz (28 gm)	18	16	14	13	11	10	9	9	80	55	33.6	22
Muenster—1 sl (1 oz) (28 gm)	26	22	19	17	16	14	13	12	103	76	48.2	27
Neufchâtel—1 oz (28 gm)	20	17	15	14	12	11	10	9	74	59	37.7	22
Parmesan, dry grated (inc Romano)—1 tbsp (5 gm)	4	4	3	3	3	2	2	2	23	14	8.6	4
Parmesan, hard (inc Romano)—1 oz (28 gm)	21	18	16	14	13	12	11	10	110	65	41.4	19
Parmesan, low-sodium—1 tbsp (5 gm)	4	4	3	3	3	2	2	2	23	14	8.6	4
Port du Salut—1 sl (1 oz) (28 gm)	24	21	18	17	15	14	13	12	98	71	42.1	34
Provolone—1 sl (1 oz) (28 gm)	22	19	17	15	13	12	11	10	98	67	43.0	19
Ricotta—½ c (123 gm)	40	35	31	28	25	23	21	19	192	116	73.2	50
Ricotta, part skim milk—½ c (124 gm)	30	26	23	21	19	17	16	14	171	88	55.0	38
Ricotta, whole milk—½ c (124 gm)	51	44	39	35	32	29	26	24	216	145	92.6	63
Romano—1 oz (28 gm)	24	21	18	16	15	13	12	11	110	68	43.7	29
Roquefort—1 oz (28 gm)	26	22	20	18	16	14	13	12	105	78	49.1	26
Swiss (inc Emmentaler, Asiago, Jarlsburg, Samsoe, Danbo, Sweitzer)—1 sl (1 oz) (28 gm)	24	21	18	16	15	13	12	11	105	69	44.8	26
Tilsit—1 sl (1 oz) (28 gm)	23	20	18	16	15	13	12	11	95	65	42.3	29

Processed, Imitation, and Mixtures

Food												
Cheese, deep-fried (inc mozzarella en carozza)—1 c (130 gm)	75	65	58	53	48	44	40	37	471	253	128.3	119
Cheese food, cold pack, American—1 oz (28 gm)	20	17	15	14	12	11	10	9	94	62	39.2	18
Cheese spread, cheddar or American cheese base, lo-fat, low-sodium (inc Velveeta)—1 tbsp (15 gm)	3	3	3	2	2	2	2	2	27	9	6.0	5
Cheese spread, cheddar or American cheese base (inc Velveeta, Cheez Whiz, Old English Smokey, Bacon, Yellow in a Jar)—1 tbsp (15 gm)	9	8	7	6	6	5	5	4	44	29	18.0	8
Cheese spread, cream cheese or Neufchâtel base (inc olive, pimiento, pineapple, onion, clam, roka, relish)—1 tbsp (15 gm)	13	11	10	9	8	7	7	6	44	39	24.3	14
Cheese spread, pressurized can—1 tbsp (14 gm)	9	7	7	6	5	5	4	4	41	27	16.8	8
Cheese straws, made w/lard—10 pieces, 5" x ⅜" x ⅜" (60 gm)	36	31	27	24	22	20	18	17	272	161	70.7	32
Cheese straws, made w/vegetable shortening—10 pieces, 5" x ⅜" x ⅜" (60 gm)	28	24	21	19	17	15	14	13	272	161	57.3	19
Cheese w/nuts (inc cheese balls)—1 tbsp (15 gm)	13	11	10	9	8	7	6	6	62	45	24.1	13
Dip, cheese base other than cream cheese (inc cheese dip)—1 tbsp (16 gm)	10	8	7	7	6	5	5	5	46	31	19.2	9
Dip, cream cheese base—1 tbsp (15 gm)	12	11	9	8	8	7	6	6	46	37	23.5	12
Imitation cheese, American or cheddar type (inc Cheez-ola, Pretend, Country Meadow)—1 tbsp (14 gm)	6	5	4	4	3	3	3	3	33	18	11.1	5

	1200	1500	1800	2100	2400	2800	3200	3600	TOT CAL	FAT CAL	S/FAT CAL	CHOL MG
Processed, Imitation, and Mixtures (cont.)												
Imitation cheese spread (inc Count Down)—1 tbsp (14 gm)	1	½	½	½	½	½	½	½	18	1	.8	1
Imitation cream cheese—1 tbsp (15 gm)	2	2	2	2	1	1	1	1	34	29	5.6	0
Pasteurized processed, pimiento—1 oz (28 gm)	26	23	20	18	16	15	13	12	106	79	50.1	27
Processed, American and Swiss blends—1 sl (.75 oz) (21 gm)	20	17	15	13	12	11	10	9	79	59	37.2	20
Processed, American, lo-fat—1 sl (.75 oz) (21 gm)	5	4	4	3	3	3	3	2	38	13	8.3	7
Processed, cheddar or American type (inc American Cheese Slices, Cheezes, Cheez Kisses)—1 sl (.75 oz) (21 gm)	20	17	15	13	12	11	10	9	79	59	37.2	20
Processed cheese food—1 sl (.75 oz) (21 gm)	15	13	11	10	9	8	7	7	69	46	29.2	13
Processed cheese food spread—1 tbsp (14 gm)	10	9	8	7	6	6	5	5	46	31	19.5	9
Processed, Swiss—1 sl (.75 oz) (21 gm)	16	14	12	11	10	9	8	8	70	47	30.3	18
Processed, Swiss, lo-fat—1 sl (.75 oz) (21 gm)	4	3	3	3	2	2	2	2	36	10	6.2	7
Processed, w/vegetables (inc pepper cheese, pimiento)—1 oz (28 gm)	26	22	20	18	16	14	13	12	105	79	49.6	26
COMPLETE DISHES **Home-Cooked or Restaurant**												
Antipasto w/ham, fish, cheese, vegetables—½ c (58 gm)	20	18	16	15	14	13	12	11	75	46	21.1	67

Food												
Baked beans, w/pork & sweet sauce—½ cup (127 gm)	7	6	6	5	5	4	4	4	191	54	9.5	19
Bean cake—1 cake (32 gm)	4	3	3	3	2	2	2	2	132	62	9.2	0
Beans & franks—¾ c (195 gm)	24	21	18	16	15	13	12	11	281	125	46.0	25
Beans, dry, cooked w/ground beef—¾ c (200 gm)	18	16	14	13	12	11	10	9	272	69	27.1	38
Beans, dry, cooked w/pork—½ c (89 gm)	2	2	2	2	1	1	1	1	120	13	4.0	3
Beef & noodles, no sauce (mixture) (inc Hamburger Helper Beef Noodle)—1 c (156 gm)	24	22	20	18	17	16	15	14	260	73	24.2	86
Beef & noodles w/cream or white sauce—1 c (249 gm)	41	36	32	29	27	25	23	21	400	188	61.8	88
Beef & noodles w/gravy—1 c (249 gm)	29	26	24	22	21	19	18	17	297	80	33.6	93
Beef & noodles w/(mushroom) soup—1 c (249 gm)	37	32	29	27	25	23	21	20	370	130	46.5	103
Beef & noodles w/tomato-base sauce (inc beef casserole)—1 c (249 gm)	24	22	20	18	17	16	15	14	272	59	22.8	89
Beef & potatoes, no sauce—1 c (190 gm)	25	22	20	19	17	16	15	14	273	74	30.7	75
Beef & potatoes w/cheese sauce—1 c (249 gm)	78	68	60	54	50	45	41	38	517	285	133.2	122
Beef & potatoes w/cream or white sauce—1 c (252 gm)	31	27	24	22	20	19	17	16	317	121	46.8	65
Beef & potatoes w/gravy in pie crust—1 piece (⅛ of 8″ sq) (216 gm)	24	21	18	17	15	14	13	12	363	137	39.2	41
Beef & potatoes w/(mushroom) soup—1 c (252 gm)	29	26	23	21	20	18	17	15	318	118	42.8	67
Beef & rice, no sauce—1 c (196 gm)	39	34	31	28	26	23	22	20	375	156	59.7	81
Beef & rice w/cream sauce—1 c (248 gm)	46	40	36	32	30	27	25	23	435	202	75.0	82
Beef & rice w/gravy—1 c (222 gm)	32	28	25	23	21	19	17	16	304	116	49.6	63
Beef & rice w/(mushroom) soup—1 c (248 gm)	40	34	31	28	25	23	21	20	398	184	66.8	65
Beef & rice w/soy-based sauce—1 c (244 gm)	39	34	30	27	25	23	21	20	347	161	60.8	74

	1200	1500	1800	2100	2400	2800	3200	3600	TOT CAL	FAT CAL	S/FAT CAL	CHOL MG
Home-Cooked or Restaurant (cont.)												
Beef & rice w/tomato-base sauce—1 c (244 gm)	23	20	18	16	15	13	12	12	288	86	35.1	45
Beef & vegetables (excluding carrots, broccoli & dark-green leafy; no potatoes), no sauce—¾ c (122 gm)	20	17	16	14	13	12	11	10	181	86	31.0	40
Beef & vegetables (excluding carrots, broccoli & dark green leafy; no potatoes), tomato-base sauce—¾ c (187 gm)	35	31	28	25	23	21	19	18	239	134	54.9	70
Beef & vegetables (excluding carrots, broccoli & dark-green leafy; (mushroom) soup—¾ c (189 gm)	37	32	28	26	24	22	20	19	272	153	57.8	70
Beef & vegetables (excluding carrots, broccoli & dark-green leafy; no potatoes), soy-base sauce—¾ c (163 gm)	12	11	10	9	8	8	7	7	199	83	17.5	30
Beef & vegetables (excluding carrots, broccoli & dark-green leafy; no potatoes), gravy—¾ c (189 gm)	13	12	11	10	10	9	8	8	152	27	10.6	54
Beef & vegetables, Hawaiian style—¾ c (189 gm)	4	4	3	3	3	3	3	2	158	22	6.0	11
Beef & vegetables (inc carrots, broccoli and/or dark-green leafy; no potatoes), no sauce—¾ c (122 gm)	20	17	16	14	13	12	11	10	183	86	31.1	40
Beef & vegetables (inc carrots, broccoli and/or dark green leafy; no potatoes), tomato-base sauce—¾ c (187 gm)	14	12	11	11	10	9	9	8	124	27	10.8	56

Food												
Beef & vegetables (inc carrots, broccoli and/or dark-green leafy; no potatoes), (mushroom) soup—¾ c (189 gm)	37	32	29	26	24	22	20	19	284	154	58.1	71
Beef & vegetables (inc carrots, broccoli and/or dark-green leafy; no potatoes), soy-base sauce—¾ c (163 gm)	13	11	10	9	9	8	7	7	122	64	17.2	33
Beef & vegetables (inc carrots, broccoli and/or dark-green leafy; no potatoes), gravy—¾ c (189 gm)	15	13	12	11	11	10	10	9	143	29	11.8	61
Beef & vegetable stew, canned—1 c (245 gm)	15	13	12	11	10	9	8	8	194	68	22.0	34
Beef & vegetable stew, home recipe, w/lean chuck, wo/salt—1 c (245 gm)	30	26	23	21	20	18	17	15	218	95	44.5	64
Beef bourguignonne—¾ c (183 gm)	17	15	14	12	11	10	10	9	194	72	24.3	41
Beef burgundy—¾ c (183 gm)	26	23	21	19	18	16	15	14	261	101	35.2	67
Beef chow mein or chop suey, no noodles—¾ c (165 gm)	17	15	14	12	11	10	10	9	204	98	25.3	39
Beef chow mein or chop suey w/noodles—1 c (220 gm)	27	23	21	19	17	16	14	13	407	188	43.6	48
Beef curry—¾ c (177 gm)	34	29	26	23	21	19	18	16	345	231	57.8	51
Beef goulash—¾ c (187 gm)	20	18	16	15	14	13	12	11	198	77	24.6	59
Beef goulash w/noodles—1 c (249 gm)	28	25	23	21	19	18	17	16	330	122	34.0	83
Beef goulash w/potatoes—1 c (242 gm)	23	20	18	17	16	14	13	12	291	108	30.6	60
Beef, ground, w/egg and onion—1 med patty (85 gm)	41	37	33	30	28	26	24	23	219	128	52.4	116
Beef loaf (inc beef meatball, w/breading no sauce)—1 med sl (108 gm)	42	37	33	31	28	26	24	22	239	137	59.8	101
Beef, noodles & vegetables (excluding carrots, broccoli & dark-green leafy), no sauce—1 c (162 gm)	16	14	13	12	12	11	10	10	181	38	14.4	62

Home-Cooked or Restaurant (cont.)

	1200	1500	1800	2100	2400	2800	3200	3600	TOT CAL	FAT CAL	S/FAT CAL	CHOL MG
Beef, noodles & vegetables (excluding carrots, broccoli & dark-green leafy), tomato-base sauce—1 c (249 gm)	17	15	14	13	12	11	11	10	223	45	16.8	61
Beef, noodles & vegetables (excluding carrots, broccoli & dark-green leafy), (mushroom) soup—1 c (249 gm)	61	53	47	42	39	35	32	30	396	229	106.7	89
Beef, noodles & vegetables (inc carrots, broccoli and/or dark-green leafy), no sauce—1 c (162 gm)	16	14	13	12	12	11	10	10	181	38	14.3	62
Beef, noodles & vegetables (inc carrots, broccoli and/or dark-green leafy), tomato-base sauce—1 c (249 gm)	15	14	12	12	11	10	9	9	208	41	15.1	54
Beef, noodles & vegetables (inc carrots, broccoli and/or dark-green leafy), (mushroom) soup—1 c (249 gm)	33	29	25	23	21	19	17	16	218	125	58.1	48
Beef, potatoes & vegetables (excluding carrots, broccoli and/or dark-green leafy), no sauce—1 c (162 gm)	13	12	11	10	10	9	8	8	163	33	13.2	48
Beef, potatoes & vegetables (excluding carrots, broccoli & dark-green leafy), (mushroom) soup—1 c (252 gm)	50	43	39	35	32	29	27	25	404	207	78.2	95
Beef, potatoes & vegetables (inc carrots, broccoli and/or dark-green leafy), no sauce—1 c (162 gm)	13	12	11	10	9	9	8	8	163	32	13.1	47

174

Food												
Beef, potatoes & vegetables (inc carrots, broccoli and/or dark-green leafy), (mushroom) soup—1 c (252 gm)	49	43	38	35	32	29	27	25	407	207	77.8	95
Beef pot pie (inc Greek meat pie, sirloin burger pie)—1 pie (8 oz, frozen) (227 gm)	46	39	34	31	28	25	23	21	554	319	90.1	37
Beef, rice & vegetables (excluding carrots, broccoli & dark-green leafy), no sauce—1 c (162 gm)	12	11	10	9	9	8	7	7	174	29	11.9	43
Beef, rice & vegetables (excluding carrots, broccoli & dark-green leafy), (mushroom) soup—1 c (249 gm)	36	32	28	26	24	22	20	18	367	153	57.5	70
Beef, rice & vegetables (excluding carrots, broccoli & dark-green leafy), soy-base sauce—1 c (217 gm)	20	18	16	14	13	12	11	10	216	87	31.1	41
Beef, rice & vegetables (excluding carrots, broccoli and/or dark-green leafy), tomato-base sauce—1 c (249 gm)	12	10	9	9	8	7	7	7	209	33	13.0	37
Beef, rice & vegetables (inc carrots, broccoli and/or dark-green leafy), no sauce—1 c (162 gm)	12	11	10	9	8	8	7	7	174	29	11.8	42
Beef, rice & vegetables (inc carrots, broccoli and/or dark-green leafy), tomato-base sauce—1 c (249 gm)	12	10	9	9	8	7	7	7	207	33	13.0	37
Beef, rice & vegetables (inc carrots, broccoli and/or dark-green leafy), (mushroom) soup—1 c (249 gm)	36	31	28	26	23	21	20	18	371	152	57.1	69
Beef, rice & vegetables (inc carrots, broccoli and/or dark-green leafy), soy-base sauce—1 c (217 gm)	20	17	15	14	13	12	11	10	214	84	30.0	40
Beef (roast) hash—1 c (190 gm)	45	38	34	31	28	25	23	21	382	222	80.2	58
Beef rolls, stuffed w/vegetables or meat mixture (inc roulades, paupiettes, bracciola)—1 beef roll (134 gm)	38	33	29	27	25	22	21	19	296	142	57.3	78

	1200	1500	1800	2100	2400	2800	3200	3600	TOT CAL	FAT CAL	S/FAT CAL	CHOL MG
Home-Cooked or Restaurant (cont.)												
Beef sloppy joe—¾ c (188 gm)	43	37	33	30	27	24	22	21	292	163	73.8	63
Beef steak w/onions, Puerto Rican style (biftec encebollado) (inc Puerto Rican stewed steak)—¾ c (134 gm)	48	41	37	33	31	28	25	24	426	295	79.9	79
Beef stew w/potatoes & vegetables (excluding carrots, broccoli & dark-green leafy), tomato-base sauce—1 c (252 gm)	19	17	15	14	13	12	11	11	226	52	21.1	60
Beef stew w/potatoes & vegetables (excluding carrots, broccoli & dark-green leafy), gravy—1 c (252 gm)	19	17	16	14	13	12	12	11	228	53	21.6	62
Beef stew w/potatoes & vegetables (inc carrots, broccoli and/or dark-green leafy), tomato-base sauce (inc meatball stew)—1 c (252 gm)	12	11	10	9	9	8	8	7	180	35	14.1	40
Beef stew w/potatoes & vegetables (inc carrots, broccoli and/or dark-green leafy), gravy—1 c (252 gm)	14	12	11	10	10	9	8	8	173	39	15.6	45
Beef stew w/potatoes, gravy—1 c (252 gm)	21	18	17	15	14	13	12	11	246	65	26.9	57
Beef stew w/potatoes, tomato-base sauce—1 c (252 gm)	21	18	17	15	14	13	12	11	249	66	27.0	57
Beef Stroganoff—¾ c (192 gm)	43	37	33	30	27	25	23	21	308	186	75.9	61
Beef Stroganoff w/noodles—1 c (256 gm)	44	38	34	30	28	25	23	21	342	182	73.4	71
Beef taco filling, beef, cheese, tomato, taco sauce—¾ c (153 gm)	47	40	36	33	30	27	25	23	259	162	77.5	78

Food												
Beef, tofu & vegetables (excluding carrots, broccoli & dark-green leafy; no potatoes), soy-base sauce—¾ c (163 gm)	14	12	11	10	9	8	8	7	213	96	23.1	26
Beef, tofu & vegetables (inc carrots, broccoli and/or dark-green leafy; no potatoes), soy-base sauce—¾ c (163 gm)	15	13	11	10	9	9	8	7	209	97	23.3	27
Beef Wellington—1 sl (116 gm)	38	33	30	27	25	23	21	20	325	164	52.9	92
Beef w/barbecue sauce—1 rib w/sauce (68 gm)	16	14	13	11	11	10	9	9	125	51	20.6	42
Beef w/cream or white sauce—¾ c (192 gm)	33	29	26	23	21	19	18	16	293	169	55.6	55
Beef w/gravy (inc country style)—1 sl w/gravy (86 gm)	14	13	11	11	10	9	8	8	116	45	18.6	39
Beef w/(mushroom) soup—¾ c (192 gm)	67	58	52	47	44	40	37	34	438	262	102.5	135
Beef w/soy-base sauce—¾ c (183 gm)	17	15	13	12	11	10	9	9	222	104	25.5	37
Beef w/sweet and sour sauce—¾ c (189 gm)	21	18	16	15	14	12	11	11	230	98	33.3	40
Beef w/tomato-base sauce (inc beef w/tomatoes, meatballs w/tomato sauce, spaghetti sauce w/beef)—¾ c (187 gm)	60	52	47	43	39	36	33	31	381	213	88.5	131
Black beans, Cuban style (habichuelas negras guisadas a la Cubana)—1 c (270 gm)	5	5	4	4	3	3	3	2	327	68	11.6	2
Bouillabaisse—1 c (227 gm)	24	22	21	19	18	17	16	15	229	75	17.9	106
Bread pudding, w/raisins—1 c (265 gm)	59	52	47	44	41	38	35	33	496	146	68.9	180
Burrito w/beans & cheese, meatless—(99 gm)	33	28	24	22	20	18	16	15	286	135	63.2	29
Burrito w/beans, meatless—(72 gm)	3	3	2	2	2	2	2	1	149	45	7.7	0
Burrito w/beef & beans—(110 gm)	24	21	19	17	15	14	13	12	291	121	38.9	43
Burrito w/beef & cheese, no beans—(132 gm)	75	65	57	52	47	43	39	36	468	275	130.8	107

	1200	1500	1800	2100	2400	2800	3200	3600	TOT CAL	FAT CAL	S/FAT CAL	CHOL MG
Home-Cooked or Restaurant *(cont.)*												
Burrito w/beef, beans & cheese— (132 gm)	52	45	39	36	33	29	27	25	388	201	91.0	72
Burrito w/beef, beans, cheese & sour cream (inc Taco Bell Burrito Supreme)—(234 gm)	91	78	69	63	57	51	47	43	678	352	163.2	119
Burrito w/beef, no beans—(110 gm)	39	34	30	27	25	23	21	20	340	155	59.7	78
Burrito w/chicken & beans—(106 gm)	12	10	9	8	8	7	7	6	236	73	15.9	29
Burrito w/chicken & cheese—(130 gm)	58	50	44	40	37	33	31	28	396	209	98.7	90
Burrito w/chicken, beans & cheese— (134 gm)	40	35	31	28	25	23	21	19	350	157	69.1	60
Cabbage w/ham hocks—¾ c (150 gm)	17	15	13	12	11	10	9	8	143	78	27.2	31
Calzone, w/cheese, meatless (inc stromboli, Pizza Hut Calizza)—½ calzone or stromboli (212 gm)	77	66	59	53	49	44	40	37	756	342	136.0	107
Calzone, w/meat & cheese (inc stromboli, Pizza Hut Calizza)—½ calzone or stromboli (212 gm)	69	60	53	48	44	40	36	34	736	341	118.0	106
Chalupas w/chicken & cheese—(150 gm)	37	32	28	25	23	21	19	18	311	154	65.2	51
Cheese fondue—½ c (108 gm)	45	39	34	31	28	25	23	21	247	131	84.7	49
Cheese soufflé—½ c (48 gm)	27	24	22	20	19	17	16	15	105	74	30.3	86
Cheese soufflé, fr home recipe— 1 portion (110 gm)	70	62	56	52	48	44	41	39	240	169	85.3	204
Chicken à la king, fr home recipe—1 c (245 gm)	80	70	63	58	53	49	45	42	468	309	114.5	186
Chicken à la king w/vegetables (excluding carrots, broccoli & dark-green leafy; no potatoes), cream or white sauce—¾ c (181 gm)	61	54	49	45	41	38	35	33	343	227	82.2	159

Food												
Chicken à la king w/vegetables (inc carrots, broccoli and/or dark-green leafy; no potatoes), cream or white sauce—¾ c (181 gm)	61	54	49	45	41	38	35	33	348	227	82.2	159
Chicken & noodles, fr home recipe—1 c (240 gm)	39	34	31	28	26	24	22	21	367	167	53.1	96
Chicken & rice, no sauce—1 c (196 gm)	18	16	14	13	12	12	11	10	283	69	19.6	58
Chicken & rice w/cream sauce—1 c (248 gm)	28	24	22	20	18	17	16	15	369	147	41.6	61
Chicken & rice w/(mushroom) soup—1 c (248 gm)	28	25	22	20	19	17	16	15	434	185	41.9	62
Chicken & rice w/tomato-base sauce—1 c (244 gm)	11	10	9	9	8	8	7	7	229	23	5.8	53
Chicken & vegetables (excluding carrots, broccoli & dark-green leafy; no potatoes), soy-base sauce—¾ c (122 gm)	25	22	20	18	17	15	14	13	220	125	34.0	62
Chicken & vegetables (inc carrots, broccoli and/or dark-green leafy; no potatoes), soy-base sauce—¾ c (122 gm)	25	22	20	18	17	15	14	13	211	125	34.0	62
Chicken cacciatore (inc chicken w/tomato sauce, chicken w/tomatoes)—1 wing w/sauce (44 gm)	8	7	7	6	6	5	5	4	83	42	10.4	23
—1 drumstick w/sauce (67 gm)	13	11	10	9	9	8	7	7	126	63	15.9	35
—1 thigh w/sauce (77 gm)	14	13	11	11	10	9	8	8	145	73	18.2	40
—½ breast w/sauce (128 gm)	24	21	19	18	16	15	14	13	241	121	30.3	67
—1 leg w/sauce (144 gm)	27	24	21	20	18	17	16	15	271	136	34.1	75
Chicken chow mein or chop suey, no noodles—¾ c (165 gm)	11	10	9	8	8	7	7	6	145	55	12.2	37
Chicken chow mein or chop suey w/noodles—1 c (220 gm)	18	16	14	13	12	11	10	10	286	117	24.3	46
Chicken cordon bleu—1 roll (½ breast w/ham & sauce) (229 gm)	83	72	64	59	54	49	45	42	437	243	132.5	155

179

	1200	1500	1800	2100	2400	2800	3200	3600	TOT CAL	FAT CAL	S/FAT CAL	CHOL MG
Home-Cooked or Restaurant (cont.)												
Chicken cornbread—1 piece (116 gm)	33	31	28	26	25	23	22	21	205	92	22.4	151
Chicken creole, wo/rice—¾ c (185 gm)	10	9	8	8	7	7	7	6	137	29	5.9	46
Chicken divan—¾ c (177 gm)	34	30	28	25	24	22	20	19	238	91	38.9	108
Chicken egg foo yung—1 omelet (86 gm)	43	39	36	34	33	31	29	28	121	74	22.2	210
Chicken fricassee—¾ c (183 gm)	25	22	20	18	17	16	14	13	239	122	35.3	62
Chicken fricassee, fr home recipe—1 c (240 gm)	43	38	34	31	29	26	24	23	386	201	64.3	96
Chicken fricassee, no potatoes, Puerto Rican style (sauce & potatoes reported separately)—¾ c (167 gm)	46	40	36	33	31	28	26	25	417	252	59.5	123
Chicken fricassee, Puerto Rican style (fricase de pollo), inc stewed chicken (pollo guisado)—1 c, boneless (223 gm)	45	39	35	32	29	27	25	23	479	327	70.3	89
Chicken fricassee, w/sauce, no potatoes, Puerto Rican style (potatoes reported separately) (inc pollo en salsa)—¾ c (167 gm)	33	29	26	24	22	20	19	17	375	262	50.2	72
Chicken Kiev—1 serving (1 whole breast) (258 gm)	109	96	87	80	74	68	63	59	639	292	143.0	292
Chicken, noodles & vegetables (excluding carrots, broccoli & dark-green leafy), no gravy or sauce—1 c (162 gm)	12	11	10	9	9	8	8	7	174	34	8.6	50

Food												
Chicken, noodles & vegetables (excluding carrots, broccoli & dark-green leafy), w/gravy—1 c (224 gm)	17	15	14	13	12	11	10	9	225	91	25.7	39
Chicken, noodles & vegetables (excluding carrots, broccoli & dark-green leafy), w/cheese sauce—1 c (244 gm)	48	42	38	35	32	29	27	25	398	169	70.0	110
Chicken, noodles & vegetables (inc carrots, broccoli and/or dark-green leafy), no gravy or sauce—1 c (162 gm)	12	11	10	10	9	8	8	8	169	34	8.8	53
Chicken, noodles & vegetables (inc carrots, broccoli and/or dark-green leafy), w/gravy—1 c (224 gm)	22	19	17	16	15	14	13	12	237	103	29.3	57
Chicken, noodles & vegetables (inc carrots, broccoli and/or dark-green leafy), w/cheese sauce—1 c (244 gm)	46	40	36	33	30	28	26	24	363	160	66.4	104
Chicken or turkey & noodles, no sauce—1 c (157 gm)	22	20	18	17	16	15	14	13	253	78	21.4	83
Chicken or turkey & noodles w/cream or white sauce—1 c (224 gm)	31	27	24	22	21	19	18	17	312	118	39.1	85
Chicken or turkey & noodles w/gravy—1 c (224 gm)	27	24	22	20	19	17	16	15	300	125	35.3	75
Chicken or turkey & noodles w/(mushroom) soup—1 c (224 gm)	27	24	22	20	19	17	16	15	307	110	32.6	81
Chicken or turkey & potatoes w/gravy—1 c (242 gm)	23	20	18	17	15	14	13	12	283	117	33.9	53
Chicken or turkey & vegetables (excluding carrots, broccoli & dark-green leafy; no potatoes), no gravy—¾ c (122 gm)	8	7	7	6	6	6	5	5	115	26	6.9	33
Chicken or turkey & vegetables (excluding carrots, broccoli & dark-green leafy; no potatoes), w/gravy—¾ c (189 gm)	20	17	16	14	13	12	12	11	209	90	25.1	55

THE 2-IN-1 SYSTEM

	1200	1500	1800	2100	2400	2800	3200	3600	TOT CAL	FAT CAL	S/FAT CAL	CHOL MG
Home-Cooked or Restaurant *(cont.)*												
Chicken or turkey & vegetables (inc carrots, broccoli and/or dark-green leafy; no potatoes), no sauce—¾ c (122 gm)	8	8	7	7	6	6	5	5	111	26	7.0	34
Chicken or turkey & vegetables (inc carrots, broccoli and/or dark-green leafy; no potatoes), w/gravy—¾ c (189 gm)	20	17	16	14	13	12	11	11	199	89	25.0	55
Chicken or turkey cake, patty, or croquette—1 croquette (62 gm)	15	13	12	11	10	9	8	8	162	87	23.4	30
—1 cake or patty (85 gm)	21	18	16	15	13	12	11	11	222	120	32.1	41
Chicken or turkey hash—1 c (190 gm)	17	15	13	12	11	10	10	9	239	100	22.1	45
Chicken or turkey pot pie—1 pie (8 oz, frozen) (227 gm)	48	42	37	33	30	27	25	23	498	265	87.2	62
Chicken or turkey salad—¾ c (137 gm)	29	26	23	22	20	18	17	16	314	216	38.1	80
Chicken or turkey tetrazzini—1 c (246 gm)	35	31	27	24	22	20	18	17	351	176	62.1	50
Chicken or turkey w/cream sauce—1 wing w/sauce (44 gm)	8	7	6	6	5	5	4	4	74	41	12.3	17
—1 drumstick w/sauce (68 gm)	13	11	10	9	8	8	7	7	115	63	19.0	27
—1 thigh w/sauce (78 gm)	15	13	11	10	10	9	8	7	132	73	21.8	31
—½ breast w/sauce (129 gm)	24	21	19	17	16	14	13	12	218	120	36.0	51
—1 leg w/sauce (145 gm)	27	23	21	19	18	16	15	14	245	135	40.5	57
Chicken parmigiana—1 piece w/sauce & cheese (182 gm)	46	41	37	35	32	30	28	26	318	141	49.3	154
Chicken, potatoes & vegetables (excluding carrots, broccoli & dark-green leafy), no sauce—1 c (162 gm)	16	15	14	13	12	11	11	10	199	50	13.6	67

Food												
Chicken, potatoes & vegetable (inc carrots, broccoli and/or dark-green leafy), no sauce—1 c (162 gm)	16	15	14	13	12	11	11	10	194	50	13.6	67
Chicken pot pie, fr home recipe, baked —1 piece (⅓ of pie) (232 gm)	55	47	42	38	34	31	28	26	545	282	98.4	72
Chicken, rice & vegetables (excluding carrots, broccoli & dark-green leafy), no gravy or sauce—1 c (162 gm)	10	9	8	8	7	7	6	6	186	31	8.4	41
Chicken, rice & vegetables (excluding carrots, broccoli & dark-green leafy), w/gravy—1 c (252 gm)	30	26	23	21	20	18	17	16	382	147	41.8	71
Chicken, rice & vegetables (excluding carrots, broccoli & dark-green leafy), w/soy-base sauce—1 c (217 gm)	17	15	14	12	11	10	10	9	214	88	25.3	39
Chicken, rice & vegetables (excluding carrots, broccoli & dark-green leafy), (mushroom) soup—1 c (252 gm)	27	24	21	20	18	16	15	14	374	140	39.4	62
Chicken, rice & vegetables (inc carrots, broccoli and/or dark-green leafy), no gravy or sauce—1 c (162 gm)	10	9	8	8	7	7	7	6	175	31	8.5	41
Chicken, rice & vegetables (inc carrots, broccoli and/or dark-green leafy), w/gravy—1 c (252 gm)	28	24	22	20	19	17	16	15	347	138	39.2	67
Chicken, rice & vegetables (inc carrots, broccoli and/or dark-green leafy), w/soy-base sauce—1 c (217 gm)	18	16	14	13	12	11	10	9	235	91	26.2	40
Chicken, rice & vegetables (inc carrots, broccoli and/or dark-green leafy), (mushroom) soup—1 c (252 gm)	27	24	21	19	18	16	15	14	340	140	39.2	62
Chicken soufflé—1 c (159 gm)	65	59	54	51	48	45	42	40	293	176	49.8	275
Chicken spread, canned—1 tbsp (13 gm)	3	3	2	2	2	2	2	2	25	14	4.0	7
Chicken stew w/potatoes & vegetables (excluding carrots, broccoli & dark-green leafy), gravy—1 c (252 gm)	17	15	14	13	13	12	11	11	217	35	8.8	82

Home-Cooked or Restaurant (*cont.*)

	1200	1500	1800	2100	2400	2800	3200	3600	TOT CAL	FAT CAL	S/FAT CAL	CHOL MG
Chicken stew w/potatoes & vegetables (excluding carrots, broccoli & dark-green leafy), tomato-base sauce—1 c (247 gm)	17	15	14	13	13	12	11	11	226	35	8.8	81
Chicken stew w/potatoes & vegetables (inc carrots, broccoli and/or dark-green leafy), gravy—1 c (252 gm)	29	26	24	22	20	19	17	16	288	127	35.7	86
Chicken stew w/potatoes & vegetables (inc carrots, broccoli and/or dark-green leafy), tomato-base sauce—1 c (247 gm)	16	15	14	13	13	12	11	11	217	35	8.7	81
Chicken teriyaki (chicken w/soy-base sauce)—1 wing w/sauce (47 gm)	6	6	5	5	5	4	4	4	66	12	3.0	30
—1 drumstick w/sauce (70 gm)	9	8	8	7	7	6	6	6	98	18	4.5	44
—1 thigh w/sauce (76 gm)	10	9	8	8	7	7	7	6	107	19	4.9	48
—½ breast w/sauce (128 gm)	16	15	14	13	12	12	11	10	180	32	8.2	80
—1 leg w/sauce (145 gm)	18	17	16	15	14	13	13	12	204	36	9.3	91
Chicken w/barbecue sauce—1 wing w/sauce (45 gm)	10	9	8	7	7	6	6	6	86	37	10.3	33
—1 drumstick w/sauce (68 gm)	15	13	12	11	10	10	9	8	130	57	15.5	50
—1 thigh w/sauce (74 gm)	16	14	13	12	11	10	10	9	141	62	16.9	54
—½ breast w/sauce (123 gm)	26	24	22	20	19	17	16	15	235	102	28.0	90
—1 leg w/sauce (140 gm)	30	27	25	23	21	20	18	17	268	117	31.9	102
Chicken w/cheese sauce—¾ c (181 gm)	35	31	28	26	24	22	20	19	272	110	43.8	98
Chicken w/dumplings—1 c (244 gm)	38	33	30	27	25	23	21	20	373	180	51.5	94
Chicken w/gravy—1 sl w/gravy (30 gm)	4	4	3	3	3	3	2	2	40	18	4.6	13
—1 chicken wing w/gravy (44 gm)	6	5	5	5	4	4	4	3	59	26	6.8	20

Food												
—1 chicken drumstick w/gravy (68 gm)	9	8	8	7	7	6	6	5	91	40	10.5	30
—1 chicken thigh w/gravy (78 gm)	11	10	9	8	8	7	7	6	104	46	12.1	35
—½ chicken breast w/gravy (129 gm)	18	16	14	13	12	11	11	10	172	76	19.9	57
—1 chicken leg w/gravy (145 gm)	20	18	16	15	14	13	12	11	193	85	22.4	64
Chicken w/(mushroom) soup—¾ c (192 gm)	30	27	25	23	21	19	18	17	281	124	35.4	94
Chili con carne—1 c (254 gm)	31	27	24	22	20	18	17	16	311	124	46.5	64
Chili con carne w/beans & macaroni—1 c (253 gm)	25	22	19	18	16	15	14	13	310	101	37.6	51
Chili con carne w/beans & rice—1 c (250 gm)	18	15	13	12	11	10	9	8	308	82	30.7	25
Chili con carne w/beans (inc kidney beans & hamburger w/tomato sauce)—1 c (254 gm)	31	27	24	22	20	18	17	16	311	124	46.5	64
Chili con carne w/beans, made w/pork—1 c (254 gm)	21	18	17	15	14	13	12	11	273	90	27.4	55
Chili con carne wo/beans—1 c (254 gm)	41	36	32	29	27	25	23	21	331	164	62.5	86
Chili con carne w/venison & beans—1 c (254 gm)	14	13	12	11	10	9	9	8	242	52	15.0	49
Chili dogs (frankfurters w/chili con carne, no bun)—1 frankfurter w/sauce (125 gm)	32	28	24	22	20	18	16	15	251	162	60.1	36
Chili w/beans, canned—½ c (128 gm)	15	13	12	11	10	9	8	7	144	63	27.1	22
Chimichanga w/beans & cheese, meatless—(118 gm)	22	19	17	15	14	12	11	10	219	116	43.1	19
Chimichanga w/beef & cheese—(118 gm)	38	33	29	27	24	22	20	19	282	140	67.7	54
Chimichanga w/chicken & sour cream, no cheese—(118 gm)	27	23	21	19	17	15	14	13	214	100	48.4	35
Chitterlings stewed, Puerto Rican style (cuajo guisado)—¾ c (180 gm)	94	82	74	67	62	56	52	49	564	483	142.8	198

	1200	1500	1800	2100	2400	2800	3200	3600	TOT CAL	FAT CAL	S/FAT CAL	CHOL MG
Home-Cooked or Restaurant *(cont.)*												
Chop suey w/meat, cooked fr home recipe—1 c (250 gm)	50	43	39	35	32	30	27	25	300	153	76.7	100
Chow fun noodles w/meat & vegetables—1 c (152 gm)	11	9	8	8	7	6	6	5	170	75	17.5	19
Chow fun noodles w/vegetables, meatless—1 c (152 gm)	1	1	1	1	1	1	½	½	125	18	2.3	0
Chow mein, chicken, wo/noodles, cooked fr home recipe—1 c (250 gm)	22	20	18	17	16	14	14	13	255	90	21.9	78
Chow mein or chop suey, no noodles—¾ c (165 gm)	20	17	16	14	13	12	11	10	217	110	28.6	46
Clam cake or patty (inc deviled)—1 cake or patty (120 gm)	40	36	33	30	28	26	24	23	397	172	42.4	135
Clam sauce, white—¾ c (180 gm)	23	20	18	16	15	14	12	12	330	274	38.2	40
Clams, stuffed (inc clams casino)—3 sm (12 sm in 11-oz package) (78 gm)	11	10	9	8	8	7	6	6	149	81	16.3	26
Coconut-rice pudding—1 c (204 gm)	1	1	1	½	½	½	½	½	292	3	1.7	0
Codfish ball or cake—1 cake (120 gm)	25	22	20	18	17	16	15	14	227	111	29.3	76
Codfish salad, Puerto Rican style (ensalada de bacalao)—1 c (150 gm)	19	17	16	14	14	12	12	11	223	150	21.8	62
Codfish salad, Puerto Rican style (serenata) (inc oil, vinegar, onion, olives, tomatoes)—¾ c (109 gm)	12	10	9	8	7	7	6	6	184	150	20.4	16
Codfish stewed, Puerto Rican style (bacalao guisado)—1 c (200 gm)	13	12	11	10	9	9	8	8	250	99	14.0	46
Corned beef hash—1 c (190 gm)	51	44	39	35	32	29	26	24	344	193	92.7	63
Corned beef patty—1 patty (100 gm)	27	23	20	18	17	15	14	13	181	102	48.8	33

Corned beef, potatoes & vegetables (excluding carrots, broccoli & dark-green leafy), no sauce—1 c (162 gm)	25	22	19	18	16	15	13	12	218	113	41.3	43
Corned beef, potatoes & vegetables (inc carrots, broccoli and/or dark-green leafy), no sauce—1 c (162 gm)	23	20	18	16	15	13	12	11	204	103	37.5	39
Corned beef w/tomato sauce & onion, Puerto Rican style—¾ c (176 gm)	46	40	36	33	31	28	26	24	313	166	64.2	112
Corn fritters, white—1 fritter (35 gm)	13	11	10	9	9	8	7	7	132	68	18.3	31
Corn fritters, yellow—1 fritter (35 gm)	13	11	10	9	9	8	7	7	132	68	18.3	31
Crab cake—1 cake (120 gm)	40	36	34	32	30	29	27	26	203	97	19.5	199
Crab, deviled—¾ c (131 gm)	32	29	27	25	24	22	21	20	252	117	24.2	139
Crab imperial (inc stuffed crab)—1 stuffed crab (194 gm)	58	53	49	46	44	41	39	37	281	143	34.2	276
Crab salad—¾ c (156 gm)	23	21	20	18	17	16	15	15	205	96	14.3	107
Creamed egg—1 egg (145 gm)	77	71	66	62	58	55	52	49	232	155	47.2	362
Creamed chipped or dried beef—¾ c (186 gm)	31	27	23	21	19	17	16	15	275	162	55.4	40
Creamed dried beef on toast—1 sl toast w/sauce (145 gm)	21	18	16	14	13	12	11	10	252	113	38.1	27
Crepes, filled w/meat, fish, or poultry—1 crepe w/filling, w/sauce (154 gm)	43	38	34	31	29	27	25	23	250	131	59.1	109
Croissant, filled w/broccoli & cheese (inc Sara Lee Le Sanwich)—(113 gm)	50	43	38	34	31	27	25	23	302	155	95.5	48
Croissant, filled w/chicken & broccoli (inc Sara Lee Le Sanwich)—(128 gm)	53	46	41	37	33	30	28	25	346	169	94.4	72
Croissant, filled w/ham & cheese (inc Sara Lee Le Sanwich)—(113 gm)	56	48	42	38	35	31	28	26	337	176	102.8	65
Deer loaf (inc deer meatball, w/breading, no sauce)—1 med sl 108 gm)	20	19	17	16	15	14	13	13	153	44	16.2	85
Dim sums, meat-filled (egg roll type) (inc shrimp, pork, ham)—3 dim sums (84 gm)	18	16	15	13	13	12	11	10	158	72	20.6	57

Home-Cooked or Restaurant (cont.)

	1200	1500	1800	2100	2400	2800	3200	3600	TOT CAL	FAT CAL	S/FAT CAL	CHOL MG
Dressing w/chicken & vegetables—1 c (161 gm)	32	28	26	24	22	20	19	18	346	166	36.7	98
Dressing w/oysters—1 c (161 gm)	16	14	12	11	10	9	8	8	289	133	28.8	21
Dumplings, fried, pork—1 dumpling (100 gm)	29	25	22	20	18	16	15	14	350	200	55.0	32
Dumplings, meat-filled (inc pierogi, piroshki, kreplach)—1 dumpling (97 gm)	32	28	24	22	20	18	16	15	349	207	63.2	28
Dumplings, plain—1 med (32 gm)	2	1	1	1	1	1	1	1	42	10	3.3	1
Dumplings, potato- or cheese-filled (inc pierogi)—1 dumpling (57 gm)	10	9	9	8	8	7	7	6	105	23	9.2	41
Dumplings, steamed, filled w/meat, poultry, or seafood (inc shui-mai)—1 dumpling (37 gm)	5	5	4	4	4	4	3	3	48	13	3.8	23
Egg dessert, custardlike, made w/water & sugar, Puerto Rican style (tocino del cielo, heaven's delight)—1 c (265 gm)	227	210	198	187	178	169	161	154	890	232	69.8	1257
Egg foo yung—1 omelet (86 gm)	45	41	39	37	35	33	32	30	85	46	13.7	247
Egg roll, meatless—(64 gm)	11	10	9	8	8	7	7	6	103	53	11.4	39
Egg roll, w/meat (inc Chinese rolls, spring rolls, lumpia)—(64 gm)	14	12	11	10	10	9	8	8	118	60	14.4	46
Egg roll, w/shrimp—(64 gm)	12	11	10	9	9	8	8	7	106	51	10.8	47
Egg salad—½ cup (111 gm)	95	87	81	76	72	67	64	61	358	312	58.3	446
Eggs a la Malaguena, Puerto Rican style (huevos a la Malaguena)—1 egg (123 gm)	55	51	48	45	43	41	39	37	131	60	17.9	300

Food												
Eggs Benedict—1 egg (155 gm)	70	63	58	54	51	47	45	42	291	164	58.4	281
Eggs, deviled—½ egg (31 gm)	29	27	25	24	23	22	20	20	63	47	11.3	156
Enchilada w/beans & cheese, meatless—(131 gm)	19	17	15	13	12	11	10	9	230	91	37.5	18
Enchilada w/beans, meatless—(118 gm)	2	2	2	1	1	1	1	1	164	35	4.8	0
Enchilada w/beef & beans—(116 gm)	11	9	8	8	7	6	6	5	192	63	16.7	21
Enchilada w/beef & cheese, no beans—(105 gm)	24	21	19	17	15	14	13	12	216	101	42.2	35
Enchilada w/beef, beans & cheese (inc Taco Bell enchirito)—(129 gm)	20	17	15	14	13	11	10	10	235	94	35.4	28
Enchilada w/beef, no beans—(114 gm)	19	17	15	14	13	12	11	10	220	91	28.6	43
Enchilada w/cheese, meatless, no beans—(102 gm)	30	26	23	21	19	17	15	14	223	117	58.2	30
Enchilada w/chicken & beans—(113 gm)	6	5	5	4	4	4	3	3	167	45	7.5	15
Enchilada w/chicken & cheese, no beans—(126 gm)	24	21	18	16	15	14	12	12	237	103	41.2	35
Enchilada w/chicken, beans & cheese—(126 gm)	16	14	12	11	10	9	8	8	214	79	28.2	23
Enchilada w/ham & cheese, no beans—(105 gm)	20	17	15	14	13	11	10	10	198	92	36.7	25
Enchilada w/seafood—(126 gm)	12	11	10	9	9	8	8	8	175	43	6.0	58
Fajita w/chicken—1 tortilla (223 gm)	19	17	15	14	13	12	11	10	393	135	29.1	41
Falafel—1 patty (approx 2¼" dia) (17 gm)	2	2	1	1	1	1	1	1	56	27	4.1	0
Fish a la creole, Puerto Rican style (pescado frito con mojo)—1 sl (4" x 3½" x ½") with ½ c sauce (213 gm)	25	22	20	18	17	16	15	14	349	216	27.9	79
Fish & rice w/cream sauce—1 c (248 gm)	32	28	25	23	21	19	17	16	405	173	54.2	53
Fish & rice w/(mushroom) soup—1 c (248 gm)	25	22	20	18	16	15	14	13	350	135	41.8	43
Fish & rice w/tomato-base sauce—1 c (248 gm)	19	16	15	13	12	11	10	10	318	84	27.0	42

Home-Cooked or Restaurant (cont.)	1200	1500	1800	2100	2400	2800	3200	3600	TOT CAL	FAT CAL	S/FAT CAL	CHOL MG
Fish cake (kamaboko) tempura—1 oz, ckd (28 gm)	3	3	3	3	2	2	2	2	46	15	3.6	12
Fish moochim (Korean style), dried fish w/soy sauce—1 tbsp (5 gm)	1	1	1	1	1	1	1	1	19	7	1.0	6
Fish sauce (bagoong)—1 c (272 gm)	30	27	26	24	23	22	21	20	283	29	8.3	166
Fish timbale or mousse—3/4 c (131 gm)	86	76	68	62	58	53	49	46	272	209	122.3	205
Fish, tofu & vegetables, tempura, Hawaiian—1 c (63 gm)	27	25	23	22	20	19	18	17	139	94	19.2	121
Fish w/cream or white sauce, not tuna or lobster—3/4 c (187 gm)	43	39	36	33	31	29	27	25	252	137	41.9	158
Fish w/tomato-base sauce (inc fish w/tomatoes)—3/4 c (167 gm)	14	13	12	12	11	10	10	9	152	42	8.7	68
Flavored pasta (inc Lipton Beef Flavor, Lipton Chicken Flavor)—3/4 c (139 gm)	6	6	5	4	4	4	3	3	88	57	12.1	7
Flavored rice, brown & wild—3/4 c (163 gm)	3	3	2	2	2	2	1	1	177	42	7.3	0
Flavored rice mixture (inc Rice-a-roni all flavors, Lipton Rice & Sauce, Uncle Ben's Rice Oriental)—3/4 c (164 gm)	19	16	15	13	12	11	10	9	104	61	37.1	18
Flavored rice mixture w/cheese (inc rice au gratin)—3/4 c (173 gm)	4	3	3	2	2	2	2	2	149	50	7.6	1
Flavored rice, white & wild—3/4 c (136 gm)	2	1	1	1	1	1	1	1	129	22	4.0	0
Flounder, stuffed—1 piece (210 gm)	36	33	31	29	27	26	24	23	311	93	20.5	174
Frankfurters & sauerkraut—1 frankfurter w/sauerkraut (120 gm)	23	20	17	16	14	13	12	11	158	119	43.8	23

Food												
Gefilte fish—5 balls (41 gm)	6	6	5	5	5	5	4	4	48	17	3.4	31
—1 c (227 gm)	35	32	30	28	27	25	24	23	268	94	19.0	170
Gizzards, stewed, Puerto Rican style (mollejitas guisadas)—¾ c (195 gm)	54	49	45	42	40	37	35	33	397	192	43.6	225
Gnocchi, cheese—1 c (70 gm)	22	19	17	16	15	14	13	12	125	75	28.2	59
Goat, stewed, Puerto Rican style (cabrito en fricase, chilindron de chivo) (inc Puerto Rican stewed kid) —1 piece with gravy (88 gm)	36	31	27	25	23	20	19	17	266	202	61.6	53
Grape leaves stuffed w/rice—1 roll (56 gm)	4	3	3	3	2	2	2	2	101	68	9.2	0
Ground beef, stewed, Puerto Rican style (picadillo para relleno)—¾ c (113 gm)	54	46	41	37	34	31	28	26	373	267	91.2	84
Ground beef w/tomato sauce & taco seasonings on a cornbread crust (inc Hamburger Helper Taco Bake)—1 piece (⅛ of whole pan) (182 gm)	42	37	33	31	28	26	24	23	375	145	54.1	114
Gumbo, no rice (New Orleans type w/shellfish, pork, and/or poultry, tomatoes, okra)—¾ c (183 gm)	8	7	6	6	5	5	5	4	95	40	8.3	26
Gumbo w/rice (New Orleans type w/shellfish, pork, and/or poultry, tomatoes, okra)—1 c (244 gm)	14	12	11	10	9	9	8	8	191	71	15.0	44
Haddock cake or patty—1 cake or patty (120 gm)	25	22	20	18	17	16	15	14	227	111	29.3	76
Ham & noodles, no sauce—1 c (157 gm)	18	16	15	14	13	12	11	11	237	69	18.4	64
Ham & noodles w/cream or white sauce—1 c (244 gm)	38	33	30	27	25	22	21	19	396	183	61.1	70
Ham & rice, no sauce—1 c (196 gm)	16	14	13	12	11	10	9	9	278	72	20.1	46
Ham & rice w/(mushroom) soup—1 c (248 gm)	21	18	16	15	13	12	11	10	318	116	36.1	32
Ham & vegetables (excluding carrots, broccoli & dark green leafy; no potatoes), no sauce—¾ c (122 gm)	21	18	16	14	13	12	11	10	165	93	33.0	38

	1200	1500	1800	2100	2400	2800	3200	3600	TOT CAL	FAT CAL	S/FAT CAL	CHOL MG
Home-Cooked or Restaurant (cont.)												
Ham & vegetables (inc carrots, broccoli and/or dark green leafy; no potatoes), no sauce—¾ c (122 gm)	21	18	16	14	13	12	11	10	171	93	33.0	38
Ham loaf (not luncheon meat) (inc pork meatball, w/breading, no sauce)—1 med sl (108 gm)	33	29	27	25	24	22	21	19	210	77	30.2	124
Ham or pork, noodles & vegetables (excluding carrots, broccoli & dark-green leafy), cheese sauce—1 c (241 gm)	59	51	46	41	38	34	32	29	463	225	97.1	102
Ham or pork salad—¾ c (137 gm)	26	23	20	19	17	16	15	14	298	207	38.1	58
Ham, potatoes & vegetables (excluding carrots, broccoli & dark-green leafy), no sauce—1 c (162 gm)	22	19	17	15	14	13	12	11	199	100	35.1	40
Ham, potatoes & vegetables (inc carrots, broccoli and/or dark-green leafy), no sauce—1 c (162 gm)	24	21	19	17	16	14	13	12	224	110	38.7	44
Ham pot pie—1 piece (⅛ of pie) (105 gm)	17	15	13	12	11	9	9	8	248	134	34.5	13
Ham Stroganoff (inc ham w/cream or white sauce)—¾ c (183 gm)	40	34	30	27	25	22	20	19	266	168	72.4	49
Italian pie, meatless (inc Pizza Hut Priazzo Florentine)—1 piece (⅛ of 11½" dia) (163 gm)	30	25	22	20	18	16	15	14	444	137	57.2	28
Italian pie w/meat (inc Priazzo, Pizza Hut Priazzo Roma, Pizza Hut Priazzo Milano, Pizza Hut Priazzo Verona)—1 piece (⅛ of 11½" dia) (191 gm)	37	32	28	25	23	21	19	17	566	210	69.8	40

Food												
Julienne salad (meat, cheese, eggs, vegetables), no dressing—1 c (76 gm) (16 gm)	22	20	18	17	16	15	14	13	73	39	17.5	91
Kamaboko (Japanese fish cake)—1 sl	1	1	1	1	1	1	1	1	18	1	.1	6
Knish, cheese—(60 gm)	22	20	18	17	16	14	14	13	209	108	25.7	71
Knish, meat—(50 gm)	21	19	17	16	15	13	13	12	176	95	23.8	66
Knish, potato—(61 gm)	21	18	17	16	14	13	13	12	202	104	22.1	69
Kung pao chicken—¾ c (122 gm)	22	19	17	15	14	13	12	11	308	194	32.9	45
Lamb loaf—1 med sl (108 gm)	41	36	33	30	28	25	24	22	211	119	54.2	108
Lamb or mutton & noodles w/gravy—1 c (249 gm)	30	26	24	22	20	18	17	16	347	153	38.9	79
Lamb or mutton & potatoes w/gravy—1 c (252 gm)	31	27	24	22	20	18	17	16	261	103	45.6	66
Lamb or mutton & potatoes w/tomato-base sauce—1 c (252 gm)	31	27	24	22	20	18	17	16	270	104	45.7	66
Lamb or mutton stew w/potatoes & vegetables (excluding carrots, broccoli & dark-green leafy), gravy—1 c (252 gm)	28	24	22	20	18	17	15	14	275	101	44.5	53
Lamb or mutton stew w/potatoes & vegetables (inc carrots, broccoli and/or dark-green leafy), gravy—1 c (252 gm)	28	24	22	20	18	17	15	14	283	100	44.4	53
Lamb, rice & vegetables (excluding carrots, broccoli & dark-green leafy), gravy—1 c (252 gm)	16	14	12	11	10	9	9	8	251	70	26.4	28
Lamb, rice & vegetables (inc carrots, broccoli and/or dark-green leafy), tomato-base sauce—1 c (252 gm)	15	13	12	10	10	9	8	7	233	66	24.1	27
Lamb, rice & vegetables (inc carrots, broccoli and/or dark-green leafy), gravy—1 c (252 gm)	23	20	18	17	15	14	13	12	245	83	30.9	59
Lamb w/gravy—1 sl w/gravy (65 gm)	11	10	9	8	7	7	6	6	83	29	12.7	33
Lasagna, meatless—1 piece (⅙ of 8" sq, approx 2½" x 4") (227 gm)	27	24	21	19	17	15	14	13	318	85	50.5	32

Home-Cooked or Restaurant (cont.)

	1200	1500	1800	2100	2400	2800	3200	3600	TOT CAL	FAT CAL	S/FAT CAL	CHOL MG
Lasagna, meatless, spinach noodles—1 piece (⅙ of 8″ sq, approx 2½″ x 4″) (227 gm)	27	24	21	19	17	15	14	13	287	85	50.5	32
Lasagna, meatless, whole wheat noodles—1 piece (⅙ of 8″ sq, approx 2½″ x 4″) (227 gm)	27	24	21	19	17	15	14	13	318	85	50.5	32
Lasagna, meatless, w/spinach—1 piece (⅙ of 8″ sq, approx 2½″ x 4″) (227 gm)	25	22	19	17	16	14	13	12	295	79	46.3	29
Lasagna w/meat & spinach—1 piece (⅙ of 8″ square, approx 2½″ x 4″) (206 gm)	31	27	24	21	20	18	16	15	304	103	54.0	45
Lasagna w/meat (inc baked ziti)—1 piece (⅙ of 8″ sq, approx 2½″ x 4″) (206 gm)	33	29	26	23	21	19	17	16	325	111	58.5	48
Lasagna w/meat, spinach noodles—1 piece (⅙ of 8″ sq, approx 2½″ x 4″) (206 gm)	33	29	26	23	21	19	17	16	300	111	58.5	48
Lasagna w/meat, whole wheat noodles—1 piece (⅙ of 8″ sq, approx 2½″ x 4″) (206 gm)	33	29	26	23	21	19	17	16	326	111	58.5	48
Lau lau (pork & fish wrapped in taro or spinach leaves)—1 lau lau (214 gm)	45	39	35	32	30	27	25	23	340	222	69.3	92
Liver, beef or calves & onions—1 med sl (6½″ x 2⅜″ x ⅜″) w/onions (143 gm)	70	65	61	58	55	52	50	47	216	85	23.1	385
Liver dumpling—1 c (250 gm)	211	195	182	172	163	154	146	139	791	402	95.3	1085

Food												
Liver hash (inc liver mush)—1 c (225 gm)	129	120	113	108	103	98	94	90	288	61	20.5	767
Livers, chicken, chopped, w/eggs & onion—¾ c (156 gm)	102	94	88	84	79	75	71	68	225	127	39.0	544
Lobster creole, Puerto Rican style (langosta a la criolla)—¾ c (182 gm)	25	22	20	19	17	16	15	14	320	197	26.1	83
Lobster Newburg (inc lobster thermidor, lobster w/cream or white sauce)—¾ c (183 gm)	137	120	108	99	91	83	77	72	455	342	199.5	313
Lobster salad—¾ c (137 gm)	19	17	16	15	14	13	13	12	102	48	9.0	94
Lobster sauce—1 c (234 gm)	53	48	44	41	38	36	33	32	385	226	47.8	203
Lobster, stuffed, baked—1 lobster (400 gm)	59	53	49	45	42	39	37	35	747	294	56.8	217
Lobster w/butter sauce (inc lobster Norfolk)—¾ c (141 gm)	91	79	70	64	59	53	49	45	333	248	147.5	163
Lo mein, meatless—1 c (200 gm)	5	4	4	4	3	3	3	3	124	12	2.6	22
Lo mein w/meat—1 c (200 gm)	21	19	17	15	14	13	12	12	285	116	25.8	61
Lomi salmon—¾ c (176 gm)	7	6	5	5	5	4	4	4	130	54	7.0	22
Macaroni & cheese, baked fr home recipe, w/butter—1 c hot (200 gm)	58	50	44	40	36	33	30	27	430	200	107.1	68
Macaroni & cheese, baked fr home recipe, w/reg margarine—1 c hot (200 gm)	42	36	32	29	26	23	21	19	430	200	80.1	42
Macaroni & cheese, baked fr home recipe, w/soft margarine—1 c hot (200 gm)	42	36	32	29	26	23	21	20	430	200	80.6	42
Macaroni & cheese, canned—1 c (240 gm)	21	18	16	14	13	12	11	10	228	86	38.0	24
Macaroni, creamed (inc fettuccine Alfredo, macaroni cooked in milk, noodles Romanoff)—¾ c (150 gm)	12	10	9	8	7	6	6	5	201	79	23.8	8
Macaroni or noodles w/beans or lentils & tomato sauce—¾ c (170 gm)	2	1	1	1	1	1	1	1	148	9	2.2	4
Macaroni or noodles w/cheese & beef—1 c (243 gm)	47	41	37	34	31	29	26	25	347	134	70.2	104

THE 2-IN-1 SYSTEM

Home-Cooked or Restaurant (cont.)

	1200	1500	1800	2100	2400	2800	3200	3600	TOT CAL	FAT CAL	S/FAT CAL	CHOL MG
Macaroni or noodles w/cheese & frankfurters—1 c (243 gm)	61	53	46	41	38	34	31	28	526	258	118.4	57
Macaroni or noodles w/cheese & tomato—1 c (243 gm)	7	6	5	5	4	4	4	3	246	29	14.3	6
Macaroni or noodles w/cheese & tuna—1 c (243 gm)	55	48	43	39	35	32	29	27	418	200	91.7	93
Macaroni or noodles w/cheese (inc macaroni casserole)—1 c (243 gm)	59	50	44	40	36	32	29	27	513	241	113.1	55
Macaroni salad (made w/celery, cucumber, lettuce, mushroom, olives, onion, peas, green or red pepper, pickles, radish, or relish)—¾ c (133 gm)	5	4	3	3	3	3	2	2	180	59	8.6	5
Macaroni salad w/crabmeat (made w/celery, cucumber, lettuce, mushroom, olives, onion, peas, green or red pepper, pickles, radish, or relish—¾ c (133 gm)	11	10	9	8	8	7	7	6	181	76	11.2	38
Macaroni salad w/egg (made w/celery, cucumber, lettuce, mushroom, olives, onion, peas, green or red pepper, pickles, radish, or relish)—¾ c (133 gm)	29	27	25	24	22	21	20	19	197	85	16.2	142
Macaroni salad w/shrimp (made with celery, cucumber, lettuce, mushroom, olives, onion, peas, green or red pepper, pickles, radish, or red relish)—¾ cup (133 gm)	11	10	9	9	8	7	7	7	192	80	11.7	39

Food												
Macaroni salad w/tuna & egg (made w/celery, cucumber, lettuce, mushroom, olives, onion, peas, green or red pepper, pickles, radish, or relish)—¾ c (133 gm)	29	26	25	23	22	20	19	18	209	87	19.1	132
Macaroni salad w/tuna (made w/celery, cucumber, lettuce, mushroom, olives, onion, peas, green or red pepper, pickles, radish, or relish)—¾ c (133 gm)	19	17	15	13	12	11	10	9	291	193	32.2	31
Mackerel cake or patty—1 cake or patty (120 gm)	39	34	31	29	27	24	23	21	316	184	48.0	111
Manapua, filled w/bean paste, meatless—(103 gm)	2	2	1	1	1	1	1	1	239	35	4.6	0
Manapua, filled w/meat—(93 gm)	22	19	17	15	14	13	12	11	290	120	35.8	37
Manicotti, cheese-filled, no sauce—(127 gm)	49	43	39	36	33	30	28	26	268	117	65.3	127
Manicotti, cheese-filled, w/meat sauce—(143 gm)	40	35	32	29	27	25	23	21	235	102	54.1	102
Manicotti, cheese-filled, w/tomato sauce, meatless—(143 gm)	38	33	30	28	26	23	22	20	223	92	50.8	98
Marinated fish, Puerto Rican style (ceviche)—¾ c (186 gm)	10	9	9	8	8	8	7	7	122	10	2.1	59
Matzo balls—1 matzo ball (35 gm)	9	9	8	8	7	7	6	6	50	17	4.5	47
Mussels w/tomato-base sauce—¾ c (180 gm)	18	17	15	15	14	13	12	12	218	37	8.9	90
Mutton stew w/vegetables (excluding carrots, broccoli & dark-green leafy; no potatoes), gravy—¾ c (189 gm)	32	28	25	23	21	19	18	16	242	122	51.4	62
Mutton stew w/vegetables (inc carrots, broccoli and/or dark-green leafy; no potatoes), gravy—¾ c (189 gm)	32	28	25	23	21	19	18	16	252	122	51.3	62
Nachos w/beans & cheese—4 nachos (75 gm)	22	19	17	15	14	12	11	10	198	108	43.2	20

Home-Cooked or Restaurant (cont.)

	1200	1500	1800	2100	2400	2800	3200	3600	TOT CAL	FAT CAL	S/FAT CAL	CHOL MG
Nachos w/beans, no cheese—4 nachos (56 gm)	3	3	2	2	2	2	2	1	122	52	7.5	0
Nachos w/beef & cheese—4 nachos (48 gm)	30	26	23	21	19	17	16	15	178	113	52.9	44
Nachos w/cheese, meatless, no beans—4 nachos (28 gm)	22	18	16	15	13	12	11	10	124	83	41.3	21
Nachos w/chili—4 nachos (28 gm)	4	3	3	3	2	2	2	2	55	26	7.0	4
Noodle pudding (inc kugel)—1 c (144 gm)	41	38	35	33	31	29	28	26	308	106	24.4	194
Octopus salad, Puerto Rican style (ensalada de pulpo)—¾ c (135 gm)	29	26	24	22	21	19	18	17	277	197	27.8	106
Omelet or scrambled egg—1 lg egg (62 gm)	51	47	44	42	40	37	36	34	105	71	21.5	267
Omelet or scrambled egg, w/cheese—1 egg (67 gm)	54	49	45	42	40	37	35	33	126	89	43.1	225
Omelet or scrambled egg, w/cheese & ham—1 egg (78 gm)	58	53	49	46	43	41	38	36	139	95	41.8	255
Omelet or scrambled egg, w/dark-green vegetables—1 egg (84 gm)	52	47	44	42	39	37	35	34	102	66	26.3	256
Omelet or scrambled egg, w/fish—1 egg (88 gm)	59	55	51	48	45	43	40	39	133	85	30.9	293
Omelet or scrambled egg, w/ground beef & onions—1 egg (95 gm)	60	55	51	47	45	42	40	38	149	96	38.9	273
Omelet or scrambled egg, w/ham or bacon—1 egg (70 gm)	49	45	42	39	37	35	33	31	115	76	29.5	231
Omelet or scrambled egg, w/mushrooms—1 egg (69 gm)	45	42	39	37	35	33	31	30	89	58	23.1	226
Omelet or scrambled egg, w/onions, peppers, tomatoes & mushrooms (inc Spanish omelet)—1 egg (145 gm)	32	29	27	25	24	22	21	20	121	78	22.4	140

Food												
Omelet or scrambled egg, w/peppers, onion & ham (inc western omelet)—1 egg (94 gm)	60	55	51	48	46	43	41	39	134	87	34.2	289
Omelet or scrambled egg, w/potatoes and/or onions (tortilla Espanola, traditional-style Spanish omelet)—1 egg (156 gm)	64	59	55	52	49	46	44	42	173	83	32.8	319
Omelet or scrambled egg, w/sausage & mushrooms—1 egg (95 gm)	63	58	53	50	47	44	42	40	169	119	44.6	280
Omelet or scrambled egg, w/vegetables other than dark-green vegetables—1 egg (64 gm)	38	35	32	30	29	27	26	25	75	48	19.2	187
Omelet or scrambled egg, w/chili, cheese, tomatoes & beans—1 egg (103 gm)	62	56	52	48	46	42	40	38	157	102	49.4	257
Omelet, ripe plantain, Puerto Rican style (tortilla de amarillo)—1 egg (85 gm)	44	40	38	35	33	31	30	28	209	133	26.0	209
Oyster fritter—1 fritter (40 gm)	13	11	10	10	9	8	8	7	124	56	14.2	42
Oyster pie (inc pot pie)—1 piece (⅙ of pie) (109 gm)	28	24	21	19	17	15	14	13	275	160	52.1	30
Oyster sauce (inc scalloped oysters)—1 c (256 gm)	36	31	28	25	23	21	20	18	346	201	55.6	72
Paella—1 c (240 gm)	17	15	14	13	12	11	10	9	350	101	20.9	50
Paella, Valenciana style—1 c, boneless (183 gm)	49	44	40	36	34	31	29	27	535	263	59.4	146
Pasta, meat-filled w/gravy (inc Chef Boy-ar-dee Mini Chicken Ravioli)—1 c (250 gm)	13	11	10	9	8	7	7	6	327	70	22.9	18
Pasta salad (macaroni or noodles, vegetables, dressing)—¾ c (133 gm)	7	6	5	4	4	4	3	3	198	107	15.5	0
Pasta salad w/meat (macaroni or noodles, vegetables, meat, dressing)—¾ c (133 gm)	17	15	13	12	11	9	8	8	249	153	34.9	12
Pasta w/carbonara sauce—1 c (201 gm)	34	30	28	26	24	22	21	20	372	101	36.5	115

Home-Cooked or Restaurant (cont.)

	1200	1500	1800	2100	2400	2800	3200	3600	TOT CAL	FAT CAL	S/FAT CAL	CHOL MG
Pasta w/cheese & meat sauce (inc cannelloni)—1 c (242 gm)	61	55	50	46	43	39	37	35	371	167	71.5	191
Pasta w/meat sauce (inc American chop suey, chili w/macaroni)—1 c (255 gm)	24	21	19	18	16	15	14	13	307	85	33.9	59
Pasta w/tomato sauce, meatless—1 c (248 gm)	½	½	½	½	½	0	0	0	179	7	1.1	0
Pastry, cheese-filled—(28 gm)	13	12	10	10	9	8	7	7	79	53	19.9	29
Pepper steak—¾ c (163 gm)	22	19	17	16	15	13	12	12	245	142	31.8	51
Pizza, cheese, thick crust (inc English muffin)—1 piece (⅛ of 12" dia) (71 gm)	11	10	8	7	7	6	5	5	203	60	22.7	7
Pizza, cheese, thin crust (inc Weight Watchers)—1 piece (⅛ of 12" dia) (63 gm)	12	10	9	8	7	7	6	5	163	56	23.9	9
Pizza, cheese, w/vegetables, thick crust—1 piece (⅛ of 12" dia) (78 gm)	10	9	8	7	6	6	5	5	192	59	21.1	7
Pizza, cheese, w/vegetables, thin crust—1 piece (⅛ of 12" dia) (70 gm)	11	9	8	7	6	6	5	5	148	52	21.0	8
Pizza rolls (inc Pizza Bites)—1 miniature roll (14 gm)	3	3	3	2	2	2	2	2	42	17	6.8	3
Pizza w/beans and vegetables, thick crust (inc Pizza Hut Taco Pizza)—1 piece (⅛ of 12" dia) (87 gm)	10	9	8	7	6	6	5	5	232	67	21.4	6
Pizza w/beans & vegetables, thin crust (inc Pizza Hut Taco Pizza)—1 piece (⅛ of 12" dia) (79 gm)	10	9	8	7	6	6	5	5	186	61	21.2	7
Pizza w/cheese topping, baked fr home recipe—⅛ pizza (65 gm)	10	9	8	7	6	6	5	5	153	49	19.0	12

Food												
Pizza w/cheese topping, chilled, baked—1/6 pizza (60 gm)	8	7	6	5	5	4	4	4	147	37	13.1	11
Pizza w/cheese topping, frozen, baked—1/2 pizza (57 gm)	7	6	6	5	5	4	4	4	140	36	13.1	10
Pizza w/meat & fruit, thick crust (inc ham & pineapple, Canadian bacon & pineapple)—1 piece (1/6 of 12" dia) (79 gm)	11	9	8	7	7	6	5	5	194	58	21.2	10
Pizza w/meat & fruit, thin crust (inc ham & pineapple, Canadian bacon & pineapple)—1 piece (1/6 of 12" dia) (71 gm)	11	10	8	8	7	6	6	5	152	51	20.9	12
Pizza w/meat & vegetables, thick crust—1 piece (1/6 of 12" dia) (87 gm)	16	14	12	11	10	9	8	7	234	88	31.9	13
Pizza w/meat & vegetables, thin crust—1 piece (1/6 of 12" dia) (79 gm)	17	15	13	12	11	10	9	8	193	86	33.4	16
Pizza w/meat, thick crust (inc sausage, ground beef, pepperoni, ham, bacon, salami)—1 piece (1/6 of 12" dia) (79 gm)	17	15	13	12	11	9	9	8	244	91	33.9	14
Pizza w/meat, thin crust—1 piece (1/6 of 12" dia) (71 gm)	19	17	15	13	12	11	10	9	209	93	37.3	18
Pizza w/sausage topping, no cheese, baked fr home recipe—1/6 pizza (67 gm)	9	8	7	6	6	5	5	4	157	56	16.3	13
Pizza w/sausage topping, w/cheese, baked fr home recipe—1/6 pizza (67 gm)	12	10	9	8	8	7	6	6	189	80	20.4	19
Pork & beans—1/2 c (127 gm)	7	6	6	5	5	4	4	4	155	30	9.5	19
Pork & rice w/tomato-base sauce—1 c (244 gm)	40	35	31	28	26	23	21	20	400	181	65.0	71
Pork & vegetables (excluding carrots, broccoli & dark-green leafy; no potatoes), no sauce—3/4 c (122 gm)	28	24	22	20	18	16	15	14	211	129	46.3	49

	1200	1500	1800	2100	2400	2800	3200	3600	TOT CAL	FAT CAL	S/FAT CAL	CHOL MG
Home-Cooked or Restaurant (cont.)												
Pork & vegetables (excluding carrots, broccoli & dark-green leafy; no potatoes), w/tomato-base sauce—¾ c (187 gm)	28	25	22	20	19	17	16	15	225	118	39.7	67
Pork & vegetables (excluding carrots, broccoli & dark-green leafy), w/soy-base sauce—¾ c (163 gm)	14	13	11	10	10	9	8	7	208	96	20.6	33
Pork & vegetables, Hawaiian style—¾ c (189 gm)	20	18	16	14	13	12	11	10	204	116	36.0	28
Pork & vegetables (inc carrots, broccoli and/or dark-green leafy; no potatoes), no sauce (inc chow yuk)—¾ c (122 gm)	28	24	22	20	18	16	15	14	221	128	46.2	49
Pork & vegetables (inc carrots, broccoli and/or dark-green leafy; no potatoes), w/tomato-base sauce—¾ c (187 gm)	28	24	22	20	19	17	16	15	230	117	39.5	66
Pork & vegetables (inc carrots, broccoli and/or dark-green leafy), w/soy-base sauce—¾ c (163 gm)	13	11	10	9	8	8	7	7	195	86	18.4	29
Pork & watercress w/soy-base sauce—¾ c (122 gm)	29	25	22	20	18	17	15	14	236	158	48.2	46
Pork chop stewed w/vegetables, Puerto Rican style (chuletas a la jardinera)—¾ c (176 gm)	17	15	14	13	12	11	10	10	177	63	20.8	51
Pork chow mein or chop suey, no noodles—¾ c (165 gm)	22	19	17	16	15	13	12	11	225	129	33.9	46
Pork chow mein or chop suey w/noodles—1 c (220 gm)	32	28	25	23	21	19	17	16	432	223	53.4	56

Food												
Pork egg foo yung—1 omelet (86 gm)	44	40	37	35	33	31	30	28	128	81	24.9	211
Pork hash, ground pork Hawaiian style, vegetables (excluding carrots, broccoli and dark-green leafy; no potatoes), soy sauce—¾ c (143 gm)	44	38	34	31	29	26	24	22	295	199	71.3	80
Pork or ham & potatoes w/gravy—1 c (252 gm)	25	22	20	18	17	15	14	13	258	94	33.0	64
Pork or ham w/barbecue sauce—1 sl w/sauce (62 gm)	16	14	13	11	11	10	9	9	124	61	20.6	42
—1 rib w/sauce (68 gm)	17	15	14	13	12	11	10	9	136	67	22.6	46
Pork or ham w/gravy—1 sl w/gravy (86 gm)	14	12	11	10	10	9	8	8	109	46	16.1	42
—1 pork chop w/gravy (117 gm)	19	17	15	14	13	12	11	10	148	63	21.9	57
Pork or ham w/(mushroom) soup—¾ c (183 gm)	38	33	30	27	25	23	21	20	305	169	56.2	85
Pork or ham w/soy-base sauce—¾ c (183 gm)	24	22	20	18	17	16	15	14	200	67	23.0	89
Pork or ham w/tomato-base sauce—1 pork chop w/sauce (116 gm)	20	18	16	14	13	12	11	10	167	92	30.7	42
Pork, potatoes & vegetables (excluding carrots, broccoli & dark-green leafy), no sauce—1 c (162 gm)	25	22	19	18	16	15	13	12	244	120	41.4	43
Pork, potatoes & vegetables (excluding carrots, broccoli & dark-green leafy), tomato-base sauce—1 c (252 gm)	23	20	18	16	15	14	13	12	276	103	34.3	48
Pork, potatoes & vegetables (inc carrots, broccoli and/or dark-green leafy), no sauce—1 c (162 gm)	25	22	19	18	16	15	13	12	251	120	41.4	43
Pork, potatoes & vegetables (inc carrots, broccoli and/or dark-green leafy), tomato-base sauce—1 c (252 gm)	23	20	18	16	15	14	13	12	267	104	34.7	49
Pork, tofu & vegetables (excluding carrots, broccoli & dark-green leafy; no potatoes), soy-base sauce—¾ c (163 gm)	24	20	18	16	15	13	12	11	266	157	43.4	28

THE 2-IN-1 SYSTEM

Home-Cooked or Restaurant (cont.)

	1200	1500	1800	2100	2400	2800	3200	3600	TOT CAL	FAT CAL	S/FAT CAL	CHOL MG
Pork, tofu & vegetables (inc carrots, broccoli and/or dark-green leafy; no potatoes), soy-base sauce—¾ c (163 gm)	24	20	18	16	15	13	12	11	262	157	43.4	28
Potato pancakes—2 pancakes (3¼" x ⅝" thick) (74 gm)	18	16	15	14	13	12	11	11	143	68	17.7	66
Potato pancakes, home-prep—1 pancake (76 gm)	28	25	23	21	20	18	17	16	234	113	30.8	93
Potato pudding—½ c (114 gm)	30	26	24	22	21	19	18	17	143	82	33.8	94
Potato salad—½ c (97 gm)	5	4	4	3	3	3	3	2	135	65	9.7	5
Potato salad, German style—½ c (88 gm)	3	2	2	2	2	2	1	1	79	15	5.1	3
Potato salad w/egg—½ c (97 gm)	15	14	13	12	11	10	10	9	137	69	11.8	63
Potted meat, inc potted beef, chicken & turkey—5½ oz can (156 gm)	75	65	58	52	48	44	40	37	387	270	126.4	122
Quesadilla—(54 gm)	19	16	14	13	11	10	9	9	191	97	37.2	15
Quiche lorraine (inc w/crabmeat, ham, shrimp)—1 piece (⅛ of 9" dia) (192 gm)	131	115	103	94	87	79	73	68	593	431	197.3	281
Rabbit, stewed, Puerto Rican style (fricase de conejo)—1 c, boneless (219 gm)	48	42	38	34	32	29	27	25	527	317	67.4	114
Ravioli, cheese-filled, no sauce—1 c (240 gm)	82	73	67	62	59	54	51	48	433	151	77.9	302
Ravioli, cheese-filled, w/meat sauce—1 c (250 gm)	59	53	48	44	42	38	36	34	358	149	64.2	196
Ravioli, cheese-filled, w/tomato sauce—1 c (250 gm)	56	50	46	42	40	37	34	32	338	131	56.8	196

Food												
Ravioli, meat-filled, no sauce—1 c (240 gm)	89	80	74	69	64	60	56	53	545	206	79.8	344
Ravioli, meat-filled, w/tomato sauce or meat sauce—1 c (250 gm)	56	50	46	43	40	37	35	33	386	153	53.6	206
Rice, brown, w/tomato sauce—1 c (243 gm)	3	2	2	2	2	1	1	1	242	29	6.3	0
Rice cake, patty, or croquette—(85 gm)	12	11	10	9	9	8	8	7	114	43	11.3	46
Rice casserole w/cheese (inc risotto)—1 c (204 gm)	45	39	34	31	28	25	23	21	385	133	83.4	52
Rice, creamed—3/4 c (153 gm)	6	5	5	4	4	3	3	3	170	34	12.7	5
Rice dessert or salad w/fruit (inc glorified rice)—1 c (155 gm)	45	39	34	31	28	25	23	21	267	134	82.3	54
Rice dressing (inc combined w/bread)—3/4 c (125 gm)	3	3	2	2	2	2	1	1	144	34	7.2	0
Rice, fried, meatless—3/4 c (123 gm)	11	10	9	9	8	8	7	7	179	77	11.2	41
Rice, fried, w/meat (inc pork, chicken; Chinese rice)—3/4 c (149 gm)	22	20	19	17	16	15	15	14	220	81	13.8	101
Rice, fried, w/shrimp—3/4 c (149 gm)	22	20	19	18	17	16	15	14	216	78	13.0	104
Rice pilaf (inc rice & vegetables, rice & mushrooms)—3/4 c (155 gm)	4	3	3	3	2	2	2	2	199	45	9.0	0
Rice-vegetable medley (inc rice Italian style)—3/4 c (155 gm)	6	5	4	4	3	3	3	3	215	69	13.6	0
Rice w/beans—1 c (239 gm)	1/2	1/2	1/2	1/2	1/2	0	0	0	269	7	1.0	0
Rice w/beans & chicken—1 c (239 gm)	23	20	18	17	16	14	13	12	350	108	28.7	64
Rice w/beans & pork—1 c (239 gm)	32	28	25	23	21	19	18	17	380	141	50.2	65
Rice w/beans & tomatoes—1 c (239 gm)	1/2	1/2	1/2	1/2	1/2	0	0	0	217	7	1.0	0
Rice w/chicken, Puerto Rican style (arroz con pollo)—1 c, boneless (163 gm)	28	25	22	20	18	17	15	14	465	182	45.6	53
Rice w/gravy—3/4 c (178 gm)	2	2	2	1	1	1	1	1	170	10	4.6	1
Rice w/raisins—3/4 c (139 gm)	0	0	0	0	0	0	0	0	176	2	.5	0
Salisbury steak w/gravy—1 steak w/gravy (129 gm)	33	29	26	23	21	19	18	17	212	117	52.5	62
Salmon cake or patty (inc salmon croquette)—1 cake or patty (120 gm)	25	22	20	18	17	16	15	14	272	153	35.3	63

THE 2-IN-1 SYSTEM

Home-Cooked or Restaurant (cont.)

	1200	1500	1800	2100	2400	2800	3200	3600	TOT CAL	FAT CAL	S/FAT CAL	CHOL MG
Salmon loaf—1 sl (105 gm)	33	30	27	25	24	22	21	20	223	115	26.0	135
Salmon salad—¾ c (156 gm)	45	41	37	35	33	31	29	27	321	216	36.7	183
Salmon, stewed, Puerto Rican style (salmon guisado)—1 c (212 gm)	18	16	14	13	12	11	10	10	368	183	25.3	44
Sardines w/mustard sauce—2 sardines (3" x 1" x ½") w/sauce (24 gm)	8	7	6	6	5	5	5	4	47	26	8.4	26
Sardines w/tomato-base sauce—2 sardines (3" x 1" x ½") w/sauce (24 gm)	8	7	6	6	5	5	5	4	47	26	7.8	26
Sausage & rice w/cheese sauce—1 c (244 gm)	53	45	40	36	33	30	27	25	433	228	93.7	71
Sausage & rice w/(mushroom) soup—1 c (244 gm)	44	37	33	30	27	24	22	21	424	241	79.8	52
Sausage & rice w/tomato-base sauce—1 c (244 gm)	45	38	34	31	28	25	23	22	467	223	77.5	65
Sausage, potatoes & vegetables (excluding carrots, broccoli & dark-green leafy), gravy—1 c (252 gm)	45	39	34	31	28	25	23	21	412	230	86.2	46
Sausage, potatoes & vegetables (inc carrots, broccoli and/or dark-green leafy), gravy—1 c (252 gm)	46	40	35	31	29	26	23	22	409	235	88.2	47
Sausage w/tomato-base sauce—1 link w/sauce (42 gm)	7	6	6	5	5	4	4	4	56	36	12.6	11
Scallops & noodles w/cheese sauce—1 c (224 gm)	36	31	28	26	24	22	20	19	359	140	50.4	86
Scallops w/cheese sauce—¾ c (183 gm)	24	21	19	17	16	15	14	13	203	65	31.1	63
Scrambled eggs w/jerked beef, Puerto Rican style (revoltillo de tasajo)—1 c (140 gm)	100	91	85	80	75	71	67	64	421	278	59.3	471

Food												
Seafood Newburg (inc frozen casserole, shrimp Newburg, crabmeat thermidor)—¾ c (183 gm)	138	121	108	99	91	83	77	72	456	343	200.0	313
Seafood paella, Puerto Rican style (paella a la marinera)—1 c (230 gm)	42	38	35	33	31	30	28	27	378	114	26.0	194
Seafood, restructured (inc Delicaseas, Sea Tails, Seat Stix, imitation crab meat)—1 cup, chunks or flakes (126 gm)	8	7	7	6	6	6	6	5	143	7	.8	46
Seafood salad—¾ c (156 gm)	19	18	17	16	15	15	14	13	118	17	2.5	115
Seafood soufflé—1 c (159 gm)	64	58	54	48	44	42	40	272	163	163	46.3	279
Seafood stew w/potatoes & vegetables (excluding carrots, broccoli & dark green leafy), tomato-base sauce—1 c (252 gm)	11	10	9	9	8	8	7	7	137	28	8.2	47
Seafood stew w/potatoes & vegetables (inc carrots, broccoli and/or dark-green leafy), tomato-base sauce—1 c (252 gm)	11	10	9	9	8	8	7	7	139	28	8.1	47
Seasoned shredded soup meat, Puerto Rican style (ropa vieja, sopa de carne ripiada)—¾ c (100 gm)	23	20	18	17	16	15	14	13	225	107	27.8	68
Shad creole, w/rice—1 c (249 gm)	29	26	23	21	19	18	16	15	360	151	43.0	66
Shellfish mixture & vegetables (excluding carrots, broccoli & dark-green leafy; no potatoes), soy-base sauce—¾ c (122 gm)	9	8	8	7	7	6	6	5	119	58	10.0	32
Shellfish mixture & vegetables (inc carrots, broccoli and/or dark-green leafy; no potatoes), soy-base sauce—¾ c (122 gm)	9	8	8	7	7	6	6	5	117	58	10.1	32
Shepherd's pie—1 c (243 gm)	22	19	17	15	14	13	12	11	303	112	32.9	45
Shepherd's pie w/beef—1 c (243 gm)	19	17	15	14	13	12	11	10	289	96	29.8	39
Shishkabob—1 shishkabob (198 gm)	14	13	11	11	10	9	9	8	157	40	15.7	46

	1200	1500	1800	2100	2400	2800	3200	3600	TOT CAL	FAT CAL	S/FAT CAL	CHOL MG
Home-Cooked or Restaurant (cont.)												
Shrimp & vegetables (excluding carrots, broccoli & dark-green leafy; no potatoes), soy-base sauce—¾ c (122 gm)	12	10	10	9	8	8	7	7	121	57	9.8	46
Shrimp & vegetables (inc carrots, broccoli and/or dark-green leafy; no potatoes), soy-base sauce—¾ c (122 gm)	12	10	10	9	8	8	7	7	126	57	9.8	46
Shrimp cake or patty (inc shrimp burger; shrimp stick, battered)—1 cake or patty (120 gm)	34	31	29	27	25	23	22	21	246	115	29.5	137
Shrimp chow mein or chop suey, no noodles—¾ c (184 gm)	18	16	15	15	14	13	13	12	141	35	5.2	99
Shrimp chow mein or chop suey w/noodles—1 c (220 gm)	24	22	20	19	18	17	16	15	265	87	15.1	109
Shrimp cocktail (shrimps w/cocktail sauce)—4 shrimps w/sauce (92 gm)	10	9	9	8	8	8	7	7	77	5	1.1	60
Shrimp creole, no rice—¾ c (184 gm)	32	30	28	26	25	24	22	21	228	80	15.1	165
Shrimp creole, w/rice (inc shrimp jambalaya)—1 c (243 gm)	32	29	27	26	24	23	22	21	301	79	14.9	161
Shrimp, curried—¾ c (177 gm)	34	30	28	26	24	22	21	20	232	103	33.0	121
Shrimp egg foo yung (inc tortas de carmaron)—1 omelet (86 gm)	46	42	39	37	35	33	31	30	156	109	23.5	226
Shrimp salad—¾ c (137 gm)	28	26	24	23	21	20	19	18	210	107	16.0	135
Shrimp, stuffed—5 stuffed shrimps (80 gm)	29	26	24	23	22	20	19	18	153	70	16.1	137
Shrimp teriyaki (shrimp w/soy-base sauce)—¾ c (151 gm)	31	29	27	26	25	24	23	22	174	10	1.6	193

Food												
Shrimp w/lobster sauce—¾ c (139 gm)	36	33	31	29	28	26	25	24	216	79	16.3	186
Somen salad—1 c (160 gm)	44	41	38	36	34	32	30	29	258	66	21.6	222
Spaghetti & meat balls, in tomato sauce, cooked, home recipe—1 c (248 gm)	25	22	20	18	17	16	15	14	332	105	29.9	74
Spaghetti & meat balls, in tomato sauce, canned—15 oz can (425 gm)	20	18	16	14	13	12	11	10	438	157	32.8	38
Spaghetti w/red clam sauce—1 c (248 gm)	5	4	4	4	3	3	3	3	226	63	8.8	8
Spaghetti w/tomato sauce & frankfurters—1 c (248 gm)	27	23	20	18	16	15	13	12	304	141	50.7	26
Spaghetti w/tomato sauce & meatballs; spaghetti w/meat sauce; or spaghetti w/meat sauce & meatballs—1 c (248 gm)	38	33	30	28	26	24	22	21	402	137	46.8	108
Spaghetti w/tomato sauce, meatless (inc spaghetti w/marinara sauce; spaghetti w/tomato sauce, meatless, w/cheese, canned)—1 c (248 gm)	½	½	½	½	½	0	0	0	156	7	1.0	0
Spaghetti w/white clam sauce—1 c (248 gm)	14	12	11	10	9	8	7	7	346	167	23.3	23
Spanakopitta (inc Greek spinach-cheese pie)—1 c (162 gm)	72	63	57	52	48	44	40	38	334	224	106.3	160
Spanish rice—¾ c (182 gm)	2	1	1	1	1	1	1	1	156	23	3.5	0
Spanish rice w/ground beef—¾ c (173 gm)	23	20	18	16	15	14	13	12	230	78	31.7	55
Spinach quiche, meatless (inc broccoli quiche)—1 piece (⅛ of 9" dia) (143 gm)	78	68	61	56	52	47	44	41	337	232	110.1	184
Steak tartare (raw ground beef & egg)—¾ c (168 gm)	77	68	61	56	52	48	44	41	370	240	104.5	197
Steak teriyaki w/sauce—¾ c (183 gm)	34	30	28	25	24	22	20	19	317	126	43.8	96
Stuffed cabbage rolls w/rice—1 cabbage roll (107 gm)	17	15	14	13	12	11	10	10	125	55	22.0	49

Home-Cooked or Restaurant (cont.)

	1200	1500	1800	2100	2400	2800	3200	3600	TOT CAL	FAT CAL	S/FAT CAL	CHOL MG
Stuffed chicken, drumstick or breast, Puerto Rican style (muslo de pollo o pechuga rellena)—1 breast (4" x 3" x ¾") (210 gm)	75	68	63	60	56	53	50	47	445	125	48.2	344
Stuffed green pepper, Puerto Rican style (pimiento relleno)—1 pepper (3½" x 3½" x 2") (250 gm)	44	38	34	30	28	25	23	21	438	180	76.8	63
Stuffed peppers, w/rice & meat—½ pepper w/filling (149 gm)	36	31	28	25	24	21	20	18	219	118	53.7	76
Stuffed peppers, w/rice, meatless—½ pepper w/filling (149 gm)	18	15	13	12	11	10	9	8	198	107	35.7	12
Stuffed rice, Dominican style (arroz relleno Dominicano)—1 c (200 gm)	48	43	39	37	34	32	30	28	552	198	51.8	163
Stuffed shells, cheese-filled, no sauce—1 jumbo shell (60 gm)	22	20	18	17	15	14	13	12	127	52	28.4	63
Stuffed shells, cheese-filled, w/meat sauce—1 jumbo shell (85 gm)	22	20	18	16	15	14	13	12	139	56	28.9	61
Stuffed shells, cheese-filled, w/tomato sauce, meatless—1 jumbo shell (85 gm)	21	19	17	15	14	13	12	12	128	49	26.6	59
Stuffed shells, w/chicken—1 jumbo shell (83 gm)	13	12	11	11	10	10	9	9	116	27	7.6	64
Sukiyaki—¾ c (122 gm)	30	27	25	24	22	21	20	19	131	52	20.9	134
Sushi, no vegetables, no fish—¾ c (109 gm)	0	0	0	0	0	0	0	0	199	2	.5	0
Sushi, w/vegetables & fish—¾ c (125 gm)	2	2	2	1	1	1	1	1	181	4	.8	9
Sushi, w/vegetables, no fish—¾ c (125 gm)	0	0	0	0	0	0	0	0	189	2	.5	0

Food												
Sushi, w/vegetables, rolled in seaweed—¾ c (125 gm)	0	0	0	0	0	0	0	0	130	1	.3	0
Swedish meatballs w/cream or white sauce—1 meatball w/sauce (45 gm)	14	12	11	10	9	8	8	7	76	41	19.3	33
Sweet & sour chicken—1 wing w/sauce (48 gm)	2	2	2	2	2	1	1	1	44	11	1.9	8
—1 drumstick w/sauce (72 gm)	3	3	3	3	2	2	2	2	66	16	2.9	13
—1 thigh w/sauce (78 gm)	4	3	3	3	3	2	2	2	71	17	3.1	14
—½ breast w/sauce (131 gm)	6	5	5	5	4	4	4	4	119	29	5.2	23
—1 leg w/sauce (148 gm)	7	6	6	5	5	5	4	4	135	33	5.9	26
Sweet & sour pork—¾ c (170 gm)	13	12	11	10	9	8	8	7	187	70	19.0	32
Sweet & sour shrimp—¾ c (132 gm)	24	22	20	18	17	16	15	14	358	190	27.5	78
Swiss steak—¾ c (187 gm)	17	15	14	13	12	11	10	10	192	75	21.8	50
Tabbouleh (inc tabbuli)—1 c (160 gm)	7	6	5	4	4	4	3	3	186	116	15.7	0
Taco or tostada w/beef & cheese (inc Taco Bell Grande)—(83 gm)	21	18	16	15	14	12	11	11	182	84	34.1	38
Taco or tostado w/beans & cheese, meatless—(88 gm)	9	7	7	6	5	5	4	4	144	45	17.3	7
Taco or tostado w/beans, cheese & meat (beef or chicken)—(83 gm)	15	13	11	10	9	8	8	7	157	62	24.1	25
Taco or tostado w/beans, meatless—(80 gm)	2	1	1	1	1	1	1	1	114	23	3.7	0
Taco or tostado w/beef—(76 gm)	14	12	11	10	9	8	8	7	153	63	20.6	30
Taco or tostado w/chicken—(72 gm)	7	6	6	5	5	4	4	4	120	38	7.9	22
Taco or tostado w/chicken & cheese—(79 gm)	14	12	11	10	9	8	8	7	148	59	21.3	29
Taco salad—1 c (122 gm)	24	21	19	17	15	14	13	12	202	114	41.0	38
Tamale casserole w/meat—1 c (244 gm)	19	17	15	14	13	12	11	10	222	91	29.3	41
Tamale, meatless—(72 gm)	14	12	10	9	7	7	7	6	165	77	28.3	7
Tamale w/meat—(70 gm)	18	16	14	13	12	10	9	9	183	85	32.4	25
Taquitos—(72 gm)	14	13	11	10	9	9	8	7	185	82	22.3	29
Tortellini, meat-filled, no sauce—1 c (190 gm)	69	63	58	54	51	48	45	43	378	142	50.8	297

THE 2-IN-1 SYSTEM

	1200	1500	1800	2100	2400	2800	3200	3600	TOT CAL	FAT CAL	S/FAT CAL	CHOL MG
Home-Cooked or Restaurant *(cont.)*												
Tortellini, meat-filled, w/tomato sauce—1 c (200 gm)	27	24	22	20	19	18	17	16	262	90	28.3	93
Tortellini, meatless, w/tomato sauce —1 c (250 gm)	56	50	46	42	40	37	34	32	338	131	56.8	196
Tortellini, spinach-filled, no sauce—1 c (122 gm)	45	41	38	36	34	31	30	28	235	84	30.2	202
Tortellini, spinach-filled, w/tomato sauce—1 c (200 gm)	22	20	18	17	16	15	14	13	230	73	22.6	79
Tostada salad, meatless—(232 gm)	47	40	35	31	28	26	23	21	351	196	90.5	41
Tuna & rice w/(mushroom) soup—1 c (248 gm)	24	21	19	17	16	14	13	12	337	127	39.7	42
Tuna cake or patty—1 cake or patty (120 gm)	33	29	26	24	22	20	19	17	300	158	44.6	82
Tuna casserole w/vegetables & (mushroom) soup, no noodles—¾ c (168 gm)	25	22	19	17	16	15	13	12	299	140	40.0	45
Tuna loaf—1 sl (105 gm)	39	35	32	30	28	26	24	23	250	117	34.2	150
Tuna noodle casserole w/cream or white sauce—1 c (224 gm)	36	31	28	25	24	21	20	18	425	180	52.8	78
Tuna noodle casserole w/(mushroom) soup—1 c (224 gm)	32	28	25	23	21	19	18	17	396	162	45.8	72
Tuna noodle casserole w/vegetables & (mushroom) soup—1 c (224 gm)	29	25	23	21	19	17	16	15	376	148	41.6	65
Tuna noodle casserole w/vegetables, cream or white sauce—1 c (224 gm)	29	25	22	20	19	17	16	15	368	148	44.9	56
Tuna potpie—1 piece (⅛ of pie) (128 gm)	26	23	20	18	16	15	13	12	291	159	49.5	28
Tuna salad—¾ c (156 gm)	23	20	18	17	16	14	13	12	290	130	30.5	60

Food												
Tuna salad w/egg—¾ c (156 gm)	37	33	31	28	27	25	23	22	295	143	33.4	142
Tuna w/cream or white sauce—¾ c (177 gm)	48	43	39	36	34	31	29	27	307	174	54.2	156
Turkey pot pie, frozen, commercial, unheated—4 oz (113 gm)	20	17	15	13	12	10	9	9	223	106	40.7	10
Turkey pot pie, home-baked—1 piece, ⅓ of pie (232 gm)	53	46	41	37	33	30	28	25	550	282	94.4	72
Turkey w/gravy—1 sl w/gravy (65 gm)	7	6	6	5	5	5	4	4	71	20	6.5	25
Turnover, chicken, w/gravy—(112 gm)	26	22	20	18	16	14	13	12	321	189	49.5	26
Turnover, meat- & cheese-filled, no gravy—(96 gm)	34	29	26	23	21	19	17	16	356	209	66.0	32
Turnover, meat-filled, no gravy—(88 gm)	30	26	23	20	19	17	15	14	328	195	59.5	26
Turnover, meat-filled, w/gravy—(152 gm)	36	31	27	24	22	20	18	16	383	222	70.0	30
Veal cordon bleu—1 roll (w/ham & sauce) (229 gm)	103	90	80	72	66	60	55	51	503	328	173.1	170
Veal fricassee, Puerto Rican style (ternera en fricase)—1 c (230 gm)	50	44	40	36	33	31	28	26	473	259	71.6	118
Veal goulash w/vegetables (excluding carrots, broccoli & dark-green leafy; no potatoes), tomato-base sauce (inc veal goulash)—¾ c (189 gm)	26	23	21	19	18	16	15	14	205	90	30.2	78
Veal goulash w/vegetables (inc carrots, broccoli and/or dark-green leafy; no potatoes), tomato-base sauce (inc veal marengo)—¾ c (189 gm)	26	23	21	20	18	17	16	15	211	93	31.2	80
Veal parmigiana—1 piece w/sauce & cheese (182 gm)	55	49	44	40	38	35	32	30	351	181	69.8	154
Veal scallopini (inc veal marsala)—1 sl with sauce (96 gm)	32	28	25	23	21	19	18	16	255	174	50.6	62
Veal w/cream sauce (inc veal paprikash)—¾ c (185 gm)	35	31	28	26	24	22	20	19	234	114	47.0	93
Veal w/gravy—1 sl w/gravy (86 gm)	19	16	15	14	13	12	11	10	124	56	24.3	50

	1200	1500	1800	2100	2400	2800	3200	3600	TOT CAL	FAT CAL	S/FAT CAL	CHOL MG
Home-Cooked or Restaurant (cont.)												
Vegetables in pastry (inc Pepperidge Farm, all varieties)—(103 gm)	12	10	9	8	7	6	5	5	162	87	24.6	4
Venison & noodles w/cream or white sauce (mixture)—1 c (249 gm)	36	32	28	26	24	22	20	19	356	169	53.4	80
Venison, potatoes & vegetables (excluding carrots, broccoli & dark-green leafy), gravy—1 c (252 gm)	10	10	9	8	7	7	7	6	171	27	9.3	40
Venison, potatoes & vegetables (inc carrots, broccoli and/or dark-green leafy), gravy—1 c (252 gm)	10	9	9	8	7	7	7	6	167	27	9.3	40
Venison stew w/potatoes & vegetables (excluding carrots, broccoli & dark-green leafy), tomato-base sauce—1 c (252 gm)	10	9	8	8	7	7	6	6	166	27	9.3	39
Venison stew w/potatoes & vegetables (inc carrots, broccoli and/or dark-green leafy), tomato-base sauce—1 c (252 gm)	10	9	8	8	7	7	6	6	166	27	9.3	39
Venison w/tomato-base sauce—¾ c (187 gm)	19	17	16	15	14	13	12	11	179	47	16.7	74
Vienna sausages stewed w/potatoes, Puerto Rican style (salchichas guisadas)—1 c (175 gm)	44	37	33	29	27	24	22	20	414	294	83.7	41
Welsh rarebit—½ cup (116 gm)	36	31	27	24	22	20	18	17	199	137	68.3	36
White rice w/tomato sauce—1 c (243 gm)	2	2	1	1	1	1	1	1	226	21	4.3	0
Wonton, fried, meat-filled—4 wontons (76 gm)	22	20	18	16	15	14	13	12	250	144	28.2	63

Frozen Meals

Food											
Beans & franks—1 serving (12 oz) (340 gm)	36	31	28	25	24	21	20	18	520	171	54.7	74
Beef & noodles w/meat sauce & cheese (diet frozen entree) (inc Weight Watchers Ziti Macaroni)—11¼ oz (319 gm)	38	33	29	27	24	22	20	19	417	125	60.4	70
Beef and pork cannelloni (diet) (inc Lean Cuisine Beef & Pork Cannelloni)—9⅝ oz (273 gm)	16	14	13	12	11	10	9	9	227	52	22.9	39
Beef enchilada, chili gravy, rice, refried beans (inc frozen Mexican meal)—15 oz (425 gm)	40	34	30	27	25	22	20	18	604	270	77.1	38
Beef, noodles, vegetables (inc Armour Dinner Classics Beef Burgundy)—10½ oz (297 gm)	24	21	18	16	15	14	12	11	387	122	42.6	32
Beef, oriental style, w/vegetables & rice (diet) (inc Lean Cuisine Oriental Beef)—8⅝ oz (245 gm)	17	15	13	12	11	10	10	9	272	57	20.9	47
Beef sirloin tips w/gravy, potatoes, & broccoli w/cheddar cheese sauce (inc Le Menu Beef Sirloin Tips)—11½ oz (326 gm)	38	33	29	27	25	22	21	19	398	140	59.2	74
Beef, sliced, gravy, hash brown potatoes, corn soup, apple cobbler (3 courses with dessert)—15 oz (425 gm)	51	44	39	36	33	30	27	25	575	212	85.9	83
Beef, sliced, gravy, potatoes, peas, apple slices (regular portion w/extra item)—11½ oz (326 gm)	25	23	21	19	18	16	15	14	371	97	29.4	79
Beef, sliced, gravy, potatoes, vegetables (regular portion) (inc Le Menu Yankee Pot Roast)—11 oz (312 gm)	46	40	36	32	29	27	24	23	464	234	80.2	69

	1200	1500	1800	2100	2400	2800	3200	3600	TOT CAL	FAT CAL	S/FAT CAL	CHOL MG
Frozen Meals (cont.)												
Beef, sliced, vegetables in sauce, au gratin potatoes (inc Banquet Gourmet Sliced Beef & Fresh Vegetables in Sauce w/Au Gratin Potatoes—10 oz (284 gm)	42	36	32	29	27	24	22	21	367	163	67.3	75
Beef, spaetzle or rice, vegetables (regular portion) (inc Budget Gourmet Peppersteak w/Rice, Stouffer's Green Pepper Steak, Le Menu Peppersteak— 10 oz (284 gm)	18	16	15	13	13	12	11	10	263	70	23.0	52
—1 Stouffer's meal (10½ oz) (298 gm)	19	17	15	14	13	12	11	11	276	74	24.1	55
—1 Le Menu meal (11½ oz) (326 gm)	21	19	17	16	14	13	12	12	302	80	26.3	60
Beef steak, rice, green beans (diet) (inc Classic Lite Beef Pepper Steak) —10 oz (283 gm)	22	20	18	17	15	14	13	12	288	77	25.8	69
Beef w/carrots (diet)—10 oz (284 gm)	51	44	40	36	33	30	28	26	360	191	78.7	101
Beef w/potatoes only (large meat portion) (inc Salisbury steak)—12 oz (340 gm)	89	77	68	61	56	50	46	42	715	390	164.8	104
Beef w/potatoes only (regular portion) (inc Swanson entrees: Gravy and Sliced Beef, Salisbury Steak, Meatballs)—1 Salisbury steak meal (5½ oz) (156 gm)	47	41	37	34	31	29	26	25	364	170	69.7	105
—1 sl beef meal (8 oz) (227 gm)	69	60	54	49	45	41	38	36	530	247	101.4	152

Food											
—1 meatballs meal (8½ oz) (241 gm)	73	64	57	52	48	44	41	38	562	262	107.6	162
Broiled steak, potatoes, vegetable (regular portion)—11 oz (312 gm)	37	32	28	26	24	22	20	19	361	143	58.4	69
Cannelloni, cheese-filled, w/tomato sauce (diet frozen entree) (inc Lean Cuisine Cheese Cannelloni w/Tomato Sauce)—9.125 oz (259 gm)	34	30	26	24	22	19	18	16	284	111	62.9	42
Cheese enchilada (frozen entree) (inc Old El Paso Cheese Enchiladas w/Sauce)—10 oz (284 gm)	58	50	44	39	36	32	29	27	562	225	111.8	55
Cheese enchilada w/beans & rice—(340 gm)	36	31	27	24	22	20	18	16	508	162	69.6	32
Chicken à la king w/rice (frozen entree) (inc Le Menu and Stouffer's Chicken à la King)—1 Stouffer's meal (9½ oz) (269 gm)	53	47	42	39	36	33	31	29	419	198	71.6	138
—1 Le Menu meal (10¼ oz) (290 gm)	57	51	46	42	39	36	33	31	452	213	77.2	148
Chicken & noodles, vegetable, chocolate cake (regular portion w/dessert)—10¼ oz (291 gm)	24	21	19	17	16	15	14	13	365	124	31.5	63
Chicken & orange sauce w/almond rice (diet frozen entree) (inc Lean Cuisine Chicken a l'Orange)—8 oz (227 gm)	14	13	12	11	11	10	9	9	285	43	10.7	61
Chicken & vegetable entree, oriental (diet) (inc Weight Watchers Sweet & Sour Chicken)—9 oz (255 gm)	11	10	9	9	8	8	7	7	202	37	9.8	45
Chicken & vegetable entree w/noodles & cream sauce (inc Budget Gourmet Chicken & Egg Noodles w/Broccoli)—10 oz (284 gm)	40	35	32	29	27	25	23	21	342	130	56.5	98
Chicken & vegetable entree w/noodles (diet) (inc Lean Cuisine Chicken & Vegetables w/Vermicelli)—12¾ oz (361 gm)	18	16	14	13	12	12	11	10	279	62	17.7	62

Frozen Meals *(cont.)*

	1200	1500	1800	2100	2400	2800	3200	3600	TOT CAL	FAT CAL	S/FAT CAL	CHOL MG
Chicken & vegetable entree w/noodles (inc Banquet Gourmet Chicken Cacciatore & Vermicelli)—10 oz (284 gm)	12	11	10	9	8	8	7	7	249	50	11.4	43
Chicken & vegetable entree w/rice (diet) (inc Lean Cuisine Glazed Chicken w/Vegetable Rice)—8½ oz (241 gm)	17	15	14	13	12	11	10	10	264	70	15.6	62
Chicken & vegetable entree w/rice, oriental (inc Green Giant Stir Fry Cashew Chicken, Chun King Crunchy Walnut Chicken—10 oz (283 gm)	17	15	14	13	12	11	10	9	301	73	23.8	43
Chicken & vegetables w/cream or white sauce (diet frozen entree) (inc Weight Watchers Chicken à la King)—9 oz (255 gm)	18	16	14	13	13	12	11	10	241	73	16.8	65
Chicken, boneless, gravy, potatoes, peas, peach cobbler (large meat portion w/dessert)—19 oz (539 gm)	50	44	39	36	33	30	28	26	746	316	74.3	111
Chicken cacciatore w/noodles (diet) (inc Lean Cuisine Chicken Cacciatore with Vermicelli)—10⅞ oz (308 gm)	19	17	15	14	13	12	11	11	285	86	21.7	59
Chicken chow mein w/rice (diet) (inc Lean Cuisine Chicken Chow Mein w/Rice)—11¼ oz (319 gm)	14	13	12	11	10	9	9	8	270	55	15.5	47
Chicken cordon bleu, carrots, rice (inc Le Menu Chicken Cordon Bleu)—11 oz (312 gm)	31	27	24	22	20	19	17	16	483	122	44.7	70

Food											
Chicken dinner—11 oz (279 gm)	40	35	32	29	27	24	23	21	504	228	56.8	95
Chicken divan w/broccoli (inc Stouffer's Chicken Divan)—8½ oz (241 gm)	50	44	39	35	32	29	27	25	364	200	83.1	86
Chicken enchilada (diet) (inc Weight Watchers Chicken Enchilada)—8½ oz (241 gm)	46	40	36	32	30	27	25	23	352	140	77.6	75
Chicken, fried, potatoes, corn, dessert (large meat portion w/dessert) (inc Banquet Extra Helping Fried Chicken Dinner, Swanson Fried Chicken)—1 Swanson meal (15¼ oz) (432 gm)	82	72	65	59	55	50	46	43	962	433	118.7	188
—1 Banquet meal (16½ oz) (467 gm)	89	78	70	64	59	54	50	46	1040	468	128.3	203
Chicken, fried, potatoes, corn, vegetable soup, dessert (3 courses w/soup)—15 oz (381 gm)	45	40	36	33	30	28	26	24	628	252	62.4	111
Chicken, fried, potatoes, mixed vegetables, dessert (reg portion w/dessert) (inc Morton Chicken Patty Dinner)—11½ oz (326 gm)	47	41	37	34	31	29	26	25	589	266	66.4	111
Chicken, fried, potatoes, peas & carrots, cornbread, dessert (3 courses w/bread)—15 oz (381 gm)	41	36	32	30	27	25	23	22	603	228	58.7	97
Chicken, fried, potatoes, peas & carrots, dessert (large meat portion w/dessert)—17 oz (432 gm)	61	54	48	44	41	37	34	32	828	358	89.7	138
Chicken, fried, potatoes, vegetables (reg portion) (inc Banquet Fried Chicken Dinner)—11 oz (279 gm)	41	36	33	30	28	25	24	22	485	234	57.6	102
Chicken, fried, w/potatoes only (large portion) (inc Swanson Hungry Man Fried Chicken)—12¾ oz (361 gm)	57	50	45	41	38	34	32	30	720	355	86.6	122
Chicken, fried, w/potatoes only (regular portion) (inc Swanson Fried Chicken Entree, Swanson Chicken Nibbles)—1 Chicken Nibbles (6 oz) (170 gm)	26	23	20	19	17	16	14	13	352	182	41.6	50

THE 2-IN-1 SYSTEM

	1200	1500	1800	2100	2400	2800	3200	3600	TOT CAL	FAT CAL	S/FAT CAL	CHOL MG
Frozen Meals (cont.)												
—8 oz (227 gm)	35	31	27	25	23	21	19	18	470	243	55.5	67
Chicken patty, breaded, w/tomato sauce & cheese, fettuccine Alfredo, green beans (inc Le Menu Breast of Chicken Parmigiana)—11½ oz (326 gm)	44	39	35	32	30	27	25	23	389	161	59.9	111
Chicken patty parmigiana, breaded, w/vegetables (diet) (inc Weight Watchers Breaded Chicken Patty Parmigiana w/Vegetable Medley—8 oz (226 gm)	34	29	26	24	22	20	18	17	303	151	55.7	57
Chicken teriyaki w/rice and vegetables (regular portion) (inc Budget Gourmet Sweet & Sour Chicken w/Rice)—10 oz (284 gm)	28	25	22	20	19	17	16	15	354	125	39.2	69
Chicken w/peas, onions & celery sauce (diet)—9½ oz (269 gm)	25	22	20	18	17	16	15	14	345	98	27.4	80
Cod, vegetable medley (diet) (inc Weight Watchers Oven Fried Fish)—6¾ oz (191 gm)	17	15	14	13	12	11	10	10	226	114	21.2	51
Corned beef hash, apple slices, peas—10 oz (284 gm)	32	28	25	22	21	19	17	16	375	131	51.2	58
Fish & chips, corn, potatoes, tomatoes (reg portion)—10¼ oz (291 gm)	39	34	31	28	26	24	22	21	478	192	55.9	92
Fish & chips (large portion)—15¾ oz (447 gm)	56	48	43	39	35	32	29	27	938	389	99.0	77
Fish & chips (reg portion) (inc Swanson Fish and Chips)—5½ oz (156 gm)	21	18	16	15	13	12	11	10	322	133	38.7	27

220

Food											
Fish, batter-dipped or fish cake, vegetables, potatoes, dessert (reg portion w/sweet item) (inc Swanson Fish and Chips)—10½ oz (298 gm)	35	30	27	24	22	20	18	17	588	264	61.4	49
Fish, batter-dipped or fish cake, vegetables, potatoes (inc Taste-O-Sea Haddock Dinner)—9 oz (255 gm)	40	35	31	28	26	24	22	20	443	197	60.0	84
Fish dinner (diet)—17 oz (482 gm)	31	27	25	23	21	20	18	17	380	64	34.4	98
Fish dinner—15¾ oz (447 gm)	61	53	47	42	38	34	31	29	922	382	110.8	77
Fish, lemon-butter sauce, starch item, vegetable—10 oz (284 gm)	23	20	18	16	15	14	13	12	279	97	36.8	44
Fish Parmesan—10 oz (280 gm)	67	60	54	50	47	44	41	39	576	204	69.8	229
Flounder, chopped broccoli (diet) (inc Lean Cuisine Fillet of Fish Divan)—12⅜ oz (351 gm)	20	18	17	15	14	13	13	12	258	45	21.5	70
Flounder, chopped broccoli, peas & carrots (diet)—17 oz (482 gm)	21	19	18	17	16	15	15	14	300	25	5.6	116
Haddock, chopped spinach (diet) (inc Lean Cuisine Fillet of Fish Florentine)—9 oz (255 gm)	18	17	15	14	13	12	12	11	189	35	16.1	72
Haddock, peas & onions, chopped cauliflower (diet)—17 oz (482 gm)	19	18	17	16	15	14	14	13	287	16	2.9	113
Lasagna w/cheese & meat sauce (diet) (inc Weight Watchers Lasagna w/meat, tomato sauce, & cheese)—12 oz (340 gm)	36	31	28	25	23	21	19	18	398	124	59.5	61
Lasagna w/cheese & sauce (diet) (inc Weight Watchers Italian Cheese Lasagna)—12 oz (340 gm)	31	27	24	22	20	18	17	16	307	90	47.0	63
Lasagna w/cheese, tomato sauce, green beans, dessert—(369 gm)	42	37	33	30	27	25	23	21	440	134	71.3	69
Lasagna w/mixed vegetables (inc Le Menu Vegetable Lasagna)—11 oz (312 gm)	44	37	33	29	27	24	22	20	375	194	84.5	39

Frozen Meals. (cont.)

	1200	1500	1800	2100	2400	2800	3200	3600	TOT CAL	FAT CAL	S/FAT CAL	CHOL MG
Linguini w/vegetables & seafood in white wine sauce (diet) (inc Budget Gourmet Linguini w/Scallops & Clams)—9½ oz (269 gm)	34	30	26	24	22	20	19	17	262	97	53.9	65
Livers, chicken, w/chopped broccoli (diet)—10½ oz (298 gm)	98	92	87	82	79	75	71	69	205	51	16.1	585
Macaroni & cheese, apples, vegetables (inc Swanson Macaroni and Cheese Dinner)—12¼ oz (347 gm)	35	30	26	24	22	19	18	16	396	144	67.7	33
Macaroni w/veal, cheese and sauce (diet)—1 meal (369 gm)	30	26	24	22	20	18	17	15	336	91	47.4	60
Meatballs, Swedish, in sauce, w/noodles (frozen entree) (inc Budget Gourmet Swedish Meatballs w/Noodles)—10 oz (284 gm)	68	59	52	48	44	39	36	34	490	244	114.0	112
Meat loaf dinner—11 oz (312 gm)	49	42	38	34	32	29	26	24	499	203	77.4	91
Meat loaf, tomato sauce, potatoes, vegetable—11 oz (312 gm)	34	30	27	24	22	20	19	17	402	172	55.0	64
Meat loaf, tomato sauce, vegetables, potatoes, chocolate brownie (reg portion w/dessert) (inc Swanson & Morton Meat Loaf Dinner—11 oz (312 gm)	49	42	38	34	32	29	26	24	499	203	77.4	91
Meat loaf w/tomato sauce, corn, applesauce—11 oz (312 gm)	30	26	23	21	19	18	16	15	366	142	49.4	52
Mexican frozen dinner w/fried beans—1 serving (18 oz) (510 gm)	50	43	38	34	31	28	25	23	738	333	95.9	50
Mosticolli w/meatballs, sauce, bread (reg portion)—11 oz (312 gm)	42	37	32	29	27	24	22	20	452	185	74.6	59

Food												
Perch, chopped broccoli, peas & carrots (diet)—17 oz (482 gm)	26	23	21	20	19	17	16	15	329	74	22.6	100
Pork, rice, & vegetables, w/soy-base sauce (diet) (inc Benihana Oriental Lite Roast Pork & Mushrooms)—9 oz (255 gm)	16	14	13	11	11	10	9	8	255	70	21.5	40
Pork, sliced, gravy, mashed potatoes, carrots, apple compote (reg portion w/sweet item) (inc Swanson Loin of Pork Dinner)—11¼ oz (319 gm)	37	32	29	26	24	22	20	19	481	213	57.4	74
Pork, sliced, sweet potatoes, mixed vegetables, apple compote (reg portion w/sweet item)—11¼ oz (319 gm)	28	24	22	20	19	17	16	15	419	116	38.6	68
Ravioli, cheese-filled, w/tomato sauce (diet frozen entree) (inc Weight Watchers Baked Cheese Ravioli)—8¹⁄₁₆ oz (229 gm)	25	21	19	17	15	14	13	12	255	83	45.1	29
Salisbury steak, baked, w/tomato sauce, mushrooms, and zucchini (diet) (inc Lean Cuisine Salisbury Steak)—9½ oz (269 gm)	39	34	31	28	26	24	22	21	295	133	56.0	92
Salisbury steak dinner—11 oz (312 gm)	50	44	39	36	33	30	26	26	473	190	77.6	100
Salisbury steak, gravy, mashed potatoes, carrots, apple cobbler (reg portion, w/sweet item) (inc Morton Salisbury Steak Dinner)—11¼ oz (319 gm)	45	39	35	32	29	26	24	22	452	209	75.8	74
Salisbury steak, gravy, potatoes, carrots & peas, dessert (large meat portion w/dessert) (inc Swanson Hungry Man Salisbury Steak)—16½ oz (468 gm)	98	84	74	67	61	55	50	46	918	443	176.8	124
Salisbury steak, gravy, potatoes, mixed vegetables or carrots (reg portion) (inc Armour Dinner Classics Salisbury Steak)—11 oz (312 gm)	76	66	59	54	50	45	41	39	459	240	121.0	145

Frozen Meals (cont.)

	1200	1500	1800	2100	2400	2800	3200	3600	TOT CAL	FAT CAL	S/FAT CAL	CHOL MG
Salisbury steak, gravy, potatoes, vegetable, soup or macaroni & cheese, dessert (3 courses w/dessert)—15½ oz (439 gm)	44	39	35	31	29	26	24	23	519	204	68.2	90
Salisbury steak, gravy, whipped potatoes, corn, chocolate cake (reg portion w/sweet item) (inc Swanson Salisbury Steak Dinner)—11 oz (312 gm)	48	41	37	33	30	28	25	23	495	239	82.8	72
Sausage & French toast (inc Swanson French Toast & Sausage)—6½ oz (184 gm)	78	71	65	61	57	53	50	47	444	205	67.0	312
Sausage & pancakes (inc Swanson Pancakes & Sausage)—6 oz (170 gm)	44	39	35	32	30	27	25	23	477	225	62.5	106
Scallops, potatoes, vegetables—8 oz (227 gm)	33	29	26	23	22	20	18	17	406	157	48.8	71
Scrambled eggs, sausage, hash brown potatoes (inc Swanson Scrambled Eggs & Sausage w/Hash Brown Potatoes)—1 meal (6¼ oz) (177 gm)	88	79	73	68	63	59	55	52	424	289	84.0	328
Seafood Newburg, rice, green beans (inc Dinner Classics Seafood Newburg)—10½ oz (297 gm)	30	26	23	21	20	18	17	16	277	84	42.4	70
Seafood platter w/fish cake, fish fillet, scallops, shrimp, potatoes (inc Taste-O-Sea Seafood Platter)—9 oz (255 gm)	52	46	41	38	35	32	29	28	527	235	75.7	121

Food												
Shrimp & clams in tomato-base sauce, w/noodles (inc Budget Gourmet Linguini w/Shrimp and Clams Marinara)—10 oz (284 gm)	25	22	20	19	17	16	15	14	243	62	28.3	79
Shrimp chow mein, egg rolls, pepper oriental—13 oz (369 gm)	17	15	14	13	12	11	10	9	415	103	22.8	45
Shrimp, potatoes, vegetables (inc Taste-O-Sea Shrimp Dinner)—7 oz (198 gm)	50	45	42	39	36	34	32	30	366	145	44.7	195
Sirloin, chopped, dinner—11 oz (312 gm)	72	62	55	50	46	42	38	36	504	260	118.6	123
Sirloin, chopped, gravy, mashed potatoes, vegetables (reg portion) (inc Le Menu Chopped Sirloin Beef)—12¼ oz (347 gm)	81	70	63	57	52	47	43	40	568	293	133.8	139
Sirloin, chopped, mushroom sauce, green beans, cauliflower (diet)—16 oz (454 gm)	76	65	58	52	48	43	39	36	539	333	136.8	98
Sirloin, chopped, or Swiss steak, gravy, vegetables, potatoes, dessert or muffin (reg portion w/sweet item)—11½ oz (326 gm)	38	33	29	26	24	22	20	19	439	174	63.3	63
Sirloin tips, gravy, potatoes, vegetables (reg portion) (inc Budget Gourmet Sirloin Tips w/Country Style Vegetables)—10 oz (283 gm)	44	38	34	31	28	25	23	21	350	184	76.6	65
Sole, chopped cauliflower (diet)—8½ oz (241 gm)	14	13	13	12	11	11	10	10	196	19	4.4	80
Spaghetti & meatballs dinner—12½ oz (354 gm)	16	14	12	11	10	9	8	8	320	80	26.9	26
Spaghetti & meatballs, peas & carrots, dessert (reg portion)—12½ oz (354 gm)	16	14	13	11	10	9	9	8	337	81	28.5	24
Spaghetti & meatballs, tomato sauce, sliced apples, bread (reg portion) (inc Morton Spaghetti & Meatballs Dinner)—11½ oz (326 gm)	30	26	23	21	19	18	16	15	449	121	46.7	57

Frozen Meals (cont.)

	1200	1500	1800	2100	2400	2800	3200	3600	TOT CAL	FAT CAL	S/FAT CAL	CHOL MG
Spaghetti w/meat & mushroom sauce (diet frozen entree) (inc Lean Cuisine Spaghetti w/Beef & Mushroom Sauce)—11½ oz (326 gm)	17	15	14	12	11	10	10	9	320	78	26.4	36
Spaghetti w/meat sauce (diet frozen entree) (inc Weight Watchers Spaghetti w/Meat Sauce)—10½ oz (297 gm)	17	15	13	12	11	10	9	9	305	74	24.8	37
Stuffed cabbage, w/meat & tomato sauce (diet frozen entree) (inc Lean Cuisine Stuffed Cabbage)—10¾ oz (305 gm)	25	22	19	18	16	15	14	13	240	99	37.5	52
Stuffed green pepper (frozen entree) (inc Green Giant Stuffed Green Pepper)—1 entree (14 oz) (396 gm)	32	28	25	22	21	19	17	16	388	184	54.6	51
Turbot, peas & carrots (diet)—8½ oz (241 gm)	19	17	16	14	13	12	11	11	283	117	23.2	58
Turbot, zucchini, carrots (diet)—17 oz (482 gm)	32	28	26	24	22	20	19	18	546	196	38.9	95
Turkey & vegetables, in sauce (diet) (inc Lean Cuisine Turkey Dijon)—9½ oz (269 gm)	23	20	18	16	15	14	13	12	275	93	31.4	55
Turkey breast, mushrooms, gravy, long-grain & wild rice, vegetables (inc Le Menu Sliced Turkey Breast Dinner)—11¼ oz (319 gm)	27	23	21	19	18	16	15	14	442	208	42.1	53
Turkey, carrots, chopped broccoli, stuffing (diet)—19 oz (539 gm)	29	26	24	22	21	19	18	17	443	80	25.0	112

Food item												
Turkey dinner—11 oz (312 gm)	71	48.7	222	519	17	18	20	22	23	26	29	33
Turkey, dressing, gravy, potato only (large meat portion) (inc Swanson Turkey Entree)—13¼ oz (376 gm)	97	38.0	168	427	18	19	20	22	24	26	28	32
Turkey, dressing, gravy, potato only (reg portion) (inc Swanson Turkey Entree)—8¾ oz (248 gm)	32	29.6	131	306	9	10	11	12	13	14	16	18
Turkey, gravy, dressing, potatoes, peas, cream of tomato soup, apple-cranberry cake cobbler (3 courses w/soup)—16 oz (454 gm)	249	73.1	306	807	41	44	47	51	54	58	64	71
Turkey, gravy, dressing, potatoes, vegetables, apple-cranberry cake cobbler (reg portion w/dessert) (inc Swanson and Morton—11½ oz (326 gm)	56	27.8	119	337	11	12	13	14	15	17	19	21
Turkey, gravy, dressing, potatoes, vegetables, dessert (large portion w/dessert) (inc Swanson Hungry Man Turkey Dinner)—18½ oz (524 gm)	94	38.5	176	603	18	19	20	22	24	26	28	32
Turkey, gravy, dressing, potatoes, vegetables (reg portion) (inc Banquet Turkey Dinner)—11 oz (312 gm)	74	45.2	203	509	17	18	19	21	23	25	28	32
Turkey tetrazzini (frozen entree) (inc Stouffer's Turkey Tetrazzini)—12 oz (340 gm)	86	84.2	262	514	25	27	30	33	36	39	44	51
Veal, breaded, w/spaghetti in tomato sauce only (inc Swanson Spaghetti with Breaded Veal)—8¼ oz (234 gm)	85	32.0	132	289	15	16	18	19	20	22	24	27
Veal dinner—11 oz (312 gm)	60	53.5	171	509	17	18	19	21	23	26	29	33
Veal, mixed vegetables, potato wedges (diet) (inc Classic Lite Veal Pepper Steak)—11 oz (312 gm)	76	39.4	96	283	16	17	18	20	21	23	26	29
Veal parmigiana, apple slices, peas, muffin (reg portion w/sweet item)—12¼ oz (347 gm)	91	104.2	255	543	29	32	35	38	42	46	52	60

	1200	1500	1800	2100	2400	2800	3200	3600	TOT CAL	FAT CAL	S/FAT CAL	CHOL MG
Frozen Meals *(cont.)*												
Veal parmigiana, green beans, fettuccine Alfredo, apple crumb cake (reg portion w/dessert) (inc Swanson's Veal Parmigiana Dinner)—12¾ oz (361 gm)	50	44	39	36	33	30	28	26	488	215	75.5	105
Veal parmigiana, potatoes, vegetable (reg portion)—11 oz (312 gm)	41	36	32	29	27	25	23	21	446	159	58.8	94
Veal parmigiana w/zucchini (diet) (inc Weight Watchers Veal Patty Parmigiana w/Zucchini)—9 oz (255 gm)	29	25	23	21	19	17	16	15	195	81	43.2	63
Veal w/peppers in sauce (diet)—13 oz (369 gm)	25	22	20	19	18	16	15	14	286	73	28.4	81
Zucchini lasagna (diet) (inc Lean Cuisine Zucchini Lasagna, Light and Elegant Lasagna Florentine)—1 Lean Cuisine meal (11 oz) (312 gm)	24	20	18	16	15	13	12	11	298	76	44.8	24
CONDIMENTS (Olives, Pickles, Relishes)												
Beans, string, green, pickled—½ c (68 gm)	0	0	0	0	0	0	0	0	19	1	.1	0
Beets, pickled (inc pickled beets w/onions, beet salad)—½ c, sliced (85 gm)	0	0	0	0	0	0	0	0	42	0	.1	0
Cabbage, fresh, pickled, Japanese style—¼ c (38 gm)	0	0	0	0	0	0	0	0	8	0	0	0
Cabbage, mustard, salted—¼ c (31 gm)	0	0	0	0	0	0	0	0	6	0	0	0

Food (Measure)	Calories	Fat (g)	Exchanges
Cabbage, red, pickled (inc sweet & sour red cabbage)—¼ c (38 gm)	55	0	—
Cauliflower, pickled—1 floweret (27 gm)	11	.1	—
Celery, pickled—1 c (150 gm)	23	.1	—
Chutney—2 tbsps (34 gm)	51	.4	—
Corn relish—1 tbsp (15 gm)	13	.2	—
Cranberry-orange relish, canned—½ c (138 gm)	246	.1	—
Cucumber, Kim Chee style—¼ c (38 gm)	8	.3	—
Cucumber pickles, dill—1 med (3¾" long) (65 gm)	7	.1	—
Cucumber pickles, fresh (inc bread and butter)—¼ cup (43 gm)	31	.3	—
Cucumber pickles, relish (inc Indian sweet relish)—1 tbsp (15 gm)	21	.2	—
Cucumber pickles, sour—1 med (3¾" long) (65 gm)	7	.2	—
Cucumber pickles, sweet (inc candied dill spears, semisweet, mixed pickles)—1 spear gherkin (20 gm)	29	.3	—
—1 lg gherkin (3" long) (35 gm)	51	.2	—
Eggplant, pickled—¼ c (34 gm)	12	.3	½ veg
Horseradish—1 tbsp (15 gm)	6	.6	—
Mustard (inc horseradish mustard, Chinese mustard)—1 tbsp (15 gm)	11	.1	—
Mustard pickles (inc chow chow, hotdog relish)—1 tbsp (15 gm)	17	.3	—
Mustard sauce—1 tbsp (10 gm)	20	.1	½ fat
Okra, pickled—1 pod (11 gm)	4	.6	—
Olives, green—5 lg (23 gm)	27	3.7	2 fat
Olives, green, stuffed—10 med (40 gm)	41	5.6	2 fat

229

	1200	1500	1800	2100	2400	2800	3200	3600	TOT CAL	FAT CAL	S/FAT CAL	CHOL MG
Condiments (Olives, Pickles, Relishes) (cont.)												
Olives, ripe—5 lg (23 gm)	2	2	1	1	1	1	1	1	30	29	4.1	0
Peppers, hot, sauce (inc Tabasco sauce)—1 tbsp (15 gm)	0	0	0	0	0	0	0	0	3	0	0	0
Peppers, pickled—1 med (20 gm)	0	0	0	0	0	0	0	0	8	1	.1	0
Pickles, mixed—¼ c (39 gm)	0	0	0	0	0	0	0	0	57	2	.5	0
Pimientos, canned, solid & liquid, wo/added calcium salts—2 oz jar (57 gm)	0	0	0	0	0	0	0	0	15	3	0	0
Radishes, pickled, Hawaiian style—¼ cup (38 gm)	0	0	0	0	0	0	0	0	8	1	.3	0
Seaweed, pickled—¼ c (38 gm)	0	0	0	0	0	0	0	0	57	1	.2	0
Tsukemono, Japanese pickles (inc nara zuke, takuan zuke, wasabe zuke)—¼ c (34 gm)	0	0	0	0	0	0	0	0	7	1	.1	0
Turnip, pickled—¼ c (39 gm)	0	0	0	0	0	0	0	0	16	1	.1	0
Vegetable relish—1 tbsp (9 gm)	0	0	0	0	0	0	0	0	4	0	0	0
Vegetables, pickled, Hawaiian style—¼ c (38 gm)	0	0	0	0	0	0	0	0	10	1	.3	0
Vegetables, pickled (inc giardiniera)—1 cup (163 gm)	0	0	0	0	0	0	0	0	43	3	.4	0
Vinegar—1 c (240 gm)	0	0	0	0	0	0	0	0	34	0	0	0
Zucchini, pickled—¼ c (43 gm)	0	0	0	0	0	0	0	0	13	1	.1	0
COOKIES and PASTRIES												
Cookies												
Almond—1 cookie (10 gm)	3	3	2	2	2	2	2	2	52	29	4.8	6
Applesauce (inc apple snacks, Dutch apple)—1 med (13 gm)	2	2	1	1	1	1	1	1	48	14	2.9	4

Food												
Bar, w/chocolate, nuts & graham crackers (inc Magic Cookie Bars)—1 bar (25 gm)	11	9	8	7	6	6	5	5	117	61	23.3	4
Brownies, from mix, w/egg/water/nuts, baked—1 brownie, 3" x 1" x 7/8" (20 gm)	5	4	4	3	3	3	3	2	86	36	8.0	9
Brownies, w/icing (inc w/nuts)—1 brownie (42 gm)	17	14	13	12	11	9	9	8	176	78	29.5	23
Brownies, wo/icing (inc w/nuts)—1 brownie (34 gm)	7	6	5	5	4	4	4	3	129	42	10.1	13
Brownies, w/nuts, frozen, commercial, w/chocolate icing—1 brownie, 1½" x 1¾" (25 gm)	8	7	6	6	5	5	4	4	103	45	15.3	10
Butter (inc sesame cookies, wedding cookies, Nabisco Social Tea Biscuits)—1 cookie (5 gm)	2	2	2	2	1	1	1	1	23	8	4.7	2
Butterscotch brownie—1 brownie (34 gm)	9	8	7	6	6	5	5	5	149	59	11.1	23
Butterscotch chip—1 med (approx 2" dia) (10 gm)	2	2	2	2	2	1	1	1	47	19	3.8	5
Carob—1 med (13 gm)	3	3	3	2	2	2	2	2	51	20	3.7	10
Carob & honey brownie—1 brownie (34 gm)	10	9	8	7	7	6	6	6	143	83	13.0	28
Chocolate & vanilla sandwich—1 sandwich (approx 1"-1½" dia) (11 gm)	2	2	2	2	1	1	1	1	54	22	5.5	0
Chocolate chip (inc w/ or wo/ nuts, Congo Bar)—1 med (approx 2" dia) (10 gm)	3	2	2	2	2	1	1	1	47	19	6.3	1
Chocolate chip, made fr home recipe or purchased at a bakery (inc tollhouse, made w/M & M's)—1 med (approx 2" dia) (10 gm)	4	3	3	3	2	2	2	2	49	25	6.6	5
Chocolate chip sandwich—1 sandwich (approx 1"-1½") (11 gm)	2	2	2	2	1	1	1	1	54	22	5.5	0

	1200	1500	1800	2100	2400	2800	3200	3600	TOT CAL	FAT CAL	S/FAT CAL	CHOL MG
Cookies (cont.)												
Chocolate chip, w/raisins (inc date-nut chocolate chip bar)—1 med (approx 2" dia) (10 gm)	3	2	2	2	2	1	1	1	46	18	5.9	1
Chocolate, chocolate-covered or fudge sandwich (inc Girl Scout Mint, Oreos, Twiddle Stix, Swiss Style Shortbread (Sunshine), Chocolate Fudge)—1 sandwich (approx 1"–1½" dia) (11 gm)	2	2	2	2	1	1	1	1	54	22	5.5	0
—1 Capri (15 gm)	3	3	2	2	2	2	2	1	74	30	7.5	0
—1 Nutty Bar (Little Debbie) (28 gm)	6	5	5	4	4	3	3	3	139	57	14.0	0
Chocolate-covered marshmallow (inc marshmallow puff, pinwheels)—1 cookie (17 gm)	4	3	3	3	2	2	2	2	70	20	6.9	4
Chocolate fudge, w/wo nuts (inc chocolate-jelly, Peanut Butter 'N Fudge Chocolate Chip—1 med (approx 2" dia) (13 gm)	3	2	2	2	2	2	1	1	58	18	5.3	3
Chocolate, made w/rice cereal—1 bar (2" square) (30 gm)	18	15	13	12	11	9	8	8	133	62	41.4	0
Chocolate, sandwich, w/extra filling (inc Double Stuff Oreo)—1 sandwich (14 gm)	3	2	2	2	1	1	1	1	66	24	5.8	0
Chocolate wafers (inc chocolate snaps, Pirouettes)—1 med (12 gm)	3	2	2	2	2	1	1	1	53	17	4.9	3
Coconut bars—1 cookie (approx 2½" x 1¼") (9 gm)	4	3	3	3	2	2	2	2	44	20	8.8	0

Food											
Cone shell, ice cream type—1 cone (6 gm)	0	0	0	0	0	0	0	23	1	.2	0
Cone shell, ice cream type, brown sugar—1 cone (11 gm)	1	1	1	1	1	1	½	44	5	2.6	0
Date bar (inc date-nut bar, date macaroon, date pinwheel)—1 cookie (16 gm)	2	2	2	2	2	2	1	57	8	2.4	8
Dietetic, apple pastry—1 cookie (24 gm)	14	12	10	9	8	7	6	115	63	31.1	0
Dietetic, chocolate chip—1 cookie (5 gm)	2	2	1	1	1	1	1	25	12	3.9	1
Dietetic, chocolate-flavored—1 wafer (6 gm)	3	3	2	2	2	2	1	30	14	7.0	0
Dietetic, fruit types—1 cookie (11 gm)	3	2	2	2	2	2	1	51	24	5.7	0
Dietetic—1 cookie (6 gm)	3	3	3	2	2	2	1	29	16	7.8	0
Dietetic, oatmeal w/raisins—1 cookie (7 gm)	1	1	1	1	1	1	1	30	11	2.8	0
Dietetic, sandwich type—1 cookie (11 gm)	3	2	2	2	2	2	1	50	25	6.0	0
Dietetic, sugar or plain—1 lemon thin (6 gm)	2	2	1	1	1	1	1	28	13	2.6	4
Fig bars—1 cookie (16 gm)	2	2	2	2	2	2	1	57	8	2.4	8
Fortune—1 cookie (8 gm)	2	2	1	1	1	1	1	39	14	2.8	5
Fruit-filled bars (inc trail mix cookies, fruitcake cookies)—1 med (16 gm)	2	2	2	2	2	2	1	57	8	2.4	8
Gingersnaps—1 med (12 gm)	2	1	1	1	1	1	1	50	10	2.5	3
Granola—1 cookie (13 gm)	7	6	5	5	4	3	3	60	20	16.8	0
Ladyfingers (inc anisette sponge, Stella D'Oro Egg Jumbo)—1 egg jumbo (9 gm)	1	1	1	1	1	1	1	32	6	2.7	1
—1 anisette sponge (12 gm)	2	1	1	1	1	1	1	43	8	3.6	1
—1 ladyfinger (14 gm)	2	1	1	1	1	1	1	50	10	4.2	1
Lebkuchen (inc honey nut bar)—1 cookie (16 gm)	1	1	1	1	1	1	1	62	8	.9	4

THE 2-IN-1 SYSTEM

	1200	1500	1800	2100	2400	2800	3200	3600	TOT CAL	FAT CAL	S/FAT CAL	CHOL MG
Cookies (cont.)												
Lemon bars—1 bar (16 gm)	5	4	4	4	3	3	3	3	70	27	5.6	16
Macaroons—1 cookie (14 gm)	6	5	4	4	3	3	3	2	67	29	12.9	0
Marshmallow pie, chocolate-covered (inc Moon Pie, Sweetie Pie, Mallow Mars, Scooter Pie)—1 scooter pie (34 gm)	8	7	6	5	5	4	4	4	139	40	13.9	9
Marshmallow pie, nonchocolate coating (inc banana)—1 pie (34 gm)	8	7	6	5	5	4	4	4	139	40	13.9	9
Marshmallow, w/coconut—1 cookie (18 gm)	4	3	3	3	3	2	2	2	74	21	7.3	5
Marshmallow, w/rice cereal (inc Rice Crispy bar cookie)—1 bar (2" sq) (30 gm)	2	2	2	1	1	1	1	1	117	26	5.1	0
Meringue (inc Marguerite, angel cup)—1 cookie (4 gm)	0	0	0	0	0	0	0	0	14	0	0	0
Meringue—1 meringue (22 gm)	0	0	0	0	0	0	0	0	65	0	0	0
Molasses (inc hermit, gingerbread man)—1 med (15 gm)	2	1	1	1	1	1	1	1	63	14	4.0	0
Oatmeal—1 med (2⅝" dia) (13 gm)	4	3	3	2	2	2	2	2	54	26	6.2	5
Oatmeal sandwich, w/creme filling (inc Keebler Chipsies)—1 sandwich (approx 1"-1½" dia) (15 gm)	4	3	3	2	2	2	2	2	73	30	7.6	2
Oatmeal, w/chocolate & peanut butter—1 cookie (34 gm)	6	5	5	4	4	3	3	3	141	56	13.5	1
Oatmeal, w/chocolate chips (inc w/chocolate frosting)—1 cookie (15 gm)	5	4	3	3	3	2	2	2	68	27	9.9	2

Food													
Oatmeal, w/raisins (inc w/dates, w/nuts)—1 med (approx 2⅜" dia) (13 gm)	2	2	2	2	2	1	1	1	1	59	18	4.5	2
Peanut butter filling, chocolate-coated (inc Girl Scout Tagalongs Peanut Butter Patties)—1 cookie (13 gm)	13	11	9	8	7	7	7	6	5	68	44	28.7	0
Peanut butter (inc peanut butter wafer)—1 med (16 gm)	4	4	3	3	3	2	2	2	2	78	36	6.7	7
Peanut butter, w/chocolate (inc Pepperidge Farm Nassau)—1 Nassau (15 gm)	5	4	4	3	3	3	3	3	2	72	29	10.7	3
Peanut butter, w/oatmeal—1 cookie (16 gm)	4	3	3	3	3	2	2	2	2	69	19	3.6	13
Peanut (inc chocolate peanut bar, peanut sandwich)—1 sandwich (approx 1"–1½" dia) (11 gm)	3	2	2	2	2	1	1	1	1	52	19	5.5	2
—1 cookie (16 gm)	4	4	3	3	3	2	2	2	2	76	28	8.1	4
Pizzelles (Italian-style wafer) (inc rosette)—1 wafer or rosette (14 gm)	4	4	3	3	3	3	3	3	3	60	10	2.2	20
Pumpkin (inc carrot bar)—1 med (11 gm)	2	1	1	2	1	1	1	1	1	43	17	3.2	5
Raisin (inc date-nut, raisin sugar)—1 cookie (16 gm)	2	1	1	1	1	1	1	1	1	61	8	2.0	5
Raisin sandwich, cream-filled (inc Little Debbie Raisin Creme Pie)—1 lg sandwich (3½"-4" dia) (30 gm)	6	5	4	4	3	3	3	3	3	130	40	10.0	7
Rugelach (inc cookie made w/cottage cheese)—1 cookie (14 gm)	3	2	2	2	2	1	1	1	1	59	33	6.5	0
Sandwich-type, not chocolate or vanilla (inc banana cream-filled, Keebler Puddin' Cremes)—1 sandwich (approx 1"–1½" dia) (11 gm)	2	2	2	2	1	1	1	1	1	54	22	5.5	0
Shortbread (include Keebler Pecan Sandies, Pepperidge Farm Milano)—1 med (15 gm)	7	6	5	5	5	4	4	4	4	75	31	11.3	13

THE 2-IN-1 SYSTEM

	1200	1500	1800	2100	2400	2800	3200	3600	TOT CAL	FAT CAL	S/FAT CAL	CHOL MG
Cookies (cont.)												
Sugar, iced (inc spice & other flavors)—1 med (16 gm)	3	3	2	2	2	2	2	2	67	17	4.8	6
Sugar (inc fruit flavors, w/or wo/nuts, spice, kichel, jelly-topped)—1 med (16 gm)	3	3	2	2	2	2	2	2	71	19	4.9	6
Sugar wafer—1 med (7 gm)	2	1	1	1	1	1	1	1	34	12	2.4	4
Sugar, w/chocolate frosting—1 med (16 gm)	4	3	3	2	2	2	2	2	66	19	6.8	4
Tea, Japanese (inc senbei, plain, ginger, or seaweed)—1 med (5 gm)	0	0	0	0	0	0	0	0	19	0	0	0
Teething (biscuit)—1 cookie (7 gm)	½	½	0	0	0	0	0	0	25	2	.7	0
Vanilla sandwich (inc Keebler French Vanilla Creme, Sunshine Vanilla Cremes, Sunshine Vienna Fingers, Nabisco Cameos, Pepperidge Farm Lidos)—1 sandwich (approx 1"-1½" dia) (11 gm)	2	2	2	2	1	1	1	1	54	22	5.5	0
Vanilla wafer—1 med (4 gm)	1	1	1	1	1	1	1	½	18	6	1.4	2
Vanilla waffle creme—1 med (9 gm)	2	2	2	2	1	1	1	1	44	16	3.1	5
Vanilla w/caramel, coconut & chocolate coating (inc Girl Scout Samoas)—1 cookie (15 gm)	9	8	7	6	5	5	4	4	77	43	20.0	1
Whole wheat, dried fruit, nut—1 cookie (14 gm)	4	3	3	3	2	2	2	2	60	28	4.5	10
Cobblers, Eclairs, Other Pastries												
Baklava (inc kadayif)—1 piece (2" x 2" x 1½") (78 gm)	42	36	32	29	26	23	21	19	333	205	83.4	35

236

Food												
Blintzes, cheese-filled —1 blintz (70 gm)	22	20	18	17	16	15	14	13	139	56	19.9	86
Blintzes, fruit-filled—1 blintz (70 gm)	16	14	13	12	12	11	10	10	125	42	13.0	64
Charlotte russe, w/lady fingers, whipped cream filling—1 serving (114 gm)	68	61	55	51	48	44	41	39	326	149	74.5	225
Cheese puffs (inc cheese straws) —2 puffs or straws (12 gm)	6	5	5	4	4	4	4	3	32	21	6.8	19
Cobbler, apple (inc fruit cobbler, school lunch)—½ c (109 gm)	7	6	5	5	4	4	3	3	235	58	15.4	2
Cobbler, apricot—½ c (109 gm)	5	5	4	3	3	3	2	2	203	45	11.7	1
Cobbler, berry—½ c (109 gm)	8	7	6	5	5	4	4	3	253	63	16.9	2
Cobbler, cherry—½ c (109 gm)	7	6	5	4	4	3	3	3	215	54	14.2	2
Cobbler, peach—½ c (109 gm)	6	6	5	4	4	3	3	3	212	46	12.6	2
Cobbler, pear—½ c (109 gm)	6	5	5	4	4	3	3	3	226	49	12.9	2
Cobbler, pineapple—½ c (109 gm)	6	5	4	4	4	3	3	3	206	47	12.8	2
Cobbler, plum—½ c (109 gm)	6	5	5	4	4	3	3	3	223	50	13.1	2
Cobbler, rhubarb—½ c (109 gm)	7	6	5	5	4	4	3	3	275	66	16.2	1
Cream puffs, eclairs, custard- or cream-filled, iced (inc napoleon)—1 miniature (30 gm)	10	9	9	8	7	7	6	6	65	32	9.5	39
—1 eclair, frozen (78 gm)	27	24	22	21	19	18	17	16	169	84	24.8	102
—1 napoleon (85 gm)	29	26	24	22	21	20	18	17	184	92	27.0	111
—1 eclair (5″ x 2″ x 1¾″) (102 gm)	35	32	29	27	25	24	22	21	221	110	32.5	133
—1 cream puff (3½″ x 2″) (112 gm)	39	35	32	30	28	26	24	23	242	121	35.6	147
Cream puffs, eclairs, custard- or cream-filled, not iced (inc French horn, lady lock)—1 miniature cream puff (23 gm)	9	8	7	7	6	6	5	5	51	27	7.5	34
—1 cream or French horn (57 gm)	21	19	18	17	16	15	14	13	127	66	18.5	85
—1 eclair (5″ x 2″ x 1¾″) (90 gm)	34	31	28	26	25	23	22	21	201	105	29.3	135

	1200	1500	1800	2100	2400	2800	3200	3600	TOT CAL	FAT CAL	S/FAT CAL	CHOL MG
Cobblers, Eclairs, Other Pastries (cont.)												
—1 cream puff (3½" x 2") (100 gm)	38	34	31	29	28	26	24	23	223	117	32.5	150
Cream puffs, w/custard filling—1 cream puff (3½" dia) (130 gm)	52	46	42	39	37	34	32	30	303	163	50.6	187
Crepes, dessert type, chocolate-filled—1 crepe w/filling (56 gm)	12	11	10	9	9	8	8	7	91	30	8.1	51
Crepes, dessert type, fruit-filled—1 crepe w/filling (56 gm)	13	12	11	10	10	9	8	8	95	27	8.3	58
Crepes, dessert type, ice cream-filled—1 crepe w/filling (56 gm)	23	20	18	17	16	15	14	13	117	53	23.9	77
Crepes suzette—1 crepe w/sauce (66 gm)	33	29	27	25	23	21	20	19	163	85	37.7	104
Crisp, apple, apple dessert (inc apple betty)—½ cup (108 gm)	3	3	2	2	2	2	1	1	177	35	7.0	0
Crisp, blueberry—½ c (123 gm)	10	8	7	6	6	5	4	4	315	114	22.0	0
Crisp, cherry—½ c (123 gm)	12	10	9	8	7	6	6	5	355	128	27.9	0
Crisp, peach—½ c (123 gm)	6	5	4	4	3	3	3	2	250	69	13.4	0
Crisp, rhubarb—½ c (123 gm)	6	5	4	4	4	3	3	3	256	71	13.9	0
Eclairs, w/custard filling & chocolate icing—1 eclair (5" x 2" x 1¾") (100 gm)	39	35	32	29	27	25	24	23	239	122	39.3	136
Fritter, apple—2 fritters (48 gm)	21	18	16	15	14	13	12	11	175	104	28.0	53
Fritter, banana—2 fritters (2" long) (88 gm)	33	29	26	24	22	20	19	18	303	172	45.7	83
Fritter, berry—2 fritters (48 gm)	18	16	14	13	12	11	10	10	157	94	24.6	45
Fritters or fried puffs, air-filled wo/syrup, Puerto Rican style (bunuelos de viento) (inc wheat flour turnover)—1 turnover (57 gm)	24	22	20	19	17	16	15	14	200	98	24.5	87

Food												
Pastry, Chinese, made w/rice flour (inc nine-layer pudding)—1 oz (28 gm)	1	1	1	1	1	½	½	½	67	13	1.8	0
Pastry, flour & water only, fried (inc Taco Bell Cinnamon Crispas)—1 crispas (10 gm)	1	1	1	1	1	1	1	1	42	16	3.0	0
Pastry, fruit-filled (inc hamantaschen)—1 pastry (78 gm)	16	14	12	11	10	9	8	8	282	155	30.0	20
Pastry, Italian, w/cheese (inc cannoli)—1 pastry (85 gm)	27	23	21	19	17	15	14	13	233	96	47.1	38
Pastry, puff (inc angel wings, flaky pastry)—1 pastry (11 gm)	3	3	2	2	2	2	1	1	57	36	7.0	0
Strudel, apple—1 piece (approx 2"–2½" sq) (64 gm)	5	4	4	3	3	3	3	2	156	29	6.0	12
Strudel, berry—1 piece (approx 2"–2½" sq) (64 gm)	5	5	4	4	4	3	3	3	159	36	7.3	14
Strudel, cheese—1 piece (approx 2"–2½" sq) (64 gm)	23	20	18	17	15	14	13	12	196	75	35.4	49
Strudel, cheese & fruit—1 piece (approx 2"–2½" sq) (64 gm)	15	13	12	11	10	9	8	8	141	49	23.2	32
Strudel, cherry—1 piece (approx 2"–2½" sq) (64 gm)	5	5	4	4	4	3	3	3	179	57	8.5	11
Strudel, peach—1 piece (approx 2"–2½" sq) (64 gm)	4	4	3	3	3	3	2	2	123	27	5.4	11
Tamale, sweet (inc w/sugar & spices)—1 tamale (34 gm)	5	4	4	3	3	2	2	2	115	45	11.0	0
Tamale, sweet, w/fruit—1 tamale (49 gm)	5	4	4	3	3	3	2	2	133	46	11.2	0
Turnover, guava—1 turnover (78 gm)	14	12	10	9	8	7	7	6	228	131	32.3	0
Turnover or dumpling, apple—1 turnover (82 gm)	15	12	11	10	9	8	7	6	289	135	33.4	0
Turnover or dumpling, berry—1 turnover (78 gm)	14	12	10	9	8	7	6	6	276	126	31.0	0
Turnover or dumpling, cherry—1 turnover (78 gm)	12	10	8	7	7	6	5	5	238	106	26.0	0

	1200	1500	1800	2100	2400	2800	3200	3600	TOT CAL	FAT CAL	S/FAT CAL	CHOL MG
Cobblers, Eclairs, Other Pastries (cont.)												
Turnover or dumpling, lemon—1 turnover (78 gm)	22	19	17	16	15	14	13	12	238	113	28.4	59
Turnover or dumpling, peach—1 turnover (78 gm)	13	11	10	9	8	7	6	5	260	120	29.8	0
Wheat flour fritters, wo/syrup—1 fritter (22 gm)	15	14	12	11	11	10	9	8	99	78	18.1	46
Danish, Doughnuts, Breakfast and Snack Bars												
Breakfast bars, cakelike (inc Carnation Breakfast Bars)—1 bar (43 gm)	28	23	20	18	16	14	13	12	192	77	62.7	0
Breakfast bars, diet meal type (inc Figurines, Slender Bars)—1 bar (25 gm)	21	18	15	14	12	11	10	9	129	72	47.6	0
Breakfast tarts (inc pop tarts, toaster pastries)—1 tart (50 gm)	6	5	5	4	4	3	3	3	196	52	14.4	0
Danish pastries (inc fruit and/or spice, pecan swirls, snail Danish, pastry with icing, babka)—1 pecan swirl (28 gm)	12	10	9	8	8	7	6	6	118	59	18.1	24
—1 Danish (64 gm)	27	24	21	19	18	16	15	14	270	135	41.5	55
Danish pastries, w/cheese—1 cheese Danish (83 gm)	30	26	24	22	20	18	17	16	319	168	44.2	67
Doughnuts, cake-type, chocolate-covered, dipped in peanuts (inc peanut stick)—1 doughnut (3¼" dia) (53 gm)	13	11	10	9	8	7	7	6	190	77	22.6	17

Food												
Doughnuts, cake-type, chocolate-covered—1 med doughnut (3¼″ dia) (53 gm)	13	12	10	9	8	8	7	6	205	83	24.4	17
Doughnuts, cake-type, chocolate (inc glazed)—1 med doughnut (3¼″ dia) (42 gm)	11	9	8	7	7	6	5	5	166	66	18.9	14
Doughnuts, cake-type, chocolate, w/chocolate icing—1 med doughnut (3¼″ dia) (53 gm)	15	13	11	10	9	8	8	7	207	89	28.4	16
Doughnuts, cake-type (inc coconut, sugar-coated, glazed)—1 med doughnut (3¼″ dia) (42 gm)	10	9	8	7	7	6	5	5	164	70	17.5	16
Doughnuts, chocolate creme-filled—1 doughnut (65 gm)	30	26	23	20	18	16	15	14	239	134	59.6	24
Doughnuts, custard-filled (inc Long John, cream-filled)—1 doughnut (65 gm)	31	27	23	21	19	17	15	14	239	135	61.7	24
Doughnuts, custard-filled, w/icing—1 doughnut (70 gm)	27	23	20	18	16	14	13	12	260	116	53.0	20
Doughnuts, jelly—1 doughnut (65 gm)	29	25	22	20	18	16	14	13	253	129	58.2	22
Doughnuts, oriental (inc Okinawan)—1 doughnut (18 gm)	6	6	5	5	4	4	4	3	75	33	8.5	17
Doughnuts, raised or yeast, chocolate—1 doughnut (approx 3″ dia) (50 gm)	27	23	20	18	16	15	13	12	203	118	54.2	20
Doughnuts, raised or yeast (inc reg glazed, crullers, doughboys, doughnut holes, honey dip, malasadas)—1 doughnut (approx 3″ dia) (50 gm)	20	17	15	13	12	11	10	9	203	103	38.7	18
Doughnuts, whole wheat (inc glazed)—1 doughnut (42 gm)	11	10	9	8	7	6	6	5	171	77	19.3	17
Granola bars, chocolate-coated (inc Quaker Oats Granola Dipps)—1 Quaker Oats Dipp (28 gm)	15	13	11	10	9	8	7	6	131	55	32.0	5
Granola bars, w/nonchocolate coating (inc Quaker Oats Peanut Butter Whipps)—1 Quaker Oats Whipp (28 gm)	6	5	5	4	4	3	3	3	133	54	14.0	1

	1200	1500	1800	2100	2400	2800	3200	3600	TOT CAL	FAT CAL	S/FAT CAL	CHOL MG
Danish, Doughnuts, Breakfast and Snack Bars (cont.)												
Granola bars, oats, sugar, raisins, coconut (inc Nature Valley Chewy Granola Bars, Quaker Oats Chewy Granola Bars, New Trail) —1 New Trail (40 gm)	18	15	13	12	11	9	8	8	182	63	41.5	0
—1 bar (43 gm)	20	17	14	13	11	10	9	8	195	68	44.6	0
	23	20	17	15	14	12	11	10	207	79	52.6	0
Granola bars, peanuts, oats, sugar, wheat germ—1 bar (43 gm)	8	7	6	5	5	4	4	3	141	45	17.9	0
Granola bars, w/nougat (inc Nature Valley Granola Clusters)—1 Nature Valley Granola Cluster (34 gm)	6	5	4	4	3	3	3	2	121	42	12.7	0
Granola bar, w/rice cereal (inc Kellogg Rice Krispy bar)—1 Rice Krispy bar (28 gm)												
CRACKERS												
Animal (inc sweet crackers)—4 crackers (12 gm)	3	3	2	2	2	2	2	2	51	10	3.8	8
Anisette toast (inc almond toast)—1 piece (11 gm)	2	2	1	1	1	1	1	1	46	9	3.7	2
Butter crackers—10 round crackers (33 gm)	8	7	6	5	5	4	4	3	151	53	17.9	0
Butter (inc snack-flavored, bacon chips, Ritz, Waverly Wafers, Nabisco Dip and Chip, Club)—4 round crackers (12 gm)	6	5	5	4	4	3	3	3	55	19	11.9	6
Butter, low sodium—1 round cracker (3 gm)	1	1	1	1	1	1	1	1	14	5	3.0	1

Cheese (inc cheese sticks, Cheez-its)—4 round crackers (12 gm)	4	4	3	3	3	2	2	2	57	23	7.9	5
Cuca—1 cracker (28 gm)	7	6	6	5	5	4	4	3	121	30	12.8	10
Fiber (inc Wasa crisp bread)—1 cracker (4½" x 2½" x ⅛") (11 gm)	0	0	0	0	0	0	0	0	38	1	.1	0
Flat bread, unleavened (inc Armenian crackers, lavosh, hardtack)—1 oz (28 gm)	0	0	0	0	0	0	0	0	89	2	.4	0
———1 Norwegian flat bread cracker (4¾" x 2¾" x 1/16") (58 gm)	½	½	0	0	0	0	0	0	185	5	.7	0
Graham cracker, chocolate-covered—2 crackers (2½" x 2" x ¼") (26 gm)	6	5	4	4	3	3	3	2	124	55	13.1	0
Graham crackers, plain—1 cracker (5" x 2½") (14 gm)	1	1	1	1	1	1	1	1	55	12	2.9	0
Graham, sugar, honey-coated, cinnamon crisps—2 lg rectangular pieces or 2 squares or 4 sm rectangular pieces (28 gm)	3	3	2	2	2	2	1	1	115	29	6.8	0
Low sodium, dietetic, matzo, matzo w/wheat germ—1 matzo (30 gm)	½	½	½	½	0	0	0	0	119	5	.9	0
Low sodium, toast thins (rye, wheat, white flour) (inc Low Salt Wheat Thins)—1 cracker (2 gm)	½	½	½	½	½	0	0	0	9	3	1.0	0
Low sodium, whole wheat (inc Triscuits, Stoned Wheat Wafers)—4 wheat wafers (16 gm)	3	2	2	2	2	1	1	1	64	20	6.1	0
Matzo—1 matzo (30 gm)	0	0	0	0	0	0	0	0	120	3	.5	0
Melba toast (inc flavors, Nutrisystem Crisp Toast)—1 sl (5 gm)	½	0	0	0	0	0	0	0	19	2	.6	0
Melba toast w/wheat germ—1 sl (5 gm)	½	0	0	0	0	0	0	0	19	2	.6	0
Milk—2 crackers (22 gm)	7	6	5	5	4	4	4	3	100	32	11.0	12
Mixed grain, salt-free (inc Venus corn wafers, wheat wafers, rye wafers, & cracked wheat wafers)—1 cracker (3 gm)	1	1	½	½	½	½	½	½	14	5	1.5	0

Crackers (cont.)

	1200	1500	1800	2100	2400	2800	3200	3600	TOT CAL	FAT CAL	S/FAT CAL	CHOL MG
Oatmeal—2 crackers (22 gm)	4	3	3	2	2	2	2	2	101	34	8.6	0
Oyster (inc chowder)—4 crackers (4 gm)	½	½	½	½	½	0	0	0	18	5	1.1	0
Puffed rice cakes (inc Arden, Chico San, Spiral)—1 cake (9 gm)	½	½	½	½	½	0	0	0	35	6	1.0	0
Rice (inc rice paper)—4 crackers (12 gm)	½	½	½	½	0	0	0	0	46	6	.9	0
Rye wafers, whole grain—10 wafers (65 gm)	0	0	0	0	0	0	0	0	224	7	0	1
Saltine (inc sesame seed, Meal Mates, Sea Toast)—4 crackers (12 gm)	3	3	2	2	2	2	2	1	52	13	5.5	4
Saltines, low sodium (inc unsalted tops)—4 crackers (12 gm)	3	3	2	2	2	2	2	1	52	13	5.5	4
Sandwich-type, peanut butter or cheese (inc rye-cheese)—1 sandwich (8 gm)	2	2	2	1	1	1	1	1	39	17	4.6	1
Soda—4 squares (12 gm)	2	1	1	1	1	1	1	1	53	14	3.4	0
Teething biscuit—1 biscuit (11 gm)	½	½	½	½	½	0	0	0	43	4	1.1	0
Toast thins (rye, wheat, white flour) (inc pizza thins, pumpernickel, onion toast, Won-Ton Chips)—4 crackers (8 gm)	2	2	1	1	1	1	1	1	37	13	4.1	0
Whole rye (inc Ry Krisp, Wasa rye crispbread, Finn rye crisp)—3 crackers (3½" x 1⅞" x ¼") (21 gm)	0	0	0	0	0	0	0	0	72	2	.3	0
Whole wheat (inc table water biscuits)—4 crackers (16 gm)	3	2	2	2	2	1	1	1	70	25	6.2	0

Food												
Whole wheat (inc Triscuits, Stoned Wheat Wafers)—4 crackers (16 gm)	3	2	2	2	1	1	1	1	64	20	6.1	0
Zwieback—1 piece (7 gm)	1	1	1	1	1	1	1	1	30	6	2.5	1

Salty Snacks

Food												
Multigrain mixture, chip w/cheese crax, peanuts (inc Chex Party Mix) (45 gm)	12	10	9	8	7	6	6	5	252	168	27.3	2
Onion-flavored rings (food starch base) (inc Funyuns)—1 oz (28 gm)	4	3	3	3	2	2	2	2	139	57	8.8	0
Popcorn, plain—2 c popped (16 gm)	½	½	½	½	0	0	0	0	62	7	.9	0
Popcorn, popped, w/coconut oil & salt —1 c (9 gm)	6	5	4	4	3	3	3	2	41	18	12.9	0
Popcorn, sugar syrup- or caramel-coated —1 c (35 gm)	1	1	1	1	½	½	½	½	134	11	1.3	0
Popcorn, sugar syrup- or caramel-coated, w/nuts—1 c (35 gm)	1	1	1	1	1	1	½	½	137	18	2.3	0
Popcorn, w/butter—2 c popped (28 gm)	16	13	12	11	10	9	8	7	128	55	30.3	14
Popcorn, w/cheese (inc w/Parmesan cheese)—1 c (26 gm)	5	4	4	3	3	3	2	2	138	75	11.2	0
Potato chips, barbecued—1 c (25 gm)	9	7	6	5	5	4	4	4	134	75	19.2	0
Potato chips (inc flavored except barbecued)—1 c reg (20 gm)	7	6	5	5	4	4	3	3	105	64	16.3	0
Potato chips, made fr dried potatoes— 1 oz (28 gm)	16	13	12	10	9	8	7	7	164	118	35.7	0
Potato chips, restructured (inc Pringles, Hearty Potato Krunch Twists)—1 c, loose (26 gm)	15	12	11	9	8	7	7	6	150	108	32.7	0
Potato chips, restructured, lo-fat and sodium (inc Pringles Light)—1 c, loose (approx 13 chips) (26 gm)	9	8	7	6	5	5	4	4	135	68	20.6	0
Potato chips, unsalted—1 c (20 gm)	7	6	5	5	4	4	3	3	105	64	16.3	0
Potato chips w/salt—10 chips (20 gm)	7	6	5	5	4	4	3	3	105	64	16.3	0

	1200	1500	1800	2100	2400	2800	3200	3600	TOT CAL	FAT CAL	S/FAT CAL	CHOL MG
Salty Snacks (cont.)												
Potato chips wo/salt—10 chips (20 gm)	7	6	5	5	4	4	3	3	105	64	16.3	0
Pretzels, baby—1 pretzel (6 gm)	0	0	0	0	0	0	0	0	24	1	.4	0
Pretzels, cheese-filled (inc Combos)— 10 Combos (30 gm)	6	5	4	4	4	3	3	3	131	46	11.9	4
Pretzels, chocolate-coated—1 pretzel (11 gm)	3	2	2	2	2	1	1	1	47	12	5.9	1
Pretzels, hard (inc cheese-flavored) —1 3-ring pretzel (3 gm)	0	0	0	0	0	0	0	0	12	1	.2	0
—1 c, thin sticks (45 gm)	2	1	1	1	1	1	1	1	176	18	3.6	0
Pretzels, soft—1 pretzel (55 gm)	3	2	2	2	2	2	1	1	190	15	5.9	2
Salt sticks, regular, w/salt coating— 10 sticks, (4¼" long) (50 gm)	2	1	1	1	1	1	1	1	192	13	2.7	2
Salt sticks, Vienna bread type—1 stick (6½" long) (35 gm)	1	1	1	1	1	1	1	1	106	10	2.2	1
Salty snacks, corn or cornmeal base, corn chips, corn-cheese chips (inc plain, flavored, or barbecued chips; Fritos)—10 chips (18 gm)	4	3	3	2	2	2	2	2	97	52	8.6	0
Salty snacks, corn or cornmeal base, corn puffs & twists; corn-cheese puffs & twists (inc Cheetos, Cheez Doodles, Planters Cheese Balls)—10 pieces (15 gm)	5	4	4	3	3	3	2	2	85	49	11.7	0
Salty snacks, corn or cornmeal base, corn-toasted (inc corn nuts, Frito Lay Toasted Corn Nuggets)—10 nuts or nuggets (18 gm)	1	1	1	1	1	½	½	½	80	23	2.1	0

Food												
Salty snacks, corn or cornmeal base, tortilla chips (inc flavored or barbecued, Tostitos, Doritos)—10 chips (18 gm)	3	2	2	2	2	1	1	1	91	43	6.4	0
Shrimp chips (tapioca base)—1 c (40 gm)	13	11	10	9	8	7	7	6	217	127	22.8	19
Wheat sticks, 100% whole wheat (inc w/sesame seeds)—1 c sticks (55 gm)	7	6	5	4	4	3	3	3	252	92	15.3	0
Yogurt chips—1 oz (28 gm)	9	7	6	6	5	4	4	4	146	72	19.0	1
EGGS (other eggs dishes are listed under "Complete Dinners")												
Chicken, white, raw, fresh & frozen—1 lg egg white (33 gm)	0	0	0	0	0	0	0	0	16	0	0	0
Chicken, whole, baked, fat added in cooking—1 lg (59 gm)	53	49	46	43	41	38	36	35	135	101	26.0	267
Chicken, whole, baked, no fat added in cooking—1 lg (45 gm)	40	37	35	33	31	29	28	27	79	50	15.0	213
Chicken, whole, boiled—1 lg (50 gm)	40	37	35	33	31	29	28	27	79	50	15.0	213
Chicken, whole, fried—1 lg (46 gm)	40	36	34	32	30	28	27	26	83	58	21.7	192
Chicken, whole, omelet—1 lg egg (64 gm)	41	38	35	33	31	29	28	26	95	64	25.4	193
Chicken, whole, poached—1 lg (50 gm)	40	37	35	33	31	29	28	27	79	50	15.0	213
Chicken, whole, scrambled—1 lg (64 gm)	41	38	35	33	31	29	28	26	95	64	25.4	213
Chicken, whole, dried—1 tbsp (5 gm)	14	13	12	12	11	10	10	9	30	19	5.7	75
Chicken, whole, pickled—1 egg (47 gm)	37	35	32	31	29	28	26	25	74	47	14.1	200
Chicken, whole, raw, fresh & frozen—1 lg (50 gm)	40	37	35	33	31	29	28	27	79	50	15.0	213
Chicken, yolk, dried—1 tbsp (4 gm)	17	16	15	14	13	13	12	11	27	23	6.7	91
Chicken, yolk only, cooked—1 yolk (17 gm)	40	37	34	32	31	29	28	27	63	50	15.1	211

	1200	1500	1800	2100	2400	2800	3200	3600	TOT CAL	FAT CAL	S/FAT CAL	CHOL MG
Eggs (cont.)												
Chicken, yolk, raw, fresh—1 lg yolk (17 gm)	40	37	34	33	31	29	28	27	63	50	15.1	212
Duck, cooked—1 egg (70 gm)	106	99	93	89	85	80	77	73	129	86	23.1	616
Duck, whole, fresh, raw—1 egg (70 gm)	107	99	94	89	85	81	77	74	130	86	23.2	619
Goose, boiled—1 egg (144 gm)	211	197	185	176	168	159	152	146	266	171	46.4	1222
Goose, whole, fresh, raw—1 egg (144 gm)	212	197	186	177	169	160	153	146	267	172	46.6	1227
Quail, whole, fresh, raw—1 egg (9 gm)	13	12	12	11	10	10	9	9	14	9	2.9	76
Turkey, whole, fresh, raw—1 egg (79 gm)	126	118	111	105	101	96	91	87	135	85	25.8	737
Egg Substitutes												
Egg substitute, frozen—¼ c (60 gm)	5	4	4	3	3	2	2	2	96	60	10.4	1
Egg substitute, liquid—1.5 fl oz (47 gm)	1	1	1	1	1	1	1	1	40	14	2.8	0
Egg substitute, powder—0.35 oz (10 gm)	10	10	9	9	8	8	7	7	44	12	3.3	57
Scrambled egg, made fr cholesterol-free frozen mixture (inc Egg Beaters)—1 c, cooked (51 gm)	5	4	3	3	3	2	2	2	93	58	10.1	1
Scrambled egg, made fr frozen mixture (inc Egg Delight)—1 c, cooked (3 tbsp, raw, equivalent to 1 lg egg, makes ¼ c cooked) (140 gm)	20	18	17	16	15	14	13	13	183	71	12.9	91
Scrambled egg, made fr packaged liquid mixture (inc Second Nature)—1 c, cooked (3 tbsp, raw, equivalent to 1 lg egg, makes ⅕ c cooked) (210 gm)	7	6	5	4	4	3	3	3	201	71	14.2	2

Food												
Scrambled egg, made fr powdered mixture (inc Eggstra)—1 pkt (75 gm)	86	20.5	96	144	13	14	15	16	17	19	20	23

FAST FOODS
Arby's

Food												
Junior Roast Beef—(86 gm)	20	35.1	77	218	9	9	10	12	13	14	16	19
Regular Roast Beef—(147 gm)	39	65.7	133	353	16	18	20	22	24	27	30	35
Giant Roast Beef—(227 gm)	65	104.4	208	531	27	29	32	35	39	43	49	56
Philly Beef 'N Swiss—(197 gm)	107	89.1	256	460	28	31	33	37	40	44	49	56
King Roast Beef—(192 gm)	49	83.7	173	467	21	23	25	28	30	34	38	45
Super Roast Beef—(234 gm)	40	76.5	199	501	19	20	22	25	27	30	34	40
Beef 'N Cheddar—(197 gm)	63	68.4	241	455	20	21	23	26	28	31	35	40
Bac'n Cheddar Deluxe—(266 gm)	83	87.3	329	526	25	27	30	33	36	40	45	52
Hot Ham 'N Cheese—(156 gm)	45	42.3	123	292	13	14	15	17	18	20	22	26
Chicken Breast Sandwich—(195 gm)	83	49.5	262	509	18	20	21	23	25	28	31	35
Turkey Deluxe—(197 gm)	39	36.9	149	375	11	12	13	14	16	17	19	22
Fish Fillet Sandwich—(211 gm)	70	90.9	287	580	25	27	29	32	35	39	44	51
French fries—(71 gm)	8	44.1	87	215	9	10	11	12	14	15	18	21
Potato cakes—(85 gm)	13	62.1	116	201	13	14	16	18	20	22	25	30
Apple turnover—(85 gm)	0	62.1	165	303	11	13	14	16	18	20	23	28
Cherry turnover—(85 gm)	0	47.7	160	280	9	10	11	12	14	15	18	21
Chocolate shake—(340 gm)	36	25.2	104	451	9	9	10	11	12	13	15	17
Jamocha shake—(326 gm)	35	22.5	95	368	8	9	9	10	11	12	14	15
Vanilla shake—(312 gm)	32	35.1	104	330	10	11	12	13	14	16	18	21

Burger King

Food												
Whopper Sandwich	90	108.0	324	628	30	32	35	39	43	47	54	62
Whopper Sandwich w/Cheese	111	153.0	387	711	41	44	48	53	59	65	73	85
Hamburger	37	45.0	108	275	12	13	15	16	18	20	22	26
Hamburger w/Cheese	48	63.0	135	317	17	18	20	22	24	27	31	35
Bacon Double Cheeseburger	104	135.0	279	510	37	40	43	48	52	58	66	76

	1200	1500	1800	2100	2400	2800	3200	3600	TOT CAL	FAT CAL	S/FAT CAL	CHOL MG
Burger King *(cont.)*												
Whopper Jr Sandwich	30	26	23	21	19	17	16	15	322	153	54.0	41
Whopper Jr Sandwich w/Cheese	40	35	31	28	25	23	21	19	364	180	72.0	52
Whaler Fish Sandwich	36	31	28	26	24	22	20	19	488	243	54.0	77
Ham & Cheese Specialty Sandwich	47	41	36	32	30	27	25	23	471	216	81.0	70
Chicken Specialty Sandwich	45	39	35	32	29	26	24	22	688	360	72.0	82
Chicken Tenders (6 pieces)	15	14	12	11	11	10	9	9	204	90	18.0	47
Onion rings	12	10	9	8	7	6	5	5	274	144	27.0	0
French fries	30	26	22	20	18	16	14	13	227	117	63.0	14
Breakfast Croissan'wich	62	56	51	48	45	42	39	37	304	171	54.0	243
Breakfast Croissan'wich w/Bacon	71	63	58	54	50	47	44	41	355	216	72.0	249
Breakfast Croissan'wich w/Sausage	102	90	82	75	70	64	60	56	538	369	126.0	293
Breakfast Croissan'wich w/Ham	65	59	54	50	47	44	42	39	355	180	54.0	262
Scrambled Egg Platter	86	78	72	67	63	59	56	53	468	270	63.0	370
Scrambled Egg Platter w/Sausage	125	112	102	94	88	82	76	72	873	468	135.0	420
Scrambled Egg Platter w/Bacon	95	86	79	73	69	64	61	57	635	324	81.0	378
French Toast Sticks	31	28	25	23	21	19	18	17	499	261	45.0	74
Great Danish	93	78	68	60	54	48	43	39	500	324	207.0	6
Milk, 2% low-fat	15	13	11	10	9	8	8	7	121	45	27.0	18
Milk, whole	29	25	22	20	18	17	15	14	157	81	54.0	35
Orange juice	0	0	0	0	0	0	0	0	82	0	0	0
Coffee, regular	0	0	0	0	0	0	0	0	2	0	0	0
Medium Pepsi Cola	0	0	0	0	0	0	0	0	159	0	0	0
Medium Diet Pepsi	0	0	0	0	0	0	0	0	1	0	0	0
Medium 7-Up	0	0	0	0	0	0	0	0	144	0	0	0
Apple pie	17	14	12	11	10	9	8	7	305	108	36.0	4
Salad, typical wo/dressing	0	0	0	0	0	0	0	0	28	0	0	0
1000 Island dressing	11	9	8	7	7	6	6	5	177	108	18.0	17

Item												
Blue cheese dressing	15	13	12	11	10	9	8	7	156	144	27.0	22
House dressing	10	8	7	7	6	5	5	5	130	117	18.0	11
Reduced-calorie Italian dressing	0	0	0	0	0	0	0	0	14	9	0	0

Carl's

Item												
Famous Star Hamburger—(231 gm)	59	50	44	39	36	32	29	27	590	324	117.0	45
Super Star Hamburger—(301 gm)	103	89	79	71	65	58	53	49	770	450	189.0	125
Western Bacon Cheeseburger—(213 gm)	76	66	58	53	48	43	40	37	630	297	135.0	105
Double Western Bacon Cheeseburger—(294 gm)	122	105	93	84	76	69	63	58	890	477	225.0	145
Old Time Star Hamburger—(168 gm)	40	35	32	29	26	24	22	21	400	153	63.0	80
Happy Star Hamburger—(86 gm)	23	20	18	16	15	14	13	12	220	72	36.0	45
Charbroiler BBQ Chicken Sandwich—(178 gm)	16	14	13	12	11	10	9	9	320	45	18.0	50
Charbroiler Chicken Club Sandwich—(234 gm)	21	19	18	16	15	14	14	13	510	198	18.0	85
California Roast Beef 'n Swiss—(209 gm)	36	33	30	28	26	24	22	21	360	72	36.0	130
Fillet of Fish Sandwich—(223 gm)	58	50	45	40	37	33	31	28	550	234	99.0	90
Chicken Fried Steak Sandwich—(205 gm)	55	47	41	37	34	30	27	25	610	297	108.0	45
American Cheese—(18 gm)	14	12	11	10	9	8	7	7	63	45	27.0	16
Swiss Cheese—(18 gm)	14	12	11	10	9	8	7	7	57	36	27.0	16
Fiesta Potato—(432 gm)	42	36	32	29	26	23	21	19	550	207	81.0	40
Broccoli & Cheese Potato—(398 gm)	22	18	16	14	13	11	10	9	470	153	45.0	10
Bacon & Cheese Potato—(400 gm)	55	47	41	37	34	30	27	25	650	306	108.0	45
Sour Cream & Chive Potato—(294 gm)	22	18	16	14	13	11	10	9	350	117	45.0	10
Cheese Potato—(403 gm)	34	29	26	23	21	19	17	16	550	198	63.0	40
Lite Potato—(278 gm)	0	0	0	0	0	0	0	0	250	3	0	0
Sunrise Sandwich w/Bacon—(127 gm)	51	44	40	37	34	31	29	27	370	171	72.0	120
Sunrise Sandwich w/Sausage—(174 gm)	74	65	58	53	49	45	41	38	500	288	108.0	165

THE 2-IN-1 SYSTEM

	1200	1500	1800	2100	2400	2800	3200	3600	TOT CAL	FAT CAL	S/FAT CAL	CHOL MG
Carl's (cont.)												
French Toast Dips, w/o syrup—(132 gm)	48	42	37	33	30	27	25	23	480	225	90.0	54
Scrambled eggs—(67 gm)	54	49	46	43	41	38	36	34	120	81	36.0	245
Hot cakes, w/margarine, w/o syrup—(156 gm)	14	12	11	10	9	8	7	7	360	108	27.0	15
English muffin, w/margarine—(57 gm)	8	7	6	5	5	4	4	3	180	54	18.0	0
Sausage—1 patty (44 gm)	20	17	15	14	12	11	10	9	190	153	36.0	25
Bacon—2 strips (10 gm)	13	11	10	9	8	7	6	6	50	36	27.0	8
Hashed Brown Nuggets—(85 gm)	18	15	13	12	11	9	8	8	170	81	36.0	10
Blueberry muffin—(99 gm)	9	8	8	7	7	6	6	5	256	63	9.0	34
Bran muffin—(113 gm)	8	7	7	6	6	6	6	6	220	54	0	50
Danish—(99 gm)	12	10	9	8	7	6	5	5	300	81	27.0	0
Chocolate chip cookies—(64 gm)	4	3	3	3	2	2	2	2	330	117	9.0	0
Chocolate cake—(92 gm)	35	30	27	25	23	21	19	18	380	180	54.0	70
French fries, reg size—(170 gm)	46	39	34	30	27	24	22	20	360	153	99.0	15
Zucchini—(121 gm)	30	25	22	19	17	15	14	13	300	144	63.0	10
Onion rings—(90 gm)	30	25	22	19	17	15	14	13	310	135	63.0	10
Reduced-calorie Italian dressing—(57 gm)	16	13	12	10	9	8	7	7	80	90	36.0	0
House dressing—(57 gm)	9	8	7	6	6	5	4	4	186	153	18.0	7
Blue cheese dressing—(57 gm)	9	8	7	6	6	5	5	4	150	126	18.0	8
1000 Island dressing—(57 gm)	12	10	9	8	7	6	5	5	231	207	27.0	0
Carbonated beverage, reg size—(600 gm)	0	0	0	0	0	0	0	0	243	0	0	0
Diet carbonated beverage, reg size—(600 gm)	0	0	0	0	0	0	0	0	2	0	0	0
Iced tea, reg size—(600 gm)	0	0	0	0	0	0	0	0	2	0	0	0
Milk, 2% low-fat, 10 fl oz—(311 gm)	12	10	9	8	7	6	5	5	175	54	27.0	0

Orange juice, small size—(249 gm)	0	0	0	0	0	0	0	0	94	9	0	0
Shakes, reg size—(330 gm)	19	16	14	13	11	10	9	9	353	63	36.0	17
Cream of broccoli soup—(186 gm)	19	17	15	13	12	11	10	9	140	54	36.0	22
Boston clam chowder—(186 gm)	15	13	12	11	10	9	8	7	140	72	27.0	22
Old-Fashioned chicken noodle soup—(186 gm)	6	5	5	4	4	4	3	3	80	9	9.0	14
Lumber Jack Mix vegetable soup—(186 gm)	½	½	½	½	½	½	½	½	70	27	0	3

Jack in the Box

Supreme Crescent—(146 gm)	80	70	63	58	53	49	45	42	547	360	118.8	178
Sausage Crescent—(156 gm)	91	80	71	65	60	54	50	47	584	387	139.5	187
Canadian Crescent—(134 gm)	74	66	60	55	51	47	44	41	452	279	87.3	226
Breakfast Jack—(126 gm)	52	47	43	40	38	35	33	31	307	117	46.8	203
Pancake syrup—(42 gm)	0	0	0	0	0	0	0	0	121	0	0	0
Grape jelly—(14 gm)	0	0	0	0	0	0	0	0	38	0	0	0
Hamburger—(103 gm)	24	21	18	17	15	14	12	11	288	117	45.9	26
Cheeseburger—(113 gm)	36	31	28	25	23	20	18	17	325	153	67.5	41
Jumbo Jack—(222 gm)	55	48	42	38	35	31	29	26	573	306	99.0	73
Jumbo Jack w/Cheese—(242 gm)	72	62	55	50	45	41	37	35	665	360	126.0	102
Bacon Cheeseburger—(230 gm)	73	63	56	50	46	41	37	34	667	351	135.0	85
Swiss & Bacon Burger—(187 gm)	94	81	71	64	58	52	47	43	681	423	180.0	92
Monterey Burger—(281 gm)	135	116	103	92	84	76	69	63	865	513	252.0	152
Ham & Swiss Burger—(259 gm)	100	86	76	68	62	56	51	47	754	441	189.0	106
Mushroom Burger—(182 gm)	50	43	38	34	31	28	26	24	470	216	90.0	64
Club Pita—(179 gm)	21	18	16	15	14	13	12	11	277	72	32.4	43
Chicken Supreme—(231 gm)	67	57	50	45	41	37	33	31	575	324	128.7	62
Pizza Pocket—(162 gm)	57	49	43	38	34	31	28	25	497	252	117.9	32
Moby Jack—(137 gm)	41	35	31	28	25	23	21	19	444	225	75.6	47
Chef salad—(369 gm)	54	47	42	38	35	32	30	28	295	162	84.6	107
Taco salad—(358 gm)	57	50	44	40	37	34	31	29	377	216	93.6	102
Pasta & seafood salad—(417 gm)	23	20	18	17	15	14	13	12	394	198	36.0	48

THE 2-IN-1 SYSTEM

	1200	1500	1800	2100	2400	2800	3200	3600	TOT CAL	FAT CAL	S/FAT CAL	CHOL MG
Jack in the Box (cont.)												
Buttermilk house dressing—(35 gm)	13	11	10	9	8	7	6	6	181	162	26.1	10
Blue cheese dressing—(35 gm)	9	8	7	6	6	5	5	4	131	99	18.0	9
Thousand Island dressing—(35 gm)	12	10	9	8	7	6	6	5	156	135	22.5	11
Reduced-calorie French dressing—(35 gm)	2	2	2	2	1	1	1	1	80	36	5.4	0
Taco—(81 gm)	24	21	18	16	15	13	12	11	191	99	46.8	21
Super Taco—(135 gm)	38	32	28	26	23	21	19	17	288	153	72.0	37
Cheese Nachos—(170 gm)	56	48	42	38	34	30	28	25	571	315	114.3	37
Regular french fries—(68 gm)	21	18	16	14	13	11	10	9	221	108	45.0	8
Large french fries—(109 gm)	34	28	25	22	20	18	16	15	353	171	71.1	13
Onion rings—(108 gm)	49	41	36	32	29	26	23	21	382	207	99.9	27
Hot apple turnover—(119 gm)	45	38	34	30	27	24	22	20	410	216	97.2	15
Cheesecake—(99 gm)	46	39	35	32	29	26	24	22	309	158	81.0	63
Orange juice—(183 gm)	0	0	0	0	0	0	0	0	80	0	0	0
Low-fat milk—(244 gm)	14	12	11	10	9	8	7	7	122	45	26.1	18
Vanilla milk shake—(317 gm)	18	16	14	13	12	10	9	9	320	54	32.4	25
Chocolate milk shake—(322 gm)	21	18	16	14	13	12	11	10	330	63	38.7	25
Strawberry milk shake—(328 gm)	21	18	16	14	13	12	11	10	320	63	38.7	25
Coca-Cola classic (12 fl oz)—(12 gm)	0	0	0	0	0	0	0	0	144	0	0	0
Diet Coke (12 fl oz)—(12 gm)	0	0	0	0	0	0	0	0	1	0	0	0
Ramblin' root beer (12 fl oz)—(12 gm)	0	0	0	0	0	0	0	0	176	0	0	0
Sprite (12 fl oz)—(12 gm)	0	0	0	0	0	0	0	0	144	0	0	0
Dr Pepper (12 fl oz)—(12 gm)	0	0	0	0	0	0	0	0	144	0	0	0
Iced tea (12 fl oz)—(12 gm)	0	0	0	0	0	0	0	0	3	0	0	0
Coffee (8 fl oz)—(8 gm)	0	0	0	0	0	0	0	0	2	0	0	0
Hot Club Supreme—(213 gm)	50	44	39	35	32	29	27	25	524	252	84.6	82

Pancake Platter—(231 gm)	50	43	39	35	33	30	27	25	612	198	77.4	99
Scrambled Egg Platter—(249 gm)	123	109	99	91	85	78	73	68	662	360	153.9	354
Sirloin Steak Dinner—(281 gm)	45	39	35	31	28	26	23	22	702	243	82.8	56
Chicken Strip Dinner—(289 gm)	65	56	49	44	41	37	33	31	674	261	115.2	86
Shrimp Dinner—(258 gm)	81	70	62	56	52	47	43	40	677	297	137.7	128
Fajita Pita—(175 gm)	15	13	12	11	10	9	8	8	278	63	23.4	30
Salsa—(25 gm)	0	0	0	0	0	0	0	0	8	0	0	0
A-1 steak sauce—(25 gm)	0	0	0	0	0	0	0	0	18	0	0	0
BBQ sauce—(25 gm)	4	3	3	3	2	2	2	2	39	9	9.0	0
Egg rolls—(3 pieces) (171 gm)	33	29	25	23	20	18	17	15	405	171	64.8	30
Shrimp—(10 pieces) (84 gm)	42	37	33	30	27	25	23	21	270	144	64.8	84
Chicken Strips—(4 pieces) (125 gm)	38	33	29	27	24	22	20	19	349	126	61.2	68
Fish Supreme—(228 gm)	64	55	48	44	40	36	32	30	554	288	121.5	66
Sweet & sour sauce—(28 gm)	4	4	3	3	2	2	2	2	40	9	9.0	1
Ultimate Cheeseburger—(280 gm)	125	107	95	85	77	69	63	58	942	621	237.6	127
Supreme Nachos—(338 gm)	76	65	57	51	46	41	37	34	787	405	150.3	59
Kentucky Fried Chicken												
Original Recipe Wing*—(56 gm)	23	21	19	17	16	15	14	13	181	111	29.4	67
Original Recipe Side Breast*—(95 gm)	33	29	27	25	23	21	20	18	276	156	41.0	96
Original Recipe Center Breast*—(107 gm)	29	26	23	22	20	19	17	16	257	123	32.4	93
Original Recipe Drumstick*—(58 gm)	22	20	18	17	16	15	14	13	147	79	20.5	81
Original Recipe Thigh*—(96 gm)	40	36	32	30	28	26	24	22	278	173	47.6	122
Extra Crispy Wing*—(57 gm)	26	23	21	19	17	16	15	14	218	140	36.4	63
Extra Crispy Side Breast*—(98 gm)	34	30	27	24	22	20	19	17	354	213	54.0	66
Extra Crispy Center Breast*—(120 gm)	36	31	28	26	24	22	20	19	353	188	47.4	93
Extra Crispy Drumstick*—(60 gm)	21	19	17	16	15	14	13	12	173	98	24.8	65
Extra Crispy Thigh*—(112 gm)	46	41	37	34	31	29	27	25	371	237	61.9	121
Kentucky Nuggets (1)—(16 gm)	5	4	4	3	3	3	3	3	46	26	6.3	12
Nugget sauces, barbecue—(28 gm)	½	½	½	½	½	½	½	0	35	5	.7	1

*—edible portion

	1200	1500	1800	2100	2400	2800	3200	3600	TOT CAL	FAT CAL	S/FAT CAL	CHOL MG
Kentucky Fried Chicken (cont.)												
Nugget sauces, sweet & sour—(28 gm)	½	½	½	½	½	0	0	0	58	5	.5	1
Nugget sauces, honey—(14 gm)	0	0	0	0	0	0	0	0	49	5	.1	1
Nugget sauces, mustard—(28 gm)	½	½	½	½	½	0	0	0	36	8	.5	1
Buttermilk biscuits (1)—(75 gm)	14	12	10	9	8	7	6	6	269	122	31.4	1
Mashed potatoes w/gravy—(86 gm)	2	1	1	1	1	1	1	1	62	13	3.5	1
Mashed potatoes—(80 gm)	1	1	1	1	1	½	½	½	59	5	1.5	1
Chicken gravy—(78 gm)	4	4	3	3	2	2	2	2	59	33	8.6	2
Kentucky fries—(119 gm)	13	11	9	8	8	7	6	5	268	115	28.2	2
Corn-on-the-cob—(143 gm)	2	2	2	1	1	1	1	1	176	28	4.5	1
Cole slaw—(79 gm)	4	3	3	3	2	2	2	2	103	51	7.5	4
Potato salad—(90 gm)	7	6	6	5	5	4	4	4	141	83	12.8	11
Baked beans—(89 gm)	2	2	1	1	1	1	1	1	105	11	4.0	1
Long John Silver's												
2 Piece Fish & Fryes—(283 gm)	38	33	30	27	25	22	20	19	660	270	65.7	60
3 Piece Fish & Fryes—(358 gm)	50	43	38	35	32	29	27	25	810	342	81.9	85
Fish & More—(381 gm)	44	38	34	31	28	25	23	22	800	333	74.7	70
3 Piece Fish Dinner—(456 gm)	56	49	43	39	36	33	30	28	960	396	90.9	100
3 Piece Chicken Plank Dinner—(321 gm)	39	34	30	27	25	22	20	19	700	306	69.3	55
4 Piece Chicken Plank Dinner—(367 gm)	47	41	36	33	30	27	25	23	810	351	81.9	70
Shrimp, Fish & Chicken Dinner—(380 gm)	49	42	38	34	31	28	26	24	840	360	81.9	80
Fish & Chicken—(398 gm)	48	42	37	33	30	27	25	23	870	360	83.7	70
Seafood Platter—(400 gm)	52	45	40	36	33	30	27	25	970	414	92.7	70
Clam Dinner—(315 gm)	37	31	27	24	22	20	18	16	840	360	78.3	15
Battered Shrimp Dinner—(267 gm)	42	36	33	30	27	25	23	21	600	288	63.9	85

Item												
Cajun Shrimp Platter—(319 gm)	38	32	28	25	23	20	18	17	770	351	76.5	25
Shrimp & Fish Dinner—(348 gm)	46	40	35	32	29	27	25	23	770	333	74.7	80
Battered Fish—1 piece (75 gm)	12	10	9	9	8	7	7	6	150	72	16.2	30
Crispy Breaded Fish—1 piece (70 gm)	12	10	9	8	7	7	6	5	170	90	20.7	15
Catfish Fillet—1 piece (79 gm)	16	14	12	11	10	9	9	8	200	108	25.2	30
Chicken Plank—1 piece (46 gm)	8	7	6	6	5	5	4	4	110	54	12.6	15
Breaded Clams—1 order (66 gm)	11	9	8	7	7	6	5	5	240	108	23.4	5
Battered Shrimp—1 piece (14 gm)	4	4	3	3	3	3	2	2	40	27	6.3	10
Cajun Shrimp—1 order (136 gm)	22	18	16	14	13	11	10	9	400	207	45.0	10
Crispy Breaded Fish Sandwich—(242 gm)	30	26	23	20	18	17	15	14	600	252	56.7	30
Clam Chowder w/Cod—(198 gm)	10	9	8	7	7	6	6	5	140	54	16.2	20
Gumbo w/Cod & Shrimp Bobs—(198 gm)	12	11	10	9	8	7	7	6	120	72	18.9	25
Fryes—1 order (85 gm)	11	9	8	7	7	6	5	5	220	90	23.4	5
Hushpuppies—1 piece (24 gm)	2	2	2	2	2	1	1	1	70	18	3.6	5
Corn on the cob w/whirl—1 ear (187 gm)	11	9	8	7	7	6	5	5	270	126	23.4	5
Mixed vegetables—1 order (113 gm)	2	2	2	2	1	1	1	1	60	18	5.4	0
Cole slaw drained on fork—1 order (98 gm)	6	6	5	5	4	4	4	3	140	54	9.0	15
Seafood salad—1 scoop (128 gm)	14	13	12	11	10	10	9	9	190	45	8.1	65
Pasta salad—1 scoop (198 gm)	23	20	17	16	14	13	12	11	360	189	38.7	35
Pecan pie—1 slice (126 gm)	40	34	31	28	26	23	21	20	530	225	64.8	70
Lemon meringue pie—1 slice (118 gm)	11	9	8	7	6	6	5	5	260	63	22.5	5
Catfish Fillet Dinner—(389 gm)	53	46	40	37	33	30	28	25	900	405	92.7	75
Pasta & Seafood Salad—(364 gm)	23	20	17	16	14	13	12	11	420	207	38.7	35
Seafood Salad—(294 gm)	14	13	12	11	10	10	9	9	250	54	8.1	65
Combo Salad—(351 gm)	27	25	23	22	21	19	19	18	300	63	10.8	140
Ocean Chef Salad—(285 gm)	13	12	11	11	10	10	9	9	240	81	5.4	70
Crispy Breaded Fish Sandwich Platter—(425 gm)	46	40	35	31	28	25	23	21	960	396	88.2	45
3 Piece Crispy Breaded Fish Dinner—(441 gm)	55	47	41	37	34	30	28	25	1020	459	104.4	55

THE 2-IN-1 SYSTEM

	1200	1500	1800	2100	2400	2800	3200	3600	TOT CAL	FAT CAL	S/FAT CAL	CHOL MG
Long John Silver's *(cont.)*												
2 Piece Crispy Breaded Fish Dinner—(371 gm)	43	37	33	29	27	24	22	20	840	369	83.7	40
Children's 1 fish, fryes, 1 hushpuppy—(184 gm)	24	21	18	17	15	14	12	11	440	180	44.1	30
Children's 2 planks, fryes, 1 hushpuppy—(201 gm)	28	24	21	19	17	16	14	13	510	216	53.1	30
Children's 1 fish, 1 plank, fryes, 1 hushpuppy—(230 gm)	32	28	25	22	20	18	17	15	550	234	56.7	45
Seafood sauce—1 pkt (35 gm)	1	1	1	1	1	1	1	½	45	9	2.7	0
Tartar sauce—1 pkt (30 gm)	3	3	2	2	2	2	2	2	80	27	5.4	5
Sweet-n-sour sauce—1 pkt (33 gm)	1	1	1	1	1	1	½	½	60	9	2.7	0
Honey-mustard sauce—1 pkt (33 gm)	1	1	1	1	½	½	½	½	60	9	1.8	0
Hot honey-mustard sauce—1 pkt (29 gm)	2	1	1	1	1	1	1	1	90	18	3.6	0
Ranch dressing—1 pkt (42 gm)	3	2	2	2	2	2	1	1	140	27	4.5	5
Blue cheese dressing—1 pkt (43 gm)	3	2	2	2	2	2	1	1	120	18	4.5	5
Reduced-calorie Italian dressing—1 pkt (49 gm)	2	2	2	2	2	1	1	1	20	9	3.6	5
Sea salad dressing—1 pkt (45 gm)	12	10	9	8	7	6	6	5	140	63	25.2	5
McDonald's												
Egg McMuffin—(138 gm)	64	58	53	50	47	44	41	39	340	142	53.1	259
Hotcakes, w/butter/syrup—(214 gm)	22	20	18	16	15	13	12	12	500	93	34.2	47
Scrambled eggs—(98 gm)	100	93	86	81	77	73	69	66	180	117	45.9	514
Pork sausage—(53 gm)	34	29	26	23	21	19	17	16	210	167	63.0	39
English muffin, w/butter—(63 gm)	12	10	9	8	7	7	6	5	186	48	20.7	15
Hash brown potatoes—(55 gm)	15	13	11	10	9	8	7	7	144	77	32.4	7

Item												
Biscuit, w/ biscuit spread—(85 gm)	32	27	23	21	19	17	15	14	330	164	68.4	9
Biscuit, w/sausage—(121 gm)	54	46	41	37	33	30	27	25	467	278	105.3	48
Biscuit, w/sausage & egg—(175 gm)	102	91	82	75	70	64	60	56	585	359	130.5	285
Biscuit, w/bacon/egg/cheese—(145 gm)	78	70	64	59	55	51	48	45	483	284	83.7	263
Sausage McMuffin—(115 gm)	50	43	38	34	31	28	25	23	427	237	90.9	59
Sausage McMuffin w/Egg—(165 gm)	95	85	77	71	66	61	57	53	517	296	114.3	287
Apple Danish—(115 gm)	18	16	14	12	11	10	9	9	389	161	31.5	26
Iced cheese Danish—(110 gm)	31	27	24	22	20	18	17	15	395	196	54.0	48
Cinnamon-raisin Danish—(110 gm)	22	19	17	15	14	13	12	11	445	189	37.8	35
Raspberry Danish—(117 gm)	17	14	13	12	11	10	9	8	414	143	27.9	27
Hamburger—(100 gm)	22	19	17	15	14	13	11	11	263	102	39.6	29
Cheeseburger—(114 gm)	33	29	25	23	21	19	17	16	318	144	60.3	41
Quarter Pounder—(160 gm)	49	42	38	34	31	28	26	24	427	212	81.9	81
Quarter Pounder w/Cheese—(186 gm)	68	59	52	47	43	39	36	33	525	284	115.2	107
Big Mac—(200 gm)	59	51	45	41	37	34	31	28	570	315	103.5	83
Fillet-o-Fish—(143 gm)	30	26	23	21	19	17	16	15	435	231	50.4	47
McD.L.T.—(254 gm)	75	65	57	52	47	42	39	36	680	396	133.2	101
Chicken McNuggets—(109 gm)	30	26	24	22	20	18	17	16	323	182	45.9	63
Hot mustard sauce—(30 gm)	2	2	2	2	2	1	1	1	63	19	4.5	3
Barbecue sauce—(32 gm)	0	0	0	0	0	0	0	0	60	4	.3	0
Sweet & sour sauce—(32 gm)	0	0	0	0	0	0	0	0	64	3	.3	0
Honey sauce—(14 gm)	0	0	0	0	0	0	0	0	50	0	.1	0
French fries—(68 gm)	20	17	15	13	12	10	9	9	220	104	41.4	9
Apple pie—(85 gm)	20	17	15	13	12	10	9	9	253	126	42.3	7
Vanilla milk shake—(291 gm)	21	18	16	15	13	12	11	10	352	76	36.9	31
Chocolate milk shake—(291 gm)	21	19	16	15	14	12	11	10	383	81	37.8	30
Strawberry milk shake—(290 gm)	21	18	16	15	14	12	11	10	362	78	36.9	32
Soft serve & cones—(115 gm)	13	11	10	9	8	7	7	6	189	47	19.8	24
Strawberry sundae—(164 gm)	17	14	13	11	11	10	10	9	320	78	28.8	25
Hot fudge sundae—(164 gm)	26	22	19	18	16	14	13	12	357	97	48.6	27
Hot caramel sundae—(165 gm)	19	16	15	13	12	11	10	9	361	90	31.5	31
McDonaldland cookies—(67 gm)	18	16	14	12	11	10	9	8	308	97	37.8	10
Chocolaty chip cookies—(69 gm)	36	30	26	24	21	19	17	16	342	147	73.8	18

McDonald's (cont.)

	1200	1500	1800	2100	2400	2800	3200	3600	TOT CAL	FAT CAL	S/FAT CAL	CHOL MG
Skim milk, 0.5% butterfat (8 fl oz)—(245 gm)	2	2	2	2	2	1	1	1	90	5	3.6	5
Milk, 2% butterfat (8 fl oz)—(244 gm)	14	12	11	10	9	8	7	7	121	42	26.1	18
Orange juice (6 fl oz)—(183 gm)	0	0	0	0	0	0	0	0	80	0	0	0
Grapefruit juice (6 fl oz)—(183 gm)	0	0	0	0	0	0	0	0	80	0	0	0
Chef salad—(273 gm)	43	38	35	32	30	27	25	24	226	117	53.1	125
Shrimp salad—(264 gm)	33	31	29	27	26	25	24	23	99	23	9.0	187
Garden salad—(204 gm)	27	25	23	21	20	18	17	16	91	50	22.5	110
Chicken Salad Oriental—(280 gm)	19	17	16	15	14	13	13	12	146	35	9.9	92
Side salad—(114 gm)	11	10	9	9	8	8	7	7	48	23	10.8	42
Croutons—(11 gm)	2	2	2	1	1	1	1	1	5	20	4.5	1
Bacon bits—(3 gm)	2	1	1	1	1	1	1	1	15	9	2.7	3
Chow mein noodles—(9 gm)	2	1	1	1	1	1	1	1	45	20	2.7	2
Blue cheese dressing—(35 gm)	15	13	11	10	9	8	8	7	171	155	27.9	16
French dressing—(28 gm)	6	5	5	4	4	3	3	3	114	93	13.5	1
House dressing—(28 gm)	12	10	9	8	7	7	6	5	163	154	25.2	7
1000 Island dressing—(35 gm)	16	14	12	11	10	9	8	8	198	177	27.9	24
Lite vinaigrette dressing—(28 gm)	1	½	½	½	½	½	½	½	25	11	.9	1
Oriental dressing—(28 gm)	0	0	0	0	0	0	0	0	51	2	.2	1
Taco Bell												
Bean Burrito—(191 gm)	22	19	17	15	13	12	11	10	360	98	45.0	14
Beef Burrito—(191 gm)	42	36	32	29	27	24	22	20	402	156	73.8	59
Combo Burrito—(191 gm)	32	27	24	22	20	18	16	15	381	127	59.4	36
Burrito Supreme—(248 gm)	41	35	31	28	25	23	21	19	422	169	81.0	35

Food												
Double Beef Burrito Supreme—(262 gm)	53	46	40	36	33	30	27	25	464	205	99.0	59
Tostada—(156 gm)	24	21	18	16	15	13	12	11	243	98	48.6	18
Beefy Tostada—(198 gm)	45	39	34	31	28	25	23	21	322	176	88.2	40
Bellbeefer—(177 gm)	30	26	23	21	19	17	16	14	312	119	54.0	39
Enchirito—(213 gm)	49	42	37	33	30	27	25	23	382	181	90.0	56
Cinnamon Crispas—(47 gm)	51	43	38	33	30	26	24	21	266	143	115.2	2
Taco Light Platter—(488 gm)	150	128	112	100	90	80	72	66	1062	523	309.6	82
Burrito Supreme Platter—(452 gm)	88	75	66	59	54	48	44	40	774	333	170.1	79
Ranch dressing—(74 gm)	24	21	18	17	15	14	13	12	236	224	41.4	36
Taco Salad wo/Beans—(516 gm)	163	138	121	108	97	87	78	72	822	515	339.3	80
Taco Salad wo/Salsa—(510 gm)	172	146	127	114	103	91	83	75	931	558	357.3	85
Taco Salad w/Ranch Dressing—(586 gm)	196	167	146	130	118	105	95	87	1167	782	398.7	121
Seafood Salad w/Ranch Dressing—(435 gm)	155	132	116	104	94	84	76	70	885	598	308.7	117
Seafood Salad wo/Dressing/Shell—(291 gm)	35	31	28	25	23	21	20	18	217	103	50.4	81
Seafood Salad wo/Dressing—(362 gm)	131	112	98	87	79	71	64	58	648	374	267.3	82
Cheesearito—(116 gm)	33	28	25	22	20	18	17	15	312	115	64.8	29
Mexican Pizza—(269 gm)	138	117	102	91	83	74	67	61	714	431	281.7	81
Taco Bellgrande Platter—(488 gm)	126	107	94	84	76	68	61	56	1002	457	256.5	80
Pintos & cheese—(128 gm)	22	19	17	15	14	12	11	10	194	86	43.2	19
Nachos—(106 gm)	51	43	37	33	30	26	24	22	356	173	111.6	9
Nachos Bellgrande—(333 gm)	98	83	72	64	58	52	47	43	719	366	205.2	43
Taco—(78 gm)	31	26	23	21	19	17	15	14	184	98	57.6	32
Taco Bellgrande—(170 gm)	58	50	44	40	36	32	29	27	351	195	112.5	55
Taco Light—(170 gm)	82	70	61	55	50	44	40	37	411	261	164.7	57
Soft Taco—(92 gm)	27	23	20	18	17	15	14	13	228	106	48.6	32
Taco Salad w/Salsa—(601 gm)	172	146	128	114	103	91	83	75	949	559	357.3	86
Taco Salad w/o Shell—(530 gm)	75	65	57	51	47	42	38	35	525	289	141.3	82
Jalapeño peppers—(100 gm)	½	½	½	½	0	0	0	0	20	2	.9	0
Fajita steak taco—(142 gm)	22	19	16	14	13	12	11	10	235	98	44.1	14
Fajita steak taco w/sour cream—(163 gm)	33	28	24	21	19	17	16	14	281	139	68.4	14

	1200	1500	1800	2100	2400	2800	3200	3600	TOT CAL	FAT CAL	S/FAT CAL	CHOL MG
Taco Bell *(cont.)*												
Fajita steak taco w/guacamole— (163 gm)	23	20	17	15	14	12	11	10	269	119	46.8	14
FATS and OILS												
Fats												
Beef tallow—1 tbsp (13 gm)	28	24	21	18	17	15	13	12	116	116	57.6	14
Butter-margarine blend—1 tbsp (14.0 gm)	21	18	16	14	13	11	10	9	100	100	41.8	15
Butter, regular, salted (inc butter, regular; seasoned, e.g., garlic butter) —1 stick (113 gm)	266	228	200	180	164	146	133	122	810	810	513.5	247
—1 tbsp (14.2 gm)	33	29	25	23	21	18	17	15	102	102	64.5	31
Butter, regular, unsalted—1 stick (113 gm)	266	228	200	180	164	146	133	122	810	810	513.5	247
—1 tbsp (14.2 gm)	33	29	25	23	21	18	17	15	102	102	64.5	31
Butter replacement, fat-free powder (inc Butter Buds)—1 tbsp (5 gm)	0	0	0	0	0	0	0	0	19	0	.3	0
Butter, whipped, salted—1 tbsp (9 gm)	21	18	16	14	13	12	11	10	65	65	40.9	20
Butter, whipped, unsalted—1 tbsp (9 gm)	21	18	16	14	13	12	11	10	65	65	40.9	20
Chicken—1 tbsp (13 gm)	17	14	13	11	10	9	8	8	115	115	34.2	11
Duck—1 tbsp (13 gm)	19	16	14	13	12	10	9	9	115	115	38.7	13
Ghee (clarified butter) (inc butter oil) —1 tbsp (12.0 gm)	35	30	26	23	21	19	17	16	105	105	66.9	31
Goose—1 tbsp (13 gm)	16	14	12	11	10	9	8	7	115	115	31.5	13
Lard—1 tbsp (12 gm)	20	17	15	14	12	11	10	9	108	108	42.3	11
Margarine, diet, salted or unsalted—1 stick (116 gm)	36	30	26	23	21	18	16	15	400	400	80.4	0

—1 tbsp (14.5 gm)	4	3	3	3	2	2			50	50	10.0	0
	4	3	2	2	2	2			51	51	8.1	0
Margarine, imitation (approximately 40% fat, corn (hydrogenated & reg)—1 tbsp (14 gm)	4	3	2	2	2	2			51	51	8.1	0
Margarine, imitation (approximately 40% fat, soybean (hydrogenated)—1 tbsp (14 gm)	5	4	3	3	2	2			51	51	10.8	0
Margarine, imitation (approximately 40% fat, soybean (hydrogenated) & cottonseed—1 tbsp (14 gm)	4	3	2	2	2	2			51	51	8.1	0
Margarine, imitation (approximately 40% fat, soybean (hydrogenated) & cottonseed (hydrogenated)—1 tbsp (14 gm)	6	5	4	4	3	3			51	51	13.5	0
Margarine, imitation (approximately 40% fat, soybean (hydrogenated) & palm (hydrogenated & reg)—1 tbsp (14 gm)	8	7	6	5	4	4			78	78	18.9	0
Margarinelike spread (approximately 60% fat), stick, soybean (hydrogenated) & palm (hydrogenated)—1 tbsp (14 gm)	7	6	5	5	4	3			78	78	16.2	0
Margarinelike spread (approximately 60% fat), tub, soybean (hydrogenated) & cottonseed (hydrogenated)—1 tbsp (14 gm)	8	7	6	5	4	4			78	78	18.9	0
Margarinelike spread (approximately 60% fat), tub, soybean (hydrogenated) & palm (hydrogenated & reg)—1 tbsp (14 gm)												
Margarine, reg, hard, coconut (hydrogenated & reg) & safflower & palm (hydrogenated)—1 tbsp (14 gm)	32	27	24	21	19	17	15	13	102	102	72.9	0
Margarine, reg, hard, corn (hydrogenated)—1 tbsp (14 gm)	7	6	5	4	4	3	3		102	102	16.2	0

Fats (cont.)

	1200	1500	1800	2100	2400	2800	3200	3600	TOT CAL	FAT CAL	S/FAT CAL	CHOL MG
Margarine, reg, hard, corn (hydrogenated & reg)—1 tbsp (14 gm)	8	7	6	5	5	4	4	3	102	102	18.9	0
Margarine, reg, hard, corn & soybean (hydrogenated) & cottonseed (hydrogenated), wo/salt—1 tbsp (14 gm)	8	7	6	5	5	4	4	3	102	102	18.9	0
Margarine, reg, hard, corn & soybean (hydrogenated) & cottonseed (hydrogenated), w/salt—1 tbsp (14 gm)	8	7	6	5	5	4	4	3	102	102	18.9	0
Margarine, reg, hard, lard (hydrogenated)—1 tbsp (14 gm)	19	16	14	12	11	10	9	8	104	104	40.5	6
Margarine, reg, hard, safflower & soybean (hydrogenated)—1 tbsp (14 gm)	7	6	5	5	4	4	3	3	102	102	16.2	0
Margarine, reg, hard, safflower & soybean (hydrogenated) & cottonseed (hydrogenated)—1 tbsp (14 gm)	7	6	5	5	4	4	3	3	102	102	16.2	0
Margarine, reg, hard, safflower & soybean (hydrogenated & reg) & cottonseed (hydrogenated)—1 tbsp (14 gm)	8	7	6	5	5	4	4	3	102	102	18.9	0
Margarine, reg, hard, sunflower & soybean (hydrogenated) & cottonseed (hydrogenated)—1 tbsp (14 gm)	7	6	5	5	4	4	3	3	102	102	16.2	0
Margarine, reg, hard, soybean (hydrogenated)—1 tbsp (14 gm)	10	8	7	6	6	5	4	4	102	102	21.6	0

Food												
Margarine, reg, hard, soybean (hydrogenated) & corn & cottonseed (hydrogenated)—1 tbsp (14 gm)	11	9	8	7	6	6	5	4	102	102	24.3	0
Margarine, reg, hard, soybean (hydrogenated) & cottonseed—1 tbsp (14 gm)	10	8	7	6	6	5	4	4	102	102	21.6	0
Margarine, reg, hard, soybean (hydrogenated) & cottonseed (hydrogenated)—1 tbsp (14 gm)	8	7	6	5	5	4	4	3	102	102	18.9	0
Margarine, reg, hard, soybean (hydrogenated) & palm (hydrogenated)—1 tbsp (14 gm)	8	7	6	5	5	4	4	3	102	102	18.9	0
Margarine, reg, hard, soybean (hydrogenated) & palm (hydrogenated & reg)—1 tbsp (14 gm)	10	8	7	6	6	5	4	4	102	102	21.6	0
Margarine, reg, hard, soybean (hydrogenated & reg)—1 tbsp (14 gm)	7	6	5	5	4	4	3	3	102	102	16.2	0
Margarine, reg, hard, soybean (hydrogenated & reg) & cottonseed (hydrogenated)—1 tbsp (14 gm)	8	7	6	5	5	4	4	3	102	102	18.9	0
Margarine, reg, liquid, soybean (hydrogenated & reg) & cottonseed—1 tbsp (14 gm)	7	6	5	5	4	4	3	3	102	102	16.2	0
Margarine, soft, corn (hydrogenated & reg)—1 tbsp (14 gm)	8	7	6	5	5	4	4	3	102	102	18.9	0
Margarine, soft, safflower & cottonseed (hydrogenated) & peanut (hydrogenated)—1 tbsp (14 gm)	7	6	5	5	4	4	3	3	102	102	16.2	0
Margarine, soft, safflower (hydrogenated & reg)—1 tbsp (14 gm)	5	4	3	3	3	2	2	2	102	102	10.8	0
Margarine, soft, soybean (hydrogenated) & cottonseed—1 tbsp (14 gm)	10	8	7	6	6	5	4	4	102	102	21.6	0

	1200	1500	1800	2100	2400	2800	3200	3600	TOT CAL	FAT CAL	S/FAT CAL	CHOL MG
Fats (cont.)												
Margarine, soft, soybean (hydrogenated) & cottonseed (hydrogenated), wo/salt —1 tbsp (14 gm)	8	7	6	5	5	4	4	3	102	102	18.9	0
Margarine, soft, soybean (hydrogenated) & cottonseed (hydrogenated), w/salt —1 tbsp (14 gm)	8	7	6	5	5	4	4	3	102	102	18.9	0
Margarine, soft, soybean (hydrogenated) & palm (hydrogenated & reg)—1 tbsp (14 gm)	10	8	7	6	6	5	4	4	102	102	21.6	0
Margarine, soft, soybean (hydrogenated & reg) & cottonseed (hydrogenated) —1 tbsp (14 gm)	10	8	7	6	6	5	4	4	102	102	21.6	0
Margarine, soft, soybean (hydrogenated & reg), wo/salt—1 tbsp (14 gm)	7	6	5	5	4	4	3	3	102	102	16.2	0
Margarine, soft, soybean (hydrogenated & reg), w/salt—1 tbsp (14 gm)	7	6	5	5	4	4	3	3	102	102	16.2	0
Margarine, soft, soybean (hydrogenated) & safflower—1 tbsp (14 gm)	6	5	4	4	3	3	3	2	102	102	13.5	0
Margarine, whipped, salted—1 tbsp (9 gm)	6	5	4	4	3	3	3	2	65	65	12.8	0
Mutton tallow—1 tbsp (13 gm)	26	22	20	17	16	14	13	12	116	116	54.9	13
Shortening, bread, soybean (hydrogenated) & cottonseed—1 tbsp (13 gm)	11	9	8	7	6	6	5	5	113	113	25.2	0
Shortening, cake mix, soybean (hydrogenated) & cottonseed (hydrogenated)—1 tbsp (13 gm)	14	12	10	9	8	7	6	6	113	113	31.5	0

Food												
Shortening, confectionery, coconut (hydrogenated) &/or palm kernel (hydrogenated)—1 tbsp (13 gm)	47	39	34	30	27	24	21	19	113	113	105.3	0
Shortening, confectionery, fractionated palm—1 tbsp (14 gm)	36	30	26	23	21	18	16	15	120	120	80.1	0
Shortening, frying (heavy duty), beef tallow & cottonseed—1 tbsp (13 gm)	25	21	18	16	15	13	12	11	115	115	51.3	13
Shortening, frying (heavy duty), palm (hydrogenated)—1 tbsp (13 gm)	24	20	18	16	14	12	11	10	113	113	54.9	0
Shortening, frying (heavy duty), soybean (hydrogenated), linoleic (less than 1%)—1 tbsp (13 gm)	11	9	8	7	6	6	5	4	113	113	24.3	0
Shortening, household, lard & vegetable oil—1 tbsp (13 gm)	22	18	16	14	13	11	10	9	115	115	46.8	7
Shortening, household, soybean (hydrogenated) & cottonseed (hydrogenated)—1 tbsp (13 gm)	13	11	9	8	7	7	6	5	113	113	28.8	0
Shortening, household, soybean (hydrogenated) & palm—1 tbsp (13 gm)	16	13	11	10	9	8	7	6	113	113	35.1	0
Turkey—1 tbsp (13 gm)	17	15	13	12	10	9	8	8	115	115	34.2	13
Oils (Vegetable)												
Almond—1 tbsp (14 gm)	4	4	4	3	3	2	2	2	120	120	9.9	0
Apricot kernel—1 tbsp (14 gm)	4	3	3	2	2	2	2	1	120	120	8.1	0
Cocoa butter—1 tbsp (14 gm)	32	27	24	21	19	17	15	13	120	120	72.9	0
Coconut—1 tbsp (14 gm)	47	40	34	30	27	24	22	20	120	120	106.2	0
Corn, salad or cooking—1 tbsp (14 gm)	7	6	5	4	4	3	3	3	120	120	15.3	0
Cottonseed, salad or cooking—1 tbsp (14 gm)	14	12	10	9	8	7	6	6	120	120	31.5	0
Grapeseed—1 tbsp (14 gm)	5	4	4	3	3	3	2	2	120	120	11.7	0
Hazelnut—1 tbsp (14 gm)	4	3	3	3	2	2	2	2	120	120	9.0	0
Linseed—1 tbsp (14 gm)	5	4	4	3	3	3	2	2	120	120	11.7	0
Nutmeg butter—1 tbsp (14 gm)	49	41	36	31	28	25	22	20	120	120	109.8	0

THE 2-IN-1 SYSTEM

	1200	1500	1800	2100	2400	2800	3200	3600	TOT CAL	FAT CAL	S/FAT CAL	CHOL MG
Oils (Vegetable) (cont.)												
Olive, salad or cooking—1 tbsp (14 gm)	7	6	5	5	4	4	3	3	119	119	16.2	0
Palm—1 tbsp (14 gm)	27	23	20	17	15	14	12	11	120	120	60.3	0
Palm kernel—1 tbsp (14 gm)	44	37	32	29	26	23	20	18	120	120	99.9	0
Peanut, salad or cooking—1 tbsp (14 gm)	9	8	7	6	5	5	4	4	119	119	20.7	0
Poppyseed—1 tbsp (14 gm)	7	6	5	5	4	4	3	3	120	120	16.2	0
Rapeseed, high erucic acid (45% & over)—1 tbsp (14 gm)	3	2	2	2	2	1	1	1	120	120	6.3	0
Rapeseed, low erucic acid (to 30%)—1 tbsp (14 gm)	3	3	2	2	2	2	1	1	120	120	7.2	0
Rapeseed, med erucic acid (30-45%)—1 tbsp (14 gm)	3	2	2	2	2	1	1	1	120	120	6.3	0
Rapeseed, zero erucic acid—1 tbsp (14 gm)	4	3	3	2	2	2	2	1	120	120	8.1	0
Rice bran—1 tbsp (14 gm)	11	9	8	7	6	6	5	4	120	120	24.3	0
Safflower, salad or cooking, linoleic, (over 70%)—1 tbsp (14 gm)	5	4	3	3	3	2	2	2	120	120	10.8	0
Safflower, salad or cooking, oleic, (over 70%)—1 tbsp (14 gm)	3	3	2	2	2	2	1	1	120	120	7.2	0
Sesame, salad or cooking—1 tbsp (14 gm)	8	6	6	5	4	4	3	3	120	120	17.1	0
Sheanut—1 tbsp (14 gm)	25	21	18	16	15	13	12	10	120	120	56.7	0
Soybean & sunflower oil (inc Puritan)—1 tbsp (13.6 gm)	6	5	4	4	4	3	3	3	120	120	13.9	0
Soybean oil (inc Crisco, Wesson)—1 tbsp (13.6 gm)	8	7	6	5	5	4	4	3	120	120	18.2	0
Soybean, salad or cooking (hydrogenated)—1 tbsp (14 gm)	8	7	6	5	5	4	4	3	120	120	18.0	0

Food												
Soybean, salad or cooking (hydrogenated) & cottonseed—1 tbsp (14 gm)	10	8	7	6	6	5	4	4	120	120	21.6	0
Sunflower, linoleic (hydrogenated)—1 tbsp (14 gm)	7	6	5	5	4	4	3	3	120	120	16.2	0
Sunflower oil (inc Wesson, Sunlite)—1 tbsp (13.6 gm)	6	5	4	4	3	3	3	2	120	120	12.6	0
Walnut—1 tbsp (14 gm)	5	4	3	3	3	2	2	2	120	120	10.8	0
Wheat germ—1 tbsp (14 gm)	10	9	8	7	6	5	5	4	120	120	23.4	0

Salad Dressings

Food												
Bacon and tomato—2 tbsp (30 gm)	7	6	5	5	4	4	3	3	98	95	15.9	1
Bacon and tomato, lo-cal—2 tbsp (32 gm)	5	4	3	3	3	2	2	2	64	60	10.2	1
Bacon (hot)—2 tbsp (29 gm)	8	7	6	5	5	4	4	3	101	97	17.0	2
Blue or Roquefort cheese—2 tbsp (31 gm)	13	11	10	9	8	7	6	6	156	146	27.6	5
Blue or Roquefort cheese, lo-cal—2 tbsp (31 gm)	1	1	1	1	1	1	1	1	31	20	3.1	0
Buttermilk (inc Hidden Valley Ranch, Seven Seas Creamy Parmesan)—2 tbsp (29 gm)	9	8	7	6	5	5	4	4	106	96	16.2	10
Caesar—2 tbsp (29 gm)	11	9	8	7	6	5	5	4	153	151	23.5	1
Caesar, lo-cal—2 tbsp (30 gm)	1	1	1	1	1	1	1	½	33	12	1.9	1
Coleslaw—2 tbsp (31 gm)	7	6	6	5	5	4	4	3	121	93	13.7	8
Cream cheese (inc Philadelphia brand)—2 tbsp (31 gm)	14	12	10	9	8	7	7	6	127	120	28.2	8
Creamy, made w/sour cream & oil (inc French made w/sour cream, Creamy Cucumber, Seven Seas Southern w/Bacon, Wishbone Creamy Southern w/Bacon, Wishbone Creamy)—2 tbsp (29 gm)	9	8	7	6	5	5	4	4	141	137	20.2	1
Creamy, made w/sour cream & oil, low- or reduced-cal (inc Creamy Cucumber)—2 tbsp (30 gm)	5	4	3	3	3	2	2	2	62	55	10.2	0

Salad Dressings (cont.)	1200	1500	1800	2100	2400	2800	3200	3600	TOT CAL	FAT CAL	S/FAT CAL	CHOL MG
French, home recipe—1 tbsp (14 gm)	7	6	5	5	4	4	3	3	88	88	16.2	0
French (inc Catalina, Ranch Style other than Hidden Valley, homemade, Sweet 'n Saucy, Holsum's 1867, Richelieu's Western)—2 tbsp (31 gm)	15	13	11	10	9	8	7	7	133	114	26.5	18
French, lo-cal—2 tbsp (32 gm)	1	1	1	1	1	1	1	1	43	17	2.3	2
Fruit, made w/cream—2 tbsp (30 gm)	16	14	13	12	11	10	9	9	63	35	20.2	43
Fruit, made w/honey, oil & water (inc w/herbs, lemon juice)—2 tbsp (29 gm)	8	7	6	5	5	4	4	3	147	124	18.5	0
Green Goddess—2 tbsp (31 gm)	11	10	8	8	7	6	6	5	157	138	20.8	12
Italian, lo-cal—2 tbsp (30 gm)	2	2	1	1	1	1	1	1	32	26	3.5	2
Italian, made w/vinegar, oil & garlic (inc Christie's Greek, California Onion, Green Onion—2 tbsp (29 gm)	8	7	6	5	5	4	4	3	136	126	18.3	0
Mayonnaise, imitation—2 tbsp (30 gm)	5	4	4	3	3	3	3	2	69	52	8.9	7
Mayonnaise, imitation, no cholesterol—2 tbsp (30 gm)	9	8	7	6	5	5	4	4	145	129	20.3	0
Mayonnaise, lo-cal or diet—2 tbsp	5	4	4	4	3	3	3	2	72	54	9.2	7
Mayonnaise, made w/yogurt (inc Yogannaise)—2 tbsp (28 gm)	5	4	4	4	3	3	3	3	27	13	7.1	12
Mayonnaise, reg—2 tbsp (28 gm)	16	14	12	11	10	9	8	7	201	200	29.7	17
Mayonnaise, soybean oil, w/salt—1 tbsp (14 gm)	8	7	6	5	5	4	4	4	99	99	14.4	8

Food												
Mayonnaise, soybean & safflower oil, w/salt—1 tbsp (14 gm)	6	5	5	4	4	3	3	3	99	99	10.8	8
Mayonnaise type salad (inc Miracle Whip)—2 tbsp (30 gm)	7	6	5	4	4	4	3	3	117	90	13.2	8
Mayonnaise type salad, lo-cal or diet (inc Miracle Whip Light)—2 tbsp (28 gm)	4	4	3	3	3	2	2	2	73	48	7.0	7
Milk, vinegar & artificial sweetener—2 tbsp (31 gm)	4	3	3	3	2	2	2	2	23	11	6.6	5
Milk, vinegar & sugar—2 tbsp (33 gm)	3	3	2	2	2	2	2	2	42	10	5.8	4
Russian—2 tbsp (31 gm)	10	8	7	7	7	5	5	4	153	142	20.4	6
Russian, lo-cal—2 tbsp (32 gm)	1	1	1	1	1	1	1	1	45	12	1.7	2
Sesame seed—1 tbsp (15 gm)	4	3	3	2	2	2	2	1	68	62	8.1	0
Sour dressing, nonbutterfat, cultured, filled cream type—1 tbsp (12 gm)	6	5	5	4	4	3	3	3	21	18	14.3	1
Sweet and sour—2 tbsp (31 gm)	0	0	0	0	0	0	0	0	5	0	0	0
Thousand Island (inc McDonald's Big Mac sauce)—2 tbsp (31 gm)	9	7	7	6	5	5	4	4	117	100	16.7	8
Thousand Island, lo-cal—2 tbsp (31 gm)	3	2	2	2	2	1	1	1	49	30	4.5	5
Vinegar, cider—1 tbsp (15 gm)	0	0	0	0	0	0	0	0	2	0	0	0
Vinegar, distilled—1 tbsp (15 gm)	0	0	0	0	0	0	0	0	2	0	0	0
Vinegar, sugar & water—2 tbsp (32 gm)	0	0	0	0	0	0	0	0	16	0	0	0
Yogurt—2 tbsp (31 gm)	3	2	2	2	2	2	1	1	22	10	5.1	3

FRUITS

Food												
Acerola (West Indian cherry), raw—1 fruit wo/pit (5 gm)	0	0	0	0	0	0	0	0	2	0	0	0
Ambrosia—½ c (97 gm)	16	13	12	10	9	8	7	7	115	42	35.9	0
Apple, pickled (inc spiced)—1 apple (29 gm)	0	0	0	0	0	0	0	0	38	1	.1	0
Apple rings, fried—1 ring (19 gm)	1	1	1	1	½	½	½	½	24	10	1.9	0
Applesauce, canned, sweetened, wo/salt—½ c (128 gm)	0	0	0	0	0	0	0	0	97	2	.4	0

Fruits (cont.)

	1200	1500	1800	2100	2400	2800	3200	3600	TOT CAL	FAT CAL	S/FAT CAL	CHOL MG
Applesauce, canned, unsweetened, w/added ascorbic acid—½ c (122 gm)	0	0	0	0	0	0	0	0	53	1	.1	0
Applesauce, canned, unsweetened, wo/added ascorbic acid—½ c (122 gm)	0	0	0	0	0	0	0	0	53	1	.1	0
Apples, baked, unsweetened—1 apple w/liquid (161 gm)	½	½	½	½	0	0	0	0	102	6	.9	0
Apples, baked, w/sugar (inc scalloped apples)—1 apple w/liquid (171 gm)	½	½	½	½	0	0	0	0	163	5	.9	0
Apples, canned, sweetened, sliced, drained, heated—½ c slices (102 gm)	½	0	0	0	0	0	0	0	68	4	.6	0
Apples, dehydrated (low moisture), sulfured, stewed—½ c (97 gm)	0	0	0	0	0	0	0	0	71	1	.2	0
Apples, dried, sulfured, stewed, w/added sugar—½ c (140 gm)	0	0	0	0	0	0	0	0	116	1	.2	0
Apples, dried, sulfured, stewed, wo/added sugar—½ c (128 gm)	0	0	0	0	0	0	0	0	72	1	.2	0
Apples, dried, sulfured, uncooked—10 rings (64 gm)	0	0	0	0	0	0	0	0	155	2	.3	0
Apples, fried—1 med apple yield (117 gm)	6	5	4	4	3	3	3	2	176	65	12.6	0
Apples, frozen, unsweetened, heated—½ c slices (103 gm)	0	0	0	0	0	0	0	0	48	3	.4	0
Apples, raw, wo/skin—1 fruit wo/core (128 gm)	0	0	0	0	0	0	0	0	72	4	.5	0
Apples, wo/skin, boiled—½ c slices (86 gm)	0	0	0	0	0	0	0	0	46	3	.4	0
Apples, raw, w/skin—1 fruit wo/core (138 gm)	½	½	0	0	0	0	0	0	81	5	.7	0

Food											
Apricots, canned, heavy syrup pack, w/skin, solid & liquid—3 halves, 1¾ tbsp liq (85 gm)	0	0	0	0	0	0	0	70	1	0	0
Apricots, canned, juice pack, w/skin, solid & liquid—3 halves, 1¾ tbsp liq (84 gm)	0	0	0	0	0	0	0	40	0	0	0
Apricots, canned, light syrup pack, w/skin, solid & liquid—3 halves, 1¾ tbsp liq (85 gm)	0	0	0	0	0	0	0	54	0	0	0
Apricots, canned, water pack, w/skin, solid & liquid—3 halves, 1¾ tbsp liq (84 gm)	0	0	0	0	0	0	0	22	1	.1	0
Apricots, canned, extra-light syrup pack, w/skin, solid & liquid—3 halves, 1¾ tbsp liq (84 gm)	0	0	0	0	0	0	0	41	1	.1	0
Apricots, dehydrated (low-moisture), sulfured, stewed—½ c (124 gm)	0	0	0	0	0	0	0	156	3	.2	0
Apricots, dried, sulfured, stewed, w/added sugar—½ c halves (135 gm)	0	0	0	0	0	0	0	153	2	.1	0
Apricots, dried, sulfured, stewed, wo/added sugar—½ c halves (125 gm)	0	0	0	0	0	0	0	106	2	.1	0
Apricots, dried, sulfured, uncooked—10 halves (35 gm)	0	0	0	0	0	0	0	83	2	.1	0
Apricots, frozen, sweetened, unthawed—½ c (121 gm)	0	0	0	0	0	0	0	119	1	.1	0
Apricots, raw, pitted—3 fruits (106 gm)	0	0	0	0	0	0	0	51	4	.3	0
Avocados, raw, California—1 fruit wo/skin & seeds (173 gm)	18	15	13	12	10	9	8	7	306	270	40.3	0
Avocados, raw, Florida—1 fruit, wo/skin & seeds (304 gm)	21	18	16	14	12	11	10	9	339	242	48.1	0
Balsam pear, pods, boiled, drained, wo/salt—½ c (½" pieces) (62 gm)	—	—	—	—	—	—	—	12	1	—	0

	THE 2-IN-1 SYSTEM											
	1200	1500	1800	2100	2400	2800	3200	3600	TOT CAL	FAT CAL	S/FAT CAL	CHOL MG
Fruits *(cont.)*												
Balsam pear, pods, raw—1 c (½ " pieces) (93 gm)	—	—	—	—	—	—	—	—	16	2	—	0
Banana, Chinese, raw (inc Cavendish, dwarf, finger)—1 med (5¼" long) (63 gm)	½	½	½	½	½	0	0	0	58	3	1.0	0
Banana, red, fried—1 fruit (7¼" long) (94 gm)	7	6	5	5	4	4	3	3	175	82	16.5	0
Banana, ripe, boiled—1 med (91 gm)	1	1	½	½	½	½	½	½	84	4	1.5	0
Banana, ripe, fried—1 med (91 gm)	9	7	6	6	5	4	4	4	184	96	19.6	0
Bananas, baked—1 banana (7¼" long) (128 gm)	1	1	1	1	1	1	½	½	167	6	2.4	0
Bananas, raw—1 fruit, wo/skin & seeds (114 gm)	1	1	1	1	½	½	½	½	105	5	1.9	0
Blackberries, canned, heavy syrup, solid & liquid—½ c (128 gm)	0	0	0	0	0	0	0	0	118	2	0	0
Blackberries, frozen, unsweetened, unthawed—1 c (151 gm)	0	0	0	0	0	0	0	0	97	5	.3	0
Blackberries, raw—½ c (72 gm)	0	0	0	0	0	0	0	0	37	3	.2	0
Blueberries, canned, heavy syrup, solid & liquid—½ c (128 gm)	0	0	0	0	0	0	0	0	112	4	.4	0
Blueberries, cooked or canned, unsweetened, water pack—½ c (122 gm)	0	0	0	0	0	0	0	0	46	3	.2	0
Blueberries, cooked or canned, w/heavy syrup (inc home canned)—½ c (128 gm)	0	0	0	0	0	0	0	0	113	4	.3	0
Blueberries, frozen, sweetened, unthawed—1 c (230 gm)	—	—	—	—	—	—	—	—	187	3	—	0

Food											
Blueberries, frozen, unsweetened, unthawed—1 c (155 gm)	0	0	0	0	0	0	0	78	9	.3	0
Blueberries, raw—1 c (145 gm)	0	0	0	0	0	0	0	82	5	.4	0
Boysenberries, canned, heavy syrup—½ c (128 gm)	—	—	—	—	—	—	—	113	1	—	0
Boysenberries, frozen, unsweetened, unthawed—1 c (132 gm)	0	0	0	0	0	0	0	66	4	.1	0
Boysenberries, raw—½ c (72 gm)	0	0	0	0	0	0	0	37	3	.1	0
Breadfruit, raw—¼ small fruit (96 gm)	—	—	—	—	—	—	—	99	2	—	0
Carambola (starfruit), raw—1 fruit, wo/seeds (127 gm)	—	—	—	—	—	—	—	42	4	—	0
Carissa (natal plum), raw—1 fruit, wo/skin & seeds (20 gm)	—	—	—	—	—	—	—	12	3	—	0
Cherimoya, raw—1 fruit, wo/skin & seeds (547 gm)	—	—	—	—	—	—	—	515	20	—	0
Cherries, maraschino—1 cherry (4 gm)	0	0	0	0	0	0	0	5	0	0	0
Cherries, sour, red, canned, heavy syrup pack, solid & liquid—½ c (128 gm)	0	0	0	0	0	0	0	116	1	.3	0
Cherries, sour, red, canned, light syrup pack, solid & liquid—½ c (126 gm)	0	0	0	0	0	0	0	94	1	.3	0
Cherries, sour, red, canned, water pack, solid & liquid—½ c (122 gm)	0	0	0	0	0	0	0	43	1	.3	0
Cherries, sour, red, canned, extra-heavy syrup pack, solid & liquid—½ c (130 gm)	0	0	0	0	0	0	0	148	1	.3	0
Cherries, sour, red, frozen, unsweetened, unthawed—1 c (155 gm)	1	½	½	½	½	½	½	72	6	1.4	0
Cherries, sour, red, raw—1 c, w/pits (103 gm)	½	0	0	0	0	0	0	51	3	.6	0
Cherries, sweet, canned, heavy syrup pack, solid & liquid—½ c, wo/pits (129 gm)	0	0	0	0	0	0	0	107	2	.4	0
Cherries, sweet, canned, juice pack, solid & liquid—½ c, wo/pits (125 gm)	0	0	0	0	0	0	0	68	0	0	0

Fruits (cont.)

	1200	1500	1800	2100	2400	2800	3200	3600	TOT CAL	FAT CAL	S/FAT CAL	CHOL MG
Cherries, sweet, canned, light syrup pack, solid & liquid—½ c, wo/pits (126 gm)	0	0	0	0	0	0	0	0	85	2	.4	0
Cherries, sweet, canned, water pack, solid & liquid—½ c, wo/pits (124 gm)	0	0	0	0	0	0	0	0	57	2	.4	0
Cherries, sweet, canned, extra-heavy syrup pack, solid & liquid—½ c, wo/pits (130 gm)	0	0	0	0	0	0	0	0	133	2	.4	0
Cherries, sweet, frozen, sweetened, thawed—1 c (259 gm)	½	½	0	0	0	0	0	0	232	3	.7	0
Cherries, sweet, raw—10 fruits, wo/pits (68 gm)	1	1	½	½	½	½	½	½	49	5	1.4	0
Cherry pie filling—1 c (264 gm)	2	1	1	1	1	1	1	1	313	20	4.0	0
Cherry pie filling, lo-cal—1 c (264 gm)	2	1	1	1	1	1	1	1	211	19	4.0	0
Crabapples, raw—1 c slices, w/skin (110 gm)	0	0	0	0	0	0	0	0	83	3	.4	0
Cranberries, cooked or canned (inc cranberry sauce)—1 sl (½", approx 8 sl per can) (57 gm)	0	0	0	0	0	0	0	0	86	1	.1	0
Cranberries, raw—1 c whole (95 gm)	0	0	0	0	0	0	0	0	46	2	.2	0
Currants, European black, raw—½ c (56 gm)	0	0	0	0	0	0	0	0	36	2	.2	0
Currants, red or white, raw—½ c (56 gm)	0	0	0	0	0	0	0	0	31	1	.1	0
Elderberries, cooked or canned—½ c (128 gm)	0	0	0	0	0	0	0	0	152	4	.2	0
Elderberries, raw—1 c (145 gm)	0	0	0	0	0	0	0	0	105	6	.3	0

Food (measure)										
Figs, canned, heavy syrup pack, solid & liquid—3 fruits, 1¾ tbsp liq (85 gm)	75	1	.2	0	0	0	0	0	0	0
Figs, canned, light syrup pack, solid & liquid—3 fruits, 1¾ tbsp liq (85 gm)	58	1	.2	0	0	0	0	0	0	0
Figs, canned, water pack, solid & liquid—3 fruits, 1¾ tbsp liq (80 gm)	42	1	.2	0	0	0	0	0	0	0
Figs, canned, extra-heavy syrup pack, solid & liquid—3 fruits, 1¾ tbsp liq (85 gm)	91	1	.2	0	0	0	0	0	0	0
Figs, raw—1 med fruit, wo/stem (50 gm)	37	1	.3	0	0	0	0	0	0	0
Fruit cocktail (peach, pineapple, pear, grape & cherry), canned, heavy syrup, solid & liquid—½ c (128 gm)	93	1	.1	0	0	0	0	0	0	0
Fruit cocktail (peach, pineapple, pear, grape & cherry), canned, juice pack, solid & liquid—½ c (124 gm)	56	0	0	0	0	0	0	0	0	0
Fruit cocktail (peach, pineapple, pear, grape & cherry), canned, light syrup, solid & liquid—½ c (126 gm)	72	1	.1	0	0	0	0	0	0	0
Fruit cocktail (peach, pineapple, pear, grape & cherry), canned, water pack, solid & liquid—½ c (122 gm)	40	1	.1	0	0	0	0	0	0	0
Fruit cocktail (peach, pineapple, pear, grape & cherry), canned, extra-heavy syrup, solid & liquid—½ c (130 gm)	115	1	.1	0	0	0	0	0	0	0
Fruit cocktail (peach, pineapple, pear, grape & cherry), canned, extra-light syrup, solid & liquid—½ c (123 gm)	55	1	.1	0	0	0	0	0	0	0
Fruit, mixed (peach, sweetened cherry, sour raspberry, grape, boysenberry), frozen, sweetened, thawed—1 c (250 gm)	245	5	.5	0	0	0	0	0	0	0

	1200	1500	1800	2100	2400	2800	3200	3600	TOT CAL	FAT CAL	S/FAT CAL	CHOL MG
Fruits (cont.)												
Fruit, mixed, (peach, pear, pineapple), canned, heavy syrup, solid & liquid—½ c (128 gm)	½	0	0	0	0	0	0	0	92	1	.2	0
Fruit, mixed (prune, apricot, pear), dried—11 oz pkg, wo/pits (293 gm)	½	½	½	½	½	0	0	0	712	13	1.1	0
Fruit salad (peach, pear, apricot, pineapple & cherry), canned, heavy syrup, solid & liquid—½ c (128 gm)	0	0	0	0	0	0	0	0	94	1	.1	0
Fruit salad (peach, pear, apricot, pineapple & cherry), canned, juice pack, solid & liquid—½ c (124 gm)	0	0	0	0	0	0	0	0	62	0	0	0
Fruit salad (peach, pear, apricot, pineapple & cherry), canned, light syrup, solid & liquid—½ c (126 gm)	0	0	0	0	0	0	0	0	73	1	.1	0
Fruit salad (peach, pear, apricot, pineapple & cherry), canned, water pack, solid & liquid—½ c (122 gm)	0	0	0	0	0	0	0	0	37	1	.1	0
Fruit salad (peach, pear, apricot, pineapple & cherry), canned, extra-heavy syrup, solid & liquid—½ c (130 gm)	0	0	0	0	0	0	0	0	114	1	.1	0
Fruit salad (pineapple, papaya, banana, guava), canned, heavy syrup, solid & liquid—½ c (128 gm)	—	—	—	—	—	—	—	—	110	1	—	0
Genip, raw—1 small (5 gm)	0	0	0	0	0	0	0	0	4	0	0	0
Gooseberries, canned, light syrup pack, solid & liquid—½ c (126 gm)	0	0	0	0	0	0	0	0	93	3	.2	0
Gooseberries, raw—1 c (150 gm)	0	0	0	0	0	0	0	0	67	8	.5	0

Food									
Grapefruit & orange sections, cooked, canned, or frozen, in light syrup—½ c (127 gm)	0	0	0	0	0	76	1	.1	0
Grapefruit & orange sections, cooked, canned, or frozen, unsweetened, water pack—½ c (122 gm)	0	0	0	0	0	32	1	.1	0
Grapefruit, raw, pink & red, California & Arizona—½ fruit (3¾" dia) (123 gm)	0	0	0	0	0	46	1	.2	0
Grapefruit, raw, pink & red, Florida—½ fruit (3¾" dia) (123 gm)	0	0	0	0	0	37	1	.2	0
Grapefruit, raw, white, California—½ fruit (3¾" dia) (118 gm)	0	0	0	0	0	43	1	.2	0
Grapefruit, raw, white, Florida—½ fruit (3¾" dia) (118 gm)	0	0	0	0	0	38	1	.2	0
Grapefruit, sections, canned, juice pack, solid & liquid—½ c (124 gm)	0	0	0	0	0	46	1	.2	0
Grapefruit, sections, canned, light syrup pack, solid & liquid—½ c (127 gm)	0	0	0	0	0	76	1	.2	0
Grapefruit, sections, canned, water pack, solid & liquid—½ c (122 gm)	0	0	0	0	0	44	1	.2	0
Grapes, American type (slip skin), raw—10 fruits (24 gm)	0	0	0	0	0	15	1	.3	0
Grapes, canned, Thompson seedless, heavy syrup pack, solid & liquid—½ c (128 gm)	0	0	0	0	0	94	1	.4	0
Grapes, canned, Thompson seedless, water pack, solid & liquid—½ c (122 gm)	0	0	0	0	0	48	1	.4	0
Grapes, European type (adherent skin), raw—10 fruits, (50 gm)	½	½	0	0	0	36	3	.9	0
Groundcherries (cape gooseberries or poha), raw—½ c (70 gm)	—	—	0	0	—	37	5	—	0
Guava sauce, cooked—½ c (119 gm)	0	0	0	0	0	43	2	.4	0

THE 2-IN-1 SYSTEM

	1200	1500	1800	2100	2400	2800	3200	3600	TOT CAL	FAT CAL	S/FAT CAL	CHOL MG
Fruits (cont.)												
Guavas, common, raw—1 fruit, wo/pits (90 gm)	1	1	½	½	½	½	½	½	45	5	1.4	0
Guava shells (canned in heavy syrup) —1 c (310 gm)	1	1	1	1	1	1	1	1	307	12	3.3	0
Guavas, strawberry, raw—1 fruit, wo/pits (6 gm)	0	0	0	0	0	0	0	0	4	1	.1	0
Java plum (jambolan), raw—3 fruits (9 gm)	—	—	—	—	—	—	—	—	5	0	—	0
Juneberry, raw—1 c (165 gm)	0	0	0	0	0	0	0	0	92	6	.4	0
Kiwifruit, (Chinese gooseberry), fresh, raw—1 med fruit, wo/skin (76 gm)	½	½	0	0	0	0	0	0	46	3	.7	0
Kumquats, cooked or canned, w/syrup—1 kumquat (14 gm)	0	0	0	0	0	0	0	0	12	0	0	0
Kumquats, raw—1 fruit, wo/pits (19 gm)	—	—	—	—	—	—	—	—	12	0	—	0
Lemon peel, raw—1 tsp (2 gm)	0	0	0	0	0	0	0	0	0	0	0	0
Lemons, raw, wo/peel—1 med fruit, wo/pits (58 gm)	0	0	0	0	0	0	0	0	17	2	.2	0
Lemons, raw, w/peel—1 med fruit, wo/pits (108 gm)	0	0	0	0	0	0	0	0	22	3	.4	0
Limes, raw—1 fruit (2" dia) (67 gm)	0	0	0	0	0	0	0	0	20	1	.2	0
Loganberries, cooked or canned—½ c (128 gm)	0	0	0	0	0	0	0	0	113	1	.1	0
Loganberries, frozen, unthawed—1 c (147 gm)	0	0	0	0	0	0	0	0	80	5	.1	0
Loganberries, raw—½ c (72 gm)	0	0	0	0	0	0	0	0	37	3	.1	0
Loquats, raw—1 fruit, wo/pits (10 gm)	0	0	0	0	0	0	0	0	5	0	0	0

Food	C1	C2	C3	C4	C5	C6	C7	C8	C9	C10
Lychee, cooked or canned, w/sugar or syrup—1 lychee w/liquid (21 gm)	0	.2	1	19	0	0	0	0	0	0
Lychee, raw (inc frozen)—1 lychee (10 gm)	0	.1	0	7	0	0	0	0	0	0
Mammy apple (mamey), raw—1 fruit wo/core (846 gm)	0	—	38	431	—	—	—	—	—	—
Mango, cooked—1 oz (28 gm)	0	.2	1	18	0	0	0	0	0	0
Mango, pickled—1 sl (28 gm)	0	.1	0	37	0	0	0	0	0	0
Mango, raw—1 fruit, wo/pits (207 gm)	0	1.3	5	135	½	½	½	½	1	1
Melon balls, frozen, unthawed—1 c (173 gm)	0	2.0	4	58	½	1	1	1	1	1
Melons, cantaloupe, raw—½ fruit (5" dia) (267 gm)	0	—	6	94	—	—	—	—	—	—
Melons, casaba, raw—⅒ fruit (7¾" x 2") (164 gm)	0	—	2	43	—	—	—	—	—	—
Melons, honeydew, raw—⅒ fruit (7" x 2") (129 gm)	0	.5	1	46	0	0	0	0	0	0
Mulberries, raw—10 fruits (15 gm)	0	0	1	7	0	0	0	0	0	½
Nectarine, cooked—½ c (131 gm)	0	.4	4	128	0	0	0	0	0	0
Nectarine, raw—1 fruit, (136 gm)	0	.6	5	67	0	0	0	0	0	0
Oheloberries, raw—10 fruits (11 gm)	0	0	0	3	0	0	0	0	0	0
Orange, mandarin, canned, juice pack—½ c (125 gm)	0	0	0	46	0	0	0	0	0	0
Orange, mandarin, canned, light syrup pack—½ c (126 gm)	0	.2	1	76	0	0	0	0	0	0
Orange peel, raw—1 tsp (2 gm)	0	0	0	0	0	0	0	0	0	0
Orange, sections, canned, juice pack—½ c (102 gm)	0	.1	1	46	0	0	0	0	0	0
Oranges, raw, California, navels—1 fruit (2⅞" dia) (140 gm)	0	.2	1	65	0	0	0	0	0	0
Oranges, raw, California, Valencias—1 fruit (2⅝" dia) (121 gm)	0	.4	4	59	0	0	0	0	0	0
Oranges, raw, Florida—1 fruit (2⅝" dia) (151 gm)	0	.4	3	69	0	0	0	0	0	0

Fruits (cont.)

	1200	1500	1800	2100	2400	2800	3200	3600	TOT CAL	FAT CAL	S/FAT CAL	CHOL MG
Oranges, raw, w/peel—1 fruit, w/seed (159 gm)	0	0	0	0	0	0	0	0	64	5	.5	0
Papaya, cooked or canned, w/sugar or syrup—1 c (132 gm)	0	0	0	0	0	0	0	0	101	1	.3	0
Papaya, green, cooked—1 c (132 gm)	1	½	½	½	½	0	0	0	24	1	.3	0
Papayas, raw—1 fruit (3½" dia) (304 gm)	0	0	0	0	½	½	0	0	117	4	1.2	0
Passion fruit (granadilla), purple, raw—1 fruit, wo/pit (18 gm)	0	0	0	0	0	0	0	0	18	1	0	0
Peaches, canned, heavy syrup pack, solid & liquid—1 half, 1¾ tbsp liq (81 gm)	0	0	0	0	0	0	0	0	60	1	.1	0
Peaches, canned, juice pack, solid & liquid—1 half, 1⅔ tbsp liq (77 gm)	0	0	0	0	0	0	0	0	34	0	0	0
Peaches, canned, light syrup pack, solid & liquid—1 half, 1¾ tbsp liq (81 gm)	0	0	0	0	0	0	0	0	44	0	0	0
Peaches, canned, water pack, solid & liquid—1 half, 1⅔ tbsp liq (77 gm)	0	0	0	0	0	0	0	0	18	0	0	0
Peaches, canned, extra-heavy syrup pack, solid & liquid—1 half, 1¾ tbsp liq (81 gm)	0	0	0	0	0	0	0	0	77	0	0	0
Peaches, canned, extra-light syrup, solid & liquid—1 half, 1⅔ tbsp liq (77 gm)	0	0	0	0	0	0	0	0	32	1	.1	0
Peaches, dehydrated (low-moisture), sulfured, stewed—½ c (121 gm)	0	0	0	0	0	0	0	0	161	5	.4	0
Peaches, dried, sulfured, stewed, w/added sugar—½ c halves (135 gm)	0	0	0	0	0	0	0	0	139	3	.3	0

Food											
Peaches, dried, sulfured, stewed, wo/added sugar—½ c halves (129 gm)	0	.3	3	99	0	0	0	0	0	0	½
Peaches, dried, sulfured, uncooked—10 halves (130 gm)	0	1.0	9	311	0	0	½	½	½	½	0
Peaches, frozen, unsweetened—½ c, sliced (125 gm)	0	.1	1	54	0	0	0	0	0	0	0
Peaches, frozen, w/sugar—½ c, sliced (125 gm)	0	.2	1	118	0	0	0	0	0	0	0
Peaches, pickled—1 fruit (88 gm)	0	0	0	110	0	0	0	0	0	0	0
Peaches, raw—1 fruit (87 gm)	0	.1	1	37	0	0	0	0	0	0	0
Peaches, spiced—1 peach half (30 gm)	0	0	0	23	0	0	0	0	0	0	0
Peaches, spiced, canned, heavy syrup pack, solid & liquid—1 fruit, 1 tbsp liq (88 gm)	0	.1	1	66	0	0	0	0	0	0	0
Pear, Japanese, raw—1 fruit (307 gm)	0	.3	3	123	0	0	0	0	0	0	0
Pears, canned, heavy syrup pack, solid & liquid—1 half, 1¾ tbsp liq (79 gm)	0	.1	1	58	0	0	0	0	0	0	0
Pears, canned, juice pack, solid & liquid—1 half, 1⅔ tbsp liq (77 gm)	0	0	1	38	0	0	0	0	0	0	0
Pears, canned, light syrup pack, solid & liquid—1 half, 1¾ tbsp liq (79 gm)	0	0	0	45	0	0	0	0	0	0	0
Pears, canned, water pack, solid & liquid—1 half, 1⅔ tbsp liq (77 gm)	0	0	0	22	0	0	0	0	0	0	0
Pears, canned, extra-heavy syrup pack, solid & liquid—1 half, 1¾ tbsp liq (79 gm)	0	.1	1	77	0	0	0	0	0	0	0
Pears, canned, extra-light syrup pack, solid & liquid—1 half, 1⅔ tbsp liq (77 gm)	0	0	1	36	0	0	0	0	0	0	0
Pears, dried, sulfured, stewed, w/added sugar—½ c halves (140 gm)	0	.2	4	196	0	0	0	0	0	0	0
Pears, dried, sulfured, stewed, wo/added sugar—½ c halves (128 gm)	0	.2	4	163	0	0	0	0	0	0	0

	1200	1500	1800	2100	2400	2800	3200	3600	TOT CAL	FAT CAL	S/FAT CAL	CHOL MG
Fruits (cont.)												
Pears, dried, sulfured, uncooked—10 halves (175 gm)	0	0	0	0	0	0	0	0	459	10	.5	0
Pears, raw—1 fruit (166 gm)	0	0	0	0	0	0	0	0	98	6	.4	0
Pears, spiced, drained solids—1 pear (45 gm)	0	0	0	0	0	0	0	0	41	1	.1	0
Persimmons, Japanese, dried—1 fruit wo/seeds (34 gm)	0	0	0	0	0	0	0	0	93	2	.4	0
Persimmons, Japanese, raw—1 fruit (2½" dia) (168 gm)	—	—	—	—	—	0	0	—	118	3	—	0
Persimmons, native, raw—1 fruit, wo/seeds (25 gm)	—	—	—	0	0	0	0	—	32	1	—	0
Pineapple, canned, heavy syrup pack, solid & liquid—1 sl, 1¼ tbsp liq (58 gm)	0	0	0	0	0	0	0	0	45	1	0	0
Pineapple, canned, juice pack, solid & liquid—1 sl, 1¼ tbsp liq (58 gm)	0	0	0	0	0	0	0	0	35	1	0	0
Pineapple, canned, light syrup pack, solid & liquid—1 sl, 1¼ tbsp liq (58 gm)	0	0	0	0	0	0	0	0	30	1	0	0
Pineapple, canned, water pack, solid & liquid—1 sl, 1¼ tbsp liq (58 gm)	0	0	0	0	0	0	0	0	19	1	0	0
Pineapple, canned, extra-heavy syrup pack, solid & liquid—1 sl, 1¼ tbsp liq (58 gm)	0	0	0	0	0	0	0	0	48	1	0	0
Pineapple, cooked or canned, in light syrup—½ c (126 gm)	0	0	0	0	0	0	0	0	66	1	.1	0
Pineapple, cooked or canned, juice pack —½ c (125 gm)	0	0	0	0	0	0	0	0	75	1	.1	0

Food										
Pineapple, cooked or canned, unsweetened, waterpack—½ c (123 gm)	0	0	0	0	0	0	39	1	.1	0
Pineapple, frozen, chunks, sweetened, unthawed—½ c chunks (122 gm)	0	0	0	0	0	0	104	1	.1	0
Pineapple, raw—1 sl (3½" dia) (84 gm)	0	0	0	0	0	0	42	4	.3	0
Pitanga (Surinams cherry), raw—1 fruit, wo/pit (7 gm)	—	—	—	—	—	—	2	0	—	0
Plantain chips—1 oz (28 gm)	5	4	4	3	3	2	100	14	11.4	0
Plantain, fried green, Puerto Rican style (tostones)—½ plantain (79 gm)	5	4	4	3	3	2	188	84	11.4	0
Plantain, fried ripe, Puerto Rican style (platano maduro frito)—½ plantain (2 sl, 4" x ¾") (76 gm)	7	5	4	4	3	3	194	94	14.7	0
Plantain, boiled green—1 med (223 gm)	1	½	½	½	½	½	259	4	1.2	0
Plantain, boiled ripe (inc baked ripe plantain)—1 med (240 gm)	1	½	½	½	½	½	278	4	1.3	0
Plantain, ripe, raw—1 med (203 gm)	1	1	1	1	1	½	248	7	2.6	0
Plantain, ripe, rolled in flour, fried—1 piece (2½" long) (45 gm)	3	3	2	2	2	1	112	50	7.8	0
Plums, canned, purple, heavy syrup pack, solid & liquid—3 fruits, 2¾ tbsp liq (133 gm)	0	0	0	0	0	0	119	1	.1	0
Plums, canned, purple, juice pack, solid & liquid—3 fruits, 2 tbsp liq (95 gm)	0	0	0	0	0	0	55	0	0	0
Plums, canned, purple, light syrup pack, solid & liquid—3 fruits, 2¼ tbsp liq (133 gm)	0	0	0	0	0	0	83	1	.1	0
Plums, canned, purple, water pack, solid & liquid—3 fruits, 2 tbsp liq (95 gm)	0	0	0	0	0	0	39	0	0	0
Plums, canned, purple, extra-heavy syrup, solid & liquid—3 fruits, 2¾ tbsp liq (133 gm)	0	0	0	0	0	0	135	1	.1	0

Fruits (cont.)

	1200	1500	1800	2100	2400	2800	3200	3600	TOT CAL	FAT CAL	S/FAT CAL	CHOL MG
Plums, pickled (inc Japanese)—1 plum (28 gm)	0	0	0	0	0	0	0	0	34	1	.1	0
Plums, raw—1 fruit (2⅛" dia) (66 gm)	0	0	0	0	0	0	0	0	36	4	.3	0
Poi—1 c (240 gm)	½	0	0	0	0	0	0	0	269	3	.6	0
Pomegranate, raw—1 fruit (3⅜" dia) (154 gm)	½	½	0	0	0	0	0	0	104	5	.7	0
Prickly pears, raw—1 fruit (103 gm)	—	—	—	—	—	—	—	—	42	5	—	0
Prunes, canned, heavy syrup pack, solid & liquid—5 fruits, 2 tbsp liq (86 gm)	0	0	0	0	0	0	0	0	90	2	.1	0
Prunes, dehydrated (low-moisture), stewed—½ c (140 gm)	0	0	0	0	0	0	0	0	158	3	.3	0
Prunes, dried, stewed, w/added sugar—½ c, wo/pits (119 gm)	0	0	0	0	0	0	0	0	147	3	.2	0
Prunes, dried, stewed, wo/added sugar—½ c, wo/pits (106 gm)	0	0	0	0	0	0	0	0	113	2	.2	0
Prunes, dried, uncooked—10 fruits wo/pits (84 gm)	0	0	0	0	0	0	0	0	201	4	.3	0
Pummelo, raw—1 fruit, wo/core (609 gm)	—	—	—	—	—	—	—	—	228	2	—	0
Quinces, raw—1 fruit, wo/core (92 gm)	0	0	0	0	0	0	0	0	53	1	.1	0
Raspberries, canned, red, heavy syrup pack, solid & liquid—½ c (128 gm)	0	0	0	0	0	0	0	0	117	1	.1	0
Raspberries, cooked or canned, unsweetened, water pack—½ c (122 gm)	0	0	0	0	0	0	0	0	40	4	.1	0
Raspberries, frozen, unsweetened—½ c (125 gm)	0	0	0	0	0	0	0	0	61	6	.2	0

Food										
Raspberries, frozen, w/sugar—½ c (125 gm)	0	0	0	0	0	0	0	129	2	.1	0
Raspberries, raw—1 c (123 gm)	0	0	0	0	0	0	0	61	6	.2	0
Rhubarb, cooked or canned, drained solids—½ c (120 gm)	0	0	0	0	0	0	0	139	1	.1	0
Rhubarb, cooked or canned, in heavy syrup (inc rhubarb sauce, home canned)—½ c (120 gm)	0	0	0	0	0	0	0	139	1	.1	0
Rhubarb, cooked or canned, in light syrup—½ c (120 gm)	0	0	0	0	0	0	0	110	1	.2	0
Rhubarb, cooked or canned, unsweetened—½ c (120 gm)	0	0	0	0	0	0	0	25	2	.4	0
Roselle, raw—1 c (57 gm)	—	—	—	—	—	—	—	28	4	—	0
Sapodilla, raw—1 fruit (3" dia) (170 gm)	1	1	1	1	1	1	½	140	17	1.4	0
Sapotes (marmalade plum), raw—1 fruit, (225 gm)	—	—	—	—	—	—	—	301	13	—	0
Soursop, raw—1 fruit (625 gm)	2	1	1	1	1	1	1	416	17	3.4	0
Strawberries, canned, heavy syrup pack, solid & liquid—½ c (127 gm)	0	0	0	0	0	0	0	117	3	.2	0
Strawberries, cooked or canned, unsweetened, water pack—½ c (121 gm)	0	0	0	0	0	0	0	24	3	.1	0
Strawberries, frozen, sweetened, sliced, unthawed—1 c (255 gm)	0	0	0	0	0	0	0	245	3	.2	0
Strawberries, frozen, sweetened, whole, unthawed—1 c (255 gm)	0	0	0	0	0	0	0	200	4	.2	0
Strawberries, frozen, unsweetened, unthawed—1 c (149 gm)	0	0	0	0	0	0	0	52	2	.1	0
Strawberries, raw—1 c (149 gm)	0	0	0	0	0	0	0	45	5	.3	0
Sugar apples (sweetsop), raw—1 fruit (2⅞" dia) (155 gm)	½	0	0	0	0	0	½	146	4	.7	0
Tamarinds, raw—1 fruit (3" x 1") (2 gm)	0	0	0	0	0	0	0	5	0	0	0
Tangerines, raw—1 fruit (2⅜" dia) (84 gm)	0	0	0	0	0	0	0	37	2	.2	0

	1200	1500	1800	2100	2400	2800	3200	3600	TOT CAL	FAT CAL	S/FAT CAL	CHOL MG
Fruits (cont.)												
Watermelon, pickled—1 sq or cube (10 gm)	0	0	0	0	0	0	0	0	10	0	0	0
Watermelon, raw—1/16 fruit (10" dia) (482 gm)	4	4	3	3	2	2	2	2	152	19	9.5	0
Fruit Mixtures												
Apple & cabbage salad w/dressing—1/2 c (80 gm)	6	5	4	4	4	3	3	3	101	72	10.7	6
Apple & fruit salad w/dressing—1/2 c (94 gm)	2	2	2	1	1	1	1	1	55	9	4.1	2
Apple salad w/dressing (inc Waldorf salad)—1/2 c (69 gm)	3	3	3	2	2	2	2	1	97	58	6.9	2
Banana whip—1/2 c (65 gm)	1/2	1/2	0	0	0	0	0	0	90	2	.7	0
Cranberry salad—1/4 c (63 gm)	1	1	1	1	1	1	1/2	1/2	86	26	2.4	0
Fruit dessert w/cream and/or pudding & nuts—1/2 c (89 gm)	20	17	15	13	12	10	9	8	225	105	45.1	0
Fruit juice bar, frozen, flavor other than orange (inc Dole Fruit 'N Juice Bar)—1 bar (2.5 fl oz) (74 gm)	0	0	0	0	0	0	0	0	70	0	0	0
Fruit juice bar, frozen, orange flavor (inc Dole Fruit 'N Juice Bar)—1 bar (2.5 fl oz) (74 gm)	0	0	0	0	0	0	0	0	68	0	0	0
Fruit salad (excluding citrus fruits) w/cream substitute—1/2 c (88 gm)	5	4	4	3	3	3	2	2	75	15	10.9	1
Fruit salad (excluding citrus fruits) w/marshmallows—1/2 c (86 gm)	2	2	1	1	1	1	1	1	94	7	4.4	0
Fruit salad (inc citrus fruits) w/cream substitute—1/2 c (88 gm)	5	4	4	3	3	3	2	2	57	14	10.9	1

Food												
Fruit salad (inc citrus fruits) w/marshmallows—½ c (86 gm)	2	2	2	1	1	1	1	1	78	6	4.4	0
Fruit salad w/cream—½ c (91 gm)	8	7	6	6	5	5	4	4	66	25	16.4	7
Fruit salad w/pudding—½ c (91 gm)	4	3	3	3	2	2	2	2	88	12	7.1	5
Fruit salad w/dressing or mayonnaise—½ c (94 gm)	4	4	3	3	3	2	2	2	73	35	9.2	2
Guacamole (inc avocado dip)—1 c (233 gm)	28	24	21	18	16	14	13	12	466	417	63.5	0
Lime soufflé (inc other citrus fruits)—½ c (60 gm)	37	34	32	30	29	27	26	25	199	88	15.0	197
Pear salad w/dressing (lettuce, ½ pear, dressing)—1 serving (112 gm)	4	3	3	2	2	2	2	2	105	46	6.5	4
Pineapple salad w/cream cheese (lettuce, 1 slice of pineapple, cream cheese)—1 serving (66 gm)	12	11	9	8	8	7	6	6	77	38	23.6	13
Pineapple salad w/dressing (lettuce, 1 cup of diced pineapple, dressing)—1 serving (184 gm)	4	3	3	2	2	2	2	2	130	47	6.5	4
Prune whip—½ c (65 gm)	0	0	0	0	0	0	0	0	97	1	.1	0
Sorbet, fruit, citrus flavor—½ c (100 gm)	0	0	0	0	0	0	0	0	92	0	0	0
Sorbet, fruit, noncitrus flavor—½ c (100 gm)	0	0	0	0	0	0	0	0	94	0	0	0
Soup, fruit—½ c (121 gm)	0	0	0	0	0	0	0	0	87	1	.2	0
Soup, sour cherry—½ c (121 gm)	15	14	12	11	11	10	9	9	104	33	17.0	49
Fruit Juices												
Acerola juice—¾ c (182 gm)	½	½	½	½	½	0	0	0	38	5	1.1	0
Apple-cherry juice—¾ c (188 gm)	0	0	0	0	0	0	0	0	88	2	.4	0
Apple cider—¾ c (186 gm)	0	0	0	0	0	0	0	0	87	2	.3	0
Apple-grape juice—¾ c (183 gm)	0	0	0	0	0	0	0	0	96	2	.3	0
Apple-grape-raspberry juice—¾ c (188 gm)	0	0	0	0	0	0	0	0	98	3	.3	0

	THE 2-IN-1 SYSTEM											
	1200	1500	1800	2100	2400	2800	3200	3600	TOT CAL	FAT CAL	S/FAT CAL	CHOL MG
Fruit Juices *(cont.)*												
Apple juice, canned or bottled, unsweetened, w/added vitamin C—¾ c (186 gm)	0	0	0	0	0	0	0	0	87	2	.3	0
Apple juice, canned or bottled, unsweetened, wo/added vitamin C—¾ c (186 gm)	0	0	0	0	0	0	0	0	87	2	.3	0
Apple juice, frozen concentrate, unsweetened, diluted w/3 cans water, w/ added vitamin C—¾ c (179 gm)	0	0	0	0	0	0	0	0	83	2	.3	0
Apple juice, frozen concentrate, unsweetened, diluted w/3 cans water, wo/added vitamin C—¾ c (179 gm)	0	0	0	0	0	0	0	0	83	2	.3	0
Apricot nectar, canned, w/added vitamin C—½ c (126 gm)	0	0	0	0	0	0	0	0	71	1	.1	0
Apricot nectar, canned, wo/added vitamin C—¾ c (188 gm)	0	0	0	0	0	0	0	0	106	1	.1	0
Apricot-orange juice—¾ c (188 gm)	0	0	0	0	0	0	0	0	92	2	.2	0
Banana nectar—½ c (125 gm)	½	½	0	0	0	0	0	0	88	2	.7	0
Cantaloupe nectar—½ c (125 gm)	½	½	0	0	0	0	0	0	75	1	.7	0
Citrus fruit juice drink, frozen concentrate, prep w/water—1 c (248 gm)	0	0	0	0	0	0	0	0	114	0	0	0
Cranberry juice cocktail, bottled—¾ c (190 gm)	—	—	—	—	—	—	—	—	108	2	—	0
Cranberry juice, unsweetened—¾ c (190 gm)	0	0	0	0	0	0	0	0	93	3	.3	0

Food									
Grapefruit & orange juice, canned, unsweetened—¾ c (185 gm)	0	.2	2	80	0	0	0	0	0
Grapefruit & orange juice, canned, w/sugar—¾ c (187 gm)	0	.2	2	95	0	0	0	0	0
Grapefruit & orange juice, fresh—¾ c (185 gm)	0	.3	2	78	0	0	0	0	0
Grapefruit & orange juice, frozen, unsweetened, reconstituted w/water—¾ c (185 gm)	0	.2	2	80	0	0	0	0	0
Grapefruit & orange juice, frozen, w/sugar, reconstituted w/water—¾ cup (186 gm)	0	.2	2	88	0	0	0	0	0
Grapefruit juice, canned, sweetened—¾ c (188 gm)	0	.2	1	87	0	0	0	0	0
Grapefruit juice, canned, unsweetened—¾ c (185 gm)	0	.2	1	70	0	0	0	0	0
Grapefruit juice, frozen concentrate, unsweetened, diluted w/3 cans water—¾ c (185 gm)	0	.3	2	77	0	0	0	0	0
Grapefruit juice, frozen, w/sugar, reconstituted w/water—¾ c (186 gm)	0	.3	2	89	0	0	0	0	0
Grapefruit juice, pink—juice fr 1 fruit (196 gm)	0	.3	2	76	0	0	0	0	0
Grapefruit juice, white—juice fr 1 fruit (196 gm)	0	.3	2	76	0	0	0	0	0
Grape juice, canned or bottled, unsweetened—¾ c (190 gm)	0	.4	1	116	0	0	0	0	0
Grape juice, frozen concentrate, sweetened, diluted w/3 cans water—¾ c (188 gm)	0	.5	1	96	0	0	0	0	0
Grape juice, lo-cal sweetener—¾ c (188 gm)	0	.4	1	115	0	0	0	0	0
Grape juice, unsweetened, w/added vitamin C—¾ c (188 gm)	0	.4	1	115	0	0	0	0	0

Fruit Juices (cont.)

	1200	1500	1800	2100	2400	2800	3200	3600	TOT CAL	FAT CAL	S/FAT CAL	CHOL MG
Grape juice, w/added vitamin C—¾ c (188 gm)	0	0	0	0	0	0	0	0	115	1	.4	0
Grape juice, w/sugar—¾ c (188 gm)	0	0	0	0	0	0	0	0	96	2	.5	0
Grape juice, w/sugar, w/added vitamin C—¾ cup (188 gm)	0	0	0	0	0	0	0	0	96	2	.5	0
Guava nectar—½ c (125 gm)	0	0	0	0	0	0	0	0	74	1	.3	0
Lemonade—1 c (8 fl oz) (248 gm)	0	0	0	0	0	0	0	0	108	0	0	0
Lemonade, frozen concentrate, pink, prep w/water—¾ c (186 gm)	0	0	0	0	0	0	0	0	75	1	.1	0
Lemonade, frozen concentrate, white, prep with water—¾ c (186 gm)	0	0	0	0	0	0	0	0	75	1	.1	0
Lemonade, lo-cal—1 c (8 fl oz) (240 gm)	0	0	0	0	0	0	0	0	43	0	0	0
Lemonade, lo-cal, w/aspartame, powder, prep w/water—2 qt & 0.42 oz pkt (1905 gm)	0	0	0	0	0	0	0	0	40	0	0	0
Lemonade, powder, prep w/water—1 c water & 2 tbsp pwd (264 gm)	0	0	0	0	0	0	0	0	102	0	0	0
Lemonade w/added vitamin C—1 c (8 fl oz) (248 gm)	0	0	0	0	0	0	0	0	108	0	0	0
Lemon juice, canned or bottled—1 tbsp (15 gm)	0	0	0	0	0	0	0	0	3	1	.1	0
Lemon juice, frozen, unsweetened, single strength—1 tbsp (15 gm)	0	0	0	0	0	0	0	0	3	1	.1	0
Lemon juice, raw—1 tbsp (15 gm)	0	0	0	0	0	0	0	0	4	0	0	0
Limeade, frozen concentrate, prep w/water—¾ c (185 gm)	0	0	0	0	0	0	0	0	77	1	0	0

Lime juice, canned or bottled, unsweetened—1 tbsp (15 gm)	0	0	0	0	0	0	0	0	3	0	0	0
Lime juice, frozen—1 c (240 gm)	0	0	0	0	0	0	0	0	65	2	.2	0
Lime juice, fresh—1 tbsp (15 gm)	0	0	0	0	0	0	0	0	4	0	0	0
Mango nectar—½ c (125 gm)	0	0	0	0	0	0	0	0	73	1	.3	0
Orange & banana juice—¾ c (188 gm)	½	½	½	½	0	0	0	0	107	3	.9	0
Orange-grapefruit juice, canned, unsweetened—¾ c (185 gm)	0	0	0	0	0	0	0	0	80	1	.2	0
Orange juice, California, chilled (inc fr concentrate)—¾ c (187 gm)	0	0	0	0	0	0	0	0	83	5	.5	0
Orange juice, canned, unsweetened—¾ c (187 gm)	0	0	0	0	0	0	0	0	78	3	.3	0
Orange juice, canned, w/sugar—¾ c (188 gm)	0	0	0	0	0	0	0	0	98	2	.3	0
Orange juice, frozen concentrate, unsweetened, diluted w/3 cans water—¾ c (187 gm)	0	0	0	0	0	0	0	0	84	1	.1	0
Orange juice, frozen, w/sugar, reconstituted w/water—¾ c (188 gm)	0	0	0	0	0	0	0	0	97	1	.1	0
Orange juice, raw—juice fr 1 fruit (86 gm)	0	0	0	0	0	0	0	0	39	2	.2	0
Papaya juice—¾ c (185 gm)	½	½	½	0	0	0	0	0	105	2	.8	0
Papaya nectar, canned—½ c (125 gm)	0	0	0	0	0	0	0	0	71	2	.5	0
Passion fruit juice, purple, fresh—¾ c (185 gm)	—	—	—	—	—	—	—	—	95	1	—	0
Passion fruit juice, yellow, fresh—¾ c (185 gm)	0	0	0	0	0	0	0	0	112	3	.4	0
Passion fruit nectar—½ c (125 gm)	0	0	0	0	0	0	0	0	83	0	0	0
Peach nectar, canned, w/added vitamin C—½ c (125 gm)	0	0	0	0	0	0	0	0	67	0	0	0
Peach nectar, canned, wo/added vitamin C—½ c (125 gm)	0	0	0	0	0	0	0	0	67	0	0	0
Pear nectar, canned, w/added vitamin C—½ c (125 gm)	0	0	0	0	0	0	0	0	75	0	0	0

Fruit Juices *(cont.)*

	1200	1500	1800	2100	2400	2800	3200	3600	TOT CAL	FAT CAL	S/FAT CAL	CHOL MG
Pear nectar, canned, wo/added vitamin C—½ c (125 gm)	0	0	0	0	0	0	0	0	75	0	0	0
Pineapple-grapefruit juice, frozen, reconstituted w/water—¾ c (188 gm)	0	0	0	0	0	0	0	0	87	1	.2	0
Pineapple juice, canned, unsweetened, wo/added vitamin C—¾ c (188 gm)	0	0	0	0	0	0	0	0	104	1	.1	0
Pineapple juice, canned, unsweetened, w/added vitamin C—¾ c (188 gm)	0	0	0	0	0	0	0	0	104	1	.1	0
Pineapple juice, frozen concentrate, unsweetened, diluted w/3 cans water—¾ c (188 gm)	0	0	0	0	0	0	0	0	97	1	0	0
Pineapple juice, w/sugar—¾ c (189 gm)	0	0	0	0	0	0	0	0	118	1	.1	0
Pineapple-orange juice, canned, unsweetened—¾ c (188 gm)	0	0	0	0	0	0	0	0	92	2	.2	0
Pineapple-orange juice, canned, w/sugar—¾ c (188 gm)	0	0	0	0	0	0	0	0	105	2	.2	0
Pineapple-orange juice, frozen, reconstituted w/water—¾ c (188 gm)	0	0	0	0	0	0	0	0	91	1	.1	0
Prune juice, canned—¾ c (192 gm)	0	0	0	0	0	0	0	0	136	1	.1	0
Prune juice, unsweetened—½ c (128 gm)	0	0	0	0	0	0	0	0	91	0	0	0
Prune juice, w/sugar—½ c (129 gm)	0	0	0	0	0	0	0	0	100	0	0	0
Soursop (guanabana) nectar—½ c (125 gm)	0	0	0	0	0	0	0	0	80	1	.1	0

Candied Fruit and Dried Fruit (continued from Fruit Juices)

Food											
Tangerine juice, canned, sweetened—¾ c (187 gm)	0	.2	3	94	0	0	0	0	0	0	0
Tangerine juice, canned, unsweetened—¾ c (185 gm)	0	.4	3	80	0	0	0	0	0	0	0
Tangerine juice, frozen concentrate, sweetened, diluted w/3 cans water—¾ c (181 gm)	0	.1	2	83	0	0	0	0	0	0	0
Tangerine juice, frozen, unsweetened, reconstituted w/water—¾ c (186 gm)	0	.4	3	80	0	0	0	0	0	0	0
Tangerine juice, fresh—¾ c (185 gm)	0	.4	3	80	0	0	0	0	0	0	0
Watermelon juice—½ c (119 gm)	1	2.4	5	38	½	½	1	1	1	1	1

Candied Fruit

Food											
Apple, candied (inc caramel apples)—1 med apple (184 gm)	13	29.8	46	263	6	6	7	8	9	10	11
Citron, candied—1 oz (28 gm)	0	0	1	88	0	0	0	0	0	0	0
Grapefruit, peel, candied—1 oz (28 gm)	0	0	1	88	0	0	0	0	0	0	0
Lemon peel, candied—1 oz (28 gm)	0	0	1	88	0	0	0	0	0	0	0
Orange peel, candied—1 oz (28 gm)	0	0	1	88	0	0	0	0	0	0	0
Pears, candied—4 oz (113 gm)	0	0	6	342	0	0	0	0	0	0	0
Pineapple, candied—4 oz container, approx ½ c (113 gm)	0	0	5	357	0	0	0	0	0	0	0

Dried Fruit

Food											
Apples, dried, uncooked (inc dried diet snack, plain or flavored)—1 ring (6 gm)	0	0	0	15	0	0	0	0	0	0	0
Banana chips—¼ c (23 gm)	4	10.1	13	85	2	2	2	3	3	3	4
Currants, dried—½ c (72 gm)	0	.2	2	204	0	0	0	0	0	0	0
Dates, domestic, natural & dry—10 fruits (83 gm)	1	1.4	4	228	½	½	½	½	½	½	1
Figs, dried, stewed—½ c (130 gm)	1	1.2	5	140	0	0	0	½	½	½	½

THE 2-IN-1 SYSTEM

	1200	1500	1800	2100	2400	2800	3200	3600	TOT CAL	FAT CAL	S/FAT CAL	CHOL MG
Dried Fruit (cont.)												
Figs, dried, uncooked—10 fruits, wo/pits (187 gm)	2	1	1	1	1	1	1	1	477	20	4.0	0
Fruit mixture, dried (mixture inc three or more of the following: apples, apricots, dates, papaya, peaches, pears, pineapples, prunes, raisins)—¼ c (34 gm)	0	0	0	0	0	0	0	0	83	2	.1	0
Papaya, dried—1 strip (23 gm)	½	0	0	0	0	0	0	0	59	2	.6	0
Pears, dried, cooked, unsweetened—½ c (128 gm)	0	0	0	0	0	0	0	0	163	4	.2	0
Pears, dried, cooked, w/sugar—½ c (140 gm)	0	0	0	0	0	0	0	0	196	4	.2	0
Raisins, cooked—½ c (148 gm)	½	½	½	½	0	0	0	0	324	3	.9	0
Raisins, golden seedless—1 c not packed (145 gm)	1	1	1	1	1	½	½	½	437	6	2.0	0
Raisins, seeded—1 c not packed (145 gm)	1	1	1	1	1	1	½	½	428	7	2.3	0
Raisins, seedless—1 c not packed (145 gm)	1	1	1	1	1	½	½	½	434	6	2.0	0
GRAVIES and SAUCES												
Barbecue sauce—1 c (8 fl oz) (250 gm)	3	2	2	2	2	1	1	1	188	41	6.0	0
Béarnaise sauce—2 tbsp (32 gm)	51	45	41	37	35	32	30	28	120	113	63.0	146
Béarnaise sauce, dehydrated, prep w/milk & butter—1 c (8 fl oz) (255 gm)	196	168	148	133	121	108	98	90	701	614	375.9	189
Cheese sauce—2 tbsp (30 gm)	5	4	4	3	3	3	3	2	33	17	9.0	6

Food												
Cheese sauce, dehydrated, prep w/milk—1 c (8 fl oz) (279 gm)	45	39	35	31	28	25	23	21	307	154	83.9	53
Cocktail sauce—1 tbsp (17 gm)	0	0	0	0	0	0	0	0	7	0	0	0
Curry sauce, dehydrated, prep w/milk—1 c (8 fl oz) (272 gm)	30	25	23	20	18	17	15	14	270	132	54.5	35
Duck sauce (inc chaisni sauce)—2 tbsp (30 gm)	6	5	5	4	4	3	3	3	71	37	12.4	4
Fruit sauce (inc all fruits)—2 tbsp (42 gm)	2	2	2	1	1	1	1	1	88	24	4.8	0
Gravy, au jus, canned—1 c (238 gm)	1	1	1	1	1	1	1	1	38	5	2.2	1
Gravy, au jus, dehydrated, prep w/water—1 c (8 fl oz) (246 gm)	2	1	1	1	1	1	1	1	19	7	3.6	1
Gravy, beef, canned—1 c (233 gm)	12	10	9	8	7	6	6	5	124	50	24.8	7
Gravy, brown, dehydrated, prep w/water—1 c (8 fl oz) (261 gm)	½	½	½	½	0	0	0	0	9	2	.9	0
Gravy, chicken, canned—1 c (238 gm)	14	12	10	9	8	7	7	6	189	122	30.2	5
Gravy, chicken, dehydrated, prep w/water—1 c (8 fl oz) (260 gm)	3	2	2	2	2	1	1	1	83	17	4.8	3
Gravy, giblet (inc any poultry gravy w/pieces of meat)—2 tbsp (30 gm)	3	3	3	3	2	2	2	2	22	11	2.9	13
Gravy, meat or poultry, prep w/water—2 tbsp (30 gm)	2	1	1	1	1	1	1	1	20	11	3.5	1
Gravy, meat, w/fruit (inc French sauce)—2 tbsp (30 gm)	1	1	1	1	1	1	1	1	19	8	2.6	1
Gravy, meat, w/wine—2 tbsp (30 gm)	1	1	1	1	1	1	1	1	20	8	2.5	1
Gravy, mushroom—2 tbsp (30 gm)	0	0	0	0	0	0	0	0	8	1	.5	0
Gravy, mushroom, canned—1 c (238 gm)	4	3	3	2	2	2	2	2	120	59	8.6	0
Gravy, mushroom, dehydrated, prep w/water—1 c (8 fl oz) (258 gm)	2	2	2	1	1	1	1	1	70	8	4.5	1
Gravy, onion, dehydrated, prep w/water—1 c (8 fl oz) (261 gm)	2	2	1	1	1	1	1	1	80	6	4.1	1
Gravy, pork, dehydrated, prep w/water—1 c (8 fl oz) (258 gm)	3	3	3	2	2	2	2	2	76	17	6.8	3

	1200	1500	1800	2100	2400	2800	3200	3600	TOT CAL	FAT CAL	S/FAT CAL	CHOL MG
Gravies and Sauces (cont.)												
Gravy, Swiss steak—2 tbsp (29 gm)	2	1	1	1	1	1	1	1	19	11	3.4	1
Gravy, turkey, canned—1 c (238 gm)	7	6	5	4	4	4	3	3	122	45	13.3	5
Gravy, turkey, dehydrated, prep w/water—1 c (8 fl oz) (261 gm)	3	2	2	2	2	1	1	1	87	17	4.9	3
Hollandaise sauce (inc butter sauce)—2 tbsp (32 gm)	60	53	48	44	41	38	35	33	138	133	74.1	172
Hollandaise sauce, w/butterfat, dehydrated, prep w/water—1 c (8 fl oz) (259 gm)	54	46	41	37	33	30	27	25	237	177	104.4	51
Hollandaise sauce, w/vegetable oil, dehydrated, prep w/milk & butter—1 c (8 fl oz) (255 gm)	197	168	148	133	121	108	99	91	703	615	376.8	189
Honey butter—1 tbsp (18.0 gm)	17	15	13	12	10	9	9	8	85	53	32.8	16
Lemon-butter sauce—2 tbsp (28 gm)	73	62	55	49	45	40	36	33	223	223	141.2	65
Milk gravy, quick gravy—2 tbsp (31 gm)	8	7	6	5	5	4	4	4	47	32	14.8	7
Mushroom sauce, dehydrated, prep w/milk—1 c (8 fl oz) (267 gm)	27	23	21	19	17	15	14	13	228	93	48.6	35
Orange sauce (for duck)—2 tbsp (29 gm)	0	0	0	0	0	0	0	0	18	1	.3	0
Plain dessert sauce (inc vanilla, rum sauce)—2 tbsp (33 gm)	2	2	1	1	1	1	1	1	48	23	4.5	0
Raisin sauce—2 tbsp (31 gm)	1	1	1	1	1	1	1	1	40	15	2.9	0
Sour cream sauce, dehydrated, prep w/milk—1 c (8 fl oz) (314 gm)	78	67	60	54	49	44	40	37	509	273	144.9	91
Soy sauce—1 tbsp (18 gm)	0	0	0	0	0	0	0	0	9	0	0	0
Spaghetti sauce (inc marinara sauce, cacciatore sauce, pizza sauce, spaghetti sauce w/mushrooms)—¼ c (63 gm)	1	1	1	1	1	1	1	1	47	21	3.1	0

Steak sauce, tomato-base (inc A1)—1 tbsp (16 gm)	0	0	0	0	0	0	0	0	10	1	.1	0
Stroganoff sauce, dehydrated, prep w/milk & water—1 c (8 fl oz) (296 gm)	33	28	25	22	21	18	17	15	271	96	60.9	38
Sweet & sour sauce (inc Vietnamese sauce)—2 tbsp (30 gm)	0	0	0	0	0	0	0	0	28	0	0	0
Sweet & sour sauce, dehydrated, prep w/water & vinegar—1 c (8 fl oz) (313 gm)	0	0	0	0	0	0	0	0	294	1	.1	0
Tartar sauce, dietary, lo-cal, (10 calories/teaspoon—1 tbsp (14 gm)	3	3	3	2	2	2	2	2	31	28	5.0	7
Tartar sauce—1 tbsp (14 gm)	7	6	5	5	4	4	4	3	74	73	13.9	7
Teriyaki sauce, dehydrated, prep w/water—1 c (8 fl oz) (283 gm)	1	½	½	½	½	½	0	0	131	8	1.2	0
Teriyaki sauce, ready-to-serve—1 tbsp (18 gm)	0	0	0	0	0	0	0	0	15	0	0	0
Tomato catsup—1 tbsp (17 gm)	0	0	0	0	0	0	0	0	18	1	.1	0
Tomato chili sauce (inc taco sauce, creole sauce, picante sauce)—1 tbsp (17 gm)	0	0	0	0	0	0	0	0	18	0	.3	0
Tomato paste—¼ cup (66 gm)	½	½	½	0	0	0	0	0	55	5	.8	0
Tomato sauce, canned, marinara—1 c (250 gm)	5	4	3	3	3	2	2	2	171	76	10.8	0
Tomato sauce, canned, w/herbs & cheese—¼ c (61 gm)	—	—	—	—	—	—	—	—	36	11	3.4	—
Tomato sauce, canned, w/mushrooms—¼ c (61 gm)	0	0	0	0	0	0	0	0	21	0	.1	0
Tomato sauce, canned, w/onions—¼ cup (61 gm)	0	0	0	0	0	0	0	0	26	1	.2	0
Tomato sauce, canned, w/onions, green peppers, & celery—¼ c (61 gm)	½	½	0	0	0	0	0	0	25	4	.7	0
Tomato sauce, canned, w/tomato tidbits—¼ c (61 gm)	0	0	0	0	0	0	0	0	20	2	.3	0

THE 2-IN-1 SYSTEM

	1200	1500	1800	2100	2400	2800	3200	3600	TOT CAL	FAT CAL	S/FAT CAL	CHOL MG
Gravies and Sauces (cont.)												
Tomato sauce, canned, spaghetti sauce —¼ c (62 gm)	2	1	1	1	1	1	1	1	68	27	3.8	0
Tomato puree—¼ c (63 gm)	0	0	0	0	0	0	0	0	26	1	.1	0
Tomato relish (inc tomato preserves)— 1 tbsp (20 gm)	0	0	0	0	0	0	0	0	30	1	.1	0
White sauce, dehydrated, prep w/milk— 1 c (8 fl oz) (264 gm)	31	26	23	21	19	17	16	14	241	121	57.6	34
White sauce, med thickness—1 c (250 gm)	85	73	64	58	53	48	43	40	405	282	154.4	103
White sauce, milk sauce—2 tbsp (31 gm)	5	4	4	3	3	3	3	2	49	33	10.2	4
White sauce, thick—1 c (250 gm)	105	90	80	72	65	59	54	49	495	351	192.4	125
White sauce, thin—1 c (250 gm)	59	50	45	40	37	33	30	28	303	196	107.6	70
LEGUMES												
Adzuki, mature seeds, canned, sweetened—½ c (148 gm)	0	0	0	0	0	0	0	0	351	1	.2	0
Adzuki, mature seeds, boiled, wo/salt—½ c (115 gm)	0	0	0	0	0	0	0	0	147	1	.4	0
Baked beans, w/tomato sauce (inc vegetarian baked beans)—½ c (128 gm)	½	½	0	0	0	0	0	0	156	6	.7	0
Baked beans, canned, plain or vegetarian—½ c (127 gm)	1	1	½	½	½	½	½	½	118	5	1.4	0
Baked beans, canned, w/beef—½ c (133 gm)	13	12	11	10	9	8	7	7	161	41	20.1	29
Baked beans, canned, w/franks—½ c (128 gm)	13	11	10	9	8	7	6	6	182	76	27.1	8

Food												
Baked beans, canned, w/pork—½ c (126 gm)	4	4	3	3	3	3	2	2	133	18	6.8	9
Baked beans, canned, w/pork & tomato sauce—½ c (126 gm)	3	3	3	2	2	2	2	2	123	12	4.5	9
Bean dip, made w/refried beans (inc garbanzo bean dip)—1 tbsp (17 gm)	½	½	½	½	½	0	0	0	24	8	1.0	0
Black bean sauce—1 tbsp (17 gm)	1	½	½	½	½	½	0	0	17	7	1.2	0
Black, brown, or bayo beans, dry, cooked, wo/fat—½ c (1 c dry yields 2⅞ c ckd) (86 gm)	0	0	0	0	0	0	0	0	98	4	.5	0
Black beans, mature seeds, boiled, wo/salt—½ c (86 gm)	½	½	½	½	½	0	0	0	113	5	1.1	0
Black bean turtle soup, mature seeds, canned—½ c (120 gm)	½	½	½	0	0	0	0	0	109	4	.8	0
Black bean turtle soup, mature seeds, boiled, wo/salt—½ c (92 gm)	½	½	0	0	0	0	0	0	120	3	.7	0
Boston baked beans—½ c (127 gm)	10	8	7	7	6	5	5	4	193	60	20.4	6
Broad beans (fava beans), mature seeds, canned—½ c (128 gm)	0	0	0	0	0	0	0	0	91	3	.4	0
Broad beans (fava beans), mature seeds, boiled, wo/salt—½ c (85 gm)	0	0	0	0	0	0	0	0	93	3	.5	0
Broad beans (fava beans), mature seeds, raw—½ c (75 gm)	1	1	1	½	½	½	½	½	256	10	1.7	0
Chick-peas, dry, cooked w/fat (inc garbanzos)—½ c (1 c dry yields 2¾ c ckd) (82 gm)	1	1	1	1	½	½	½	½	146	18	1.8	0
Chick-peas (garbanzo beans, bengal gram), mature seeds, canned—½ c (120 gm)	1	½	½	½	½	½	½	0	143	13	1.3	0
Chili beans, barbecue beans, ranch-style beans or Mexican-style beans (inc beans in chili sauce)—½ c (127 gm)	1	1	½	½	½	½	½	½	112	11	1.4	0
Cowpeas, dry, cooked wo/fat (inc black-eyed peas, field peas)—½ c (1 c dry yields 2⅓ c ckd) (87 gm)	½	½	.0	0	0	0	0	0	66	2	.7	0

Legumes (cont.)

	1200	1500	1800	2100	2400	2800	3200	3600	TOT CAL	FAT CAL	S/FAT CAL	CHOL MG
Cowpeas, dry, cooked w/pork (inc black-eyed peas w/pork, field peas w/pork)—½ c (90 gm)	5	5	4	4	3	3	3	3	95	20	6.6	14
Cranberry (roman), mature seeds, canned—½ c (130 gm)	½	½	½	0	0	0	0	0	108	4	.8	0
Cranberry (roman), mature seeds, boiled, wo/salt—½ c (88 gm)	½	½	½	½	½	0	0	0	120	4	1.0	0
French beans, mature seeds, boiled, wo/salt—½ c (86 gm)	½	0	0	0	0	0	0	0	111	5	.6	0
Great northern beans, mature seeds, canned—½ c (131 gm)	1	1	½	½	½	½	½	½	150	5	1.4	0
Great northern beans, mature seeds boiled, wo/salt—½ c (88 gm)	½	½	½	½	½	0	0	0	104	4	1.1	0
Green or yellow split peas, dry, cooked wo/fat—½ c (1 c dry yields 3 c ckd) (101 gm)	0	0	0	0	0	0	0	0	116	3	.5	0
Kidney beans, all types, mature seeds, canned—½ c (128 gm)	0	0	0	0	0	0	0	0	104	4	.5	0
Kidney beans, all types, mature seeds, boiled, wo/salt—½ c (88 gm)	0	0	0	0	0	0	0	0	112	4	.5	0
Kidney bean salad—½ cup (116 gm)	8	7	6	5	5	4	4	4	175	62	14.0	9
Lentils, dry, cooked wo/fat—½ c (1 c dry yields 2½ c ckd) (96 gm)	0	0	0	0	0	0	0	0	101	0	0	0
Lentils, mature seeds, boiled, wo/salt—½ c (99 gm)	0	0	0	0	0	0	0	0	115	4	.4	0
Lima beans, dry, cooked wo/fat (inc butter beans)—½ c (1 c dry yields 2⅔ c ckd) (94 gm)	½	½	½	½	½	0	0	0	129	5	1.1	0

Lima beans, immature seeds, canned, reg pack, solid & liquid—½ c (124 gm)	½	½	0	0	0	0	0	0	93	4	.7	0
Lima beans, immature seeds, canned, special dietary pack, solid & liquid—½ c (124 gm)	½	½	0	0	0	0	0	0	93	4	.7	0
Lima beans, immature seeds, frozen, baby, boiled, drained, wo/salt—10 oz pkt (311 gm)	1	1	1	1	½	½	½	½	326	8	1.9	0
Lupins, mature seeds, boiled, wo/salt—1/ c (83 gm)	1	1	1	1	1	1	1	½	98	22	2.6	0
Lupins, mature seeds, raw—½ c (90 gm)	4	4	3	3	2	2	2	2	334	79	9.4	0
Mongo beans (wo/fat)—½ c (90 gm)	½	½	½	½	½	0	0	0	94	3	1.0	0
Mung beans (wo/fat)—½ c (1 c dry yields 3⅔ c ckd) (102 gm)	1	½	½	½	½	½	0	0	112	4	1.2	0
Mung beans, dehydrated—½ c (70 gm)	0	0	0	0	0	0	0	0	246	1	.1	0
Mung beans, mature seeds, sprouted, canned, drained solids—½ c (62 gm)	0	0	0	0	0	0	0	0	8	1	.1	0
Navy beans, mature seeds, canned—½ c (131 gm)	1	1	½	½	½	½	½	½	148	5	1.4	0
Navy beans, mature seeds, boiled, wo/salt—½ c (91 gm)	1	½	½	½	½	½	0	0	129	5	1.2	0
Peas, dry, cooked w/pork—½ c (99 gm)	10	9	8	7	6	6	5	5	163	59	21.2	7
Pink beans, mature seeds, boiled, wo/salt—½ c (84 gm)	½	½	½	½	½	0	0	0	125	4	1.0	0
Pinto, calico, & red Mexican beans, dry, wo/fat (inc October beans, Shellie beans)—½ c (1 c dry yields 2⅔ c ckd) (88 gm)	0	0	0	0	0	0	0	0	102	3	.4	0
Pinto beans, immature seeds, frozen, boiled, drained, wo/salt—⅓ of 10 oz pkt (94 gm)	0	0	0	0	0	0	0	0	152	4	.4	0
Pinto beans, mature seeds, canned—½ c (120 gm)	½	½	0	0	0	0	0	0	93	4	.7	0

Legumes (cont.)

	1200	1500	1800	2100	2400	2800	3200	3600	TOT CAL	FAT CAL	S/FAT CAL	CHOL MG
Red kidney beans, dry, cooked wo/fat —½ c (1 c dry yields 2⁷/₁₀ c ckd) (88 gm)	0	0	0	0	0	0	0	0	103	4	.5	0
Refried beans—½ c (127 gm)	4	4	3	3	2	2	2	2	208	76	9.7	0
Refried beans w/cheese—½ c (127 gm)	11	9	8	7	7	6	5	5	224	92	21.9	7
Refried beans, canned—½ c (126 gm)	2	2	2	1	1	1	1	1	134	13	4.7	0
Shellie beans, canned, solid & liquid—½ c (122 gm)	0	0	0	0	0	0	0	0	37	2	.3	0
Small white beans, mature seeds, boiled, wo/salt—½ c (90 gm)	1	1	½	½	½	½	½	½	127	5	1.4	0
Snap beans, green variety, boiled, drained, wo/salt—½ c (62 gm)	0	0	0	0	0	0	0	0	22	2	.4	0
Snap beans, green variety, canned, reg pack, solid & liquid—½ c (120 gm)	0	0	0	0	0	0	0	0	18	1	.3	0
Snap beans, green variety, canned, spec dietary pack, drained solids—½ c (68 gm)	0	0	0	0	0	0	0	0	13	1	.2	0
Snap beans, green variety, frozen, all styles, boiled, drained, wo/salt—½ c (68 gm)	0	0	0	0	0	0	0	0	18	1	.2	0
Snap beans, yellow variety, boiled, drained, wo/salt—½ c (62 gm)	0	0	0	0	0	0	0	0	22	2	.4	0
Snap beans, yellow variety, canned, reg pack, drained, solids—½ c (68 gm)	0	0	0	0	0	0	0	0	13	1	.2	0
Snap beans, yellow variety, canned, spec dietary pack, drained solids—½ c (68 gm)	0	0	0	0	0	0	0	0	13	1	.2	0

Food												
Snap beans, yellow variety, frozen, all styles, boiled, drained, wo/salt—½ c (68 gm)	0	0	0	0	0	0	0	0	18	1	.2	0
Soybeans, green, boiled, drained, wo/salt—½ c (90 gm)	3	2	2	2	1	1	1	1	127	52	5.7	0
Soybeans, green, raw—½ c (128 gm)	4	3	3	2	2	2	2	2	188	78	8.5	0
Soybeans, mature, cooked, boiled, wo/salt—½ c (86 gm)	4	4	3	3	2	2	2	2	149	69	10.0	0
Soybeans, mature seeds, dry roasted—½ c (86 gm)	11	9	8	7	6	5	5	4	387	167	24.2	0
Soybeans, mature seeds, roasted—½ c (86 gm)	13	11	9	8	7	6	6	5	405	196	28.4	0
Soybeans, mature seeds, sprouted, steamed, wo/salt—½ c (47 gm)	1	1	1	1	1	½	½	½	38	19	2.1	0
Soybeans, mature seeds, sprouted, raw—10 sprouts (10 gm)	½	0	0	0	0	0	½	0	12	5	.6	0
Soybeans, sprouted seeds, boiled, drained—1 c (125 gm)	0	0	0	0	0	0	0	0	48	16	0	0
Stewed green peas, Puerto Rican style (habichuelas del pais)—½ c (130 gm)	5	4	4	3	3	3	2	2	110	39	8.8	5
White beans, dry, cooked wo/fat (inc Navy, pea, Great northern)—½ c (1 c dry yields 2¾ c ckd) (88 gm)	½	0	0	0	0	0	0	0	103	5	.6	0
White beans, mature seeds, canned—½ c (131 gm)	½	½	½	0	0	0	0	0	153	4	.9	0
Winged bean, immature seeds, boiled, drained, wo/salt—½ c (31 gm)	0	0	0	0	0	0	0	0	12	2	.5	0
Winged, mature seeds, boiled, wo/salt—½ c (86 gm)	3	2	2	2	1	1	1	1	126	45	6.4	0
Yardlong beans, mature seeds, boiled, wo/salt—½ c (86 gm)	½	½	½	0	0	0	0	0	102	4	.9	0
Yellow beans, mature seeds, boiled, wo/salt—½ c (88 gm)	1	1	1	1	1	½	½	½	126	8	2.3	0

Soybean Products

	1200	1500	1800	2100	2400	2800	3200	3600	TOT CAL	FAT CAL	S/FAT CAL	CHOL MG
Bean curd (tofu) (inc stir fried)—½ c (½" cubes) (92 gm)	2	2	2	1	1	1	1	1	66	35	5.1	0
Curd cheese—1 c (225 gm)	11	9	8	7	6	5	5	4	304	164	23.7	0
Miso—1 c (272 gm)	5	4	4	3	3	4	3	3	465	113	11.0	0
Natto—1 c (175 gm)	7	6	5	5	4	4	3	2	292	117	16.9	0
Soybean curd (tofu), breaded, fried—1 sl (2¾" x 1" x ½") (29 gm)	4	3	3	3	2	2	2	2	46	26	5.6	8
Soybean curd (tofu), deep-fried (inc aburage)—1 oz (28 gm)	3	3	2	2	2	2	2	1	74	51	7.4	0
Soybean meal—1 c (122 gm)	14	12	10	9	8	7	7	6	514	223	32.2	0
Soy flour, defatted—½ c stirred (50 gm)	½	0	0	0	0	0	0	0	165	5	.6	0
Soy flour, full-fat, raw—½ c stirred (42 gm)	5	4	4	3	3	3	2	2	183	78	11.3	0
Soy flour, full-fat, roasted—½ c stirred (42 gm)	5	4	4	3	3	3	2	2	185	83	12.0	0
Soy flour, lo-fat—½ c stirred (44 gm)	2	1	1	1	1	1	1	1	144	26	3.9	0
Soy meal, defatted, raw—½ c (61 gm)	1	1	½	½	½	½	½	½	207	14	1.4	0
Soy nuts (inc pernuts)—1 pkt (4¾ oz) (135 gm)	17	14	12	11	10	9	8	7	612	292	38.5	0
Soy sauce (inc shoyu)—1 c (288 gm)	0	0	0	0	0	0	0	0	184	0	0	0
Teriyaki sauce (inc Oriental barbecue sauce)—1 c (288 gm)	0	0	0	0	0	0	0	0	242	0	0	0
Tofu, raw, reg—½ c (124 gm)	3	3	2	2	2	2	2	1	94	53	7.7	0
Vermicelli, made fr soybeans—1 c (140 gm)	0	0	0	0	0	0	0	0	500	1	.1	0
Worcestershire sauce—1 c (272 gm)	0	0	0	0	0	0	0	0	174	0	0	0

LUNCHEON MEATS and SAUSAGES

Bacon, cooked—2 sl ckd (approx 8 oz, raw) (13 gm)	10	9	8	7	6	6	5	5	60	41	16.9	15
Bacon, formed, lean meat added, cooked (inc Sizzlean)—2 strips (22 gm)	17	14	13	12	11	10	9	8	99	68	28.4	26
Barbecue loaf—1 sl (5⅞" x 3½") (23 gm)	4	4	3	3	3	3	2	2	40	19	6.6	9
Beef, dried, chipped, uncooked—4 oz (113 gm)	24	22	20	18	17	16	15	14	230	64	26.6	80
Beef, pressed, luncheon meat—1 sl (1 oz, 4" x 4" x 3/32" thick) (28 gm)	3	3	3	2	2	2	2	2	34	8	3.3	11
Beef sausage, brown & serve, links, cooked—2 links (2 oz, raw) (26 gm, ckd)	14	12	11	10	9	8	7	7	84	69	28.0	12
Beef sausage, smoked (inc Eckrich, Hillshire Farms)—1 patty (27 gm)	15	13	11	10	9	8	7	7	86	70	28.7	14
Beef sausage, smoked, stick (inc beef jerky)—1 stick (10 gm)	9	8	7	6	6	5	5	4	53	43	17.7	9
Beef sausage w/cheese, smoked (inc Hillshire Farms)—1 link (86 gm)	51	43	38	34	31	28	25	23	278	225	97.5	48
Beerwurst (beer salami), beef—1 sl (4" dia x ⅛") (23 gm)	14	12	11	10	9	8	7	7	76	62	27.0	14
Beerwurst (beer salami), pork—1 sl (4" dia x ⅛") (23 gm)	8	7	6	5	5	5	4	4	55	39	13.0	13
Berliner—1 oz (28 gm)	9	8	7	6	5	5	4	4	65	44	15.5	13
Blood sausage (inc blutwurst, blood pudding)—1 sl, loaf shape (5" x 4⅝" x 1/16") (25 gm)	18	16	14	13	12	10	10	9	95	78	30.1	30
Bockwurst—1 stick (7/lb) (65 gm)	27	24	21	19	17	16	14	13	184	133	46.6	43
Boiled ham, luncheon meat—1 sl (1 oz, 6¼" x 4" x 1/16") (28 gm)	5	5	4	4	4	3	3	3	45	21	6.8	15
Bologna, beef (inc veal bologna)—1 med sl (4½" dia x ⅛") (28 gm)	16	13	12	11	10	9	8	7	88	71	29.4	16

Luncheon Meats and Sausages (cont.)

	1200	1500	1800	2100	2400	2800	3200	3600	TOT CAL	FAT CAL	S/FAT CAL	CHOL MG
Bologna, beef & pork—1 sl (1 oz, 4½" dia) (28 gm)	15	13	11	10	9	8	7	7	89	72	27.3	16
Bologna, Lebanon—1 med sl (4½" dia x ⅛") (28 gm)	10	8	8	7	6	6	5	5	63	37	15.5	18
Bologna, pork—1 sl (1 oz, 4½") (28 gm)	10	9	8	7	7	6	6	5	70	50	17.5	17
Bologna ring, smoked (inc Alderfers)—1 med sl (1 oz) (28 gm)	14	12	11	10	9	8	7	7	88	71	27.0	15
Bologna, turkey (inc chicken bologna, turkey salami)—1 sl (1 oz) (28 gm)	10	9	8	7	7	6	6	6	56	38	12.8	28
Bologna, w/cheese—1 med sl (28 gm)	16	13	12	11	10	9	8	7	90	72	29.2	17
Bratwurst—1 oz (28 gm)	13	11	10	9	8	7	7	6	85	66	23.8	17
Bratwurst, cooked—1 stick (3 oz) (85 gm)	40	34	30	27	25	22	20	19	256	198	71.3	51
Braunschweiger (liver sausage), smoked—1 oz (28 gm)	19	17	15	14	13	12	11	10	102	82	27.8	44
Breakfast strips—3 sl (34 gm)	26	22	20	18	16	15	14	13	153	105	43.9	40
Brotwurst—1 oz (28 gm)	14	12	11	10	9	8	7	7	92	71	25.3	18
Brown & serve, before browning—1 link (37/8 x 7/8") (21 gm)	13	11	10	9	8	7	7	6	83	68	24.6	13
Capicola—1 sl (4¼" x 4¼" x 1/16") (21 gm)	3	3	2	2	2	2	2	2	28	9	3.1	10
Cervalat, soft—1 sl (4⅜" dia x ⅛") (28 gm)	16	14	12	11	10	9	8	8	97	75	30.3	19
Cheesefurter, cheese smokie—1 oz (28 gm)	15	13	11	10	9	8	8	7	93	74	26.7	19
Chicken liver paste or pâté—1 sl (1 oz) (28 gm)	21	20	18	17	17	16	15	14	56	33	10.1	109

Food												
Chicken or turkey loaf, luncheon meat—1 sl (1 oz) (28 gm)	4	4	4	3	3	3	3	3	42	18	5.1	14
Chicken salad spread—1 tbsp (13 gm)	2	2	2	2	2	1	1	1	26	16	4.0	4
Chorizos (inc Spanish sausage, lenguica, Portuguese sausage)—6 sl (1¾" dia x ⅛") (30 gm)	21	18	16	15	13	12	11	10	137	103	38.8	26
Corned beef, brisket—4 oz (113 gm)	46	40	36	33	31	28	26	24	284	193	64.7	111
Corned beef, pressed—1 sl (1 oz) (28 gm)	5	4	4	4	3	3	3	3	46	17	6.8	12
Corned beef spread—1 tbsp (14 gm)	4	4	3	3	3	2	2	2	33	22	7.5	5
Deviled ham, canned—4½ oz can (128 gm)	72	62	55	49	45	40	37	34	449	372	134.0	83
Dutch brand loaf—2 sl (4" x 4" x ³/₃₂) (57 gm)	19	16	14	13	12	11	10	9	136	91	32.4	27
Frankfurter, bacon- & cheese-filled—1 frankfurter (10/lb) (45 gm)	24	20	18	16	15	13	12	11	147	117	42.4	31
Frankfurter, beef—1 frankfurter (10/lb) (45 gm)	25	22	19	17	16	14	13	12	150	122	49.4	22
Frankfurter, beef & pork (inc Smokie Links)—1 frankfurter (10/lb) (45 gm)	23	20	18	16	14	13	12	11	149	121	44.2	23
Frankfurter, breaded, baked—1 frankfurter (51 gm)	28	24	21	19	18	16	14	13	176	133	48.8	39
Frankfurter, cheese-filled—1 frankfurter (8/12 oz package) (43 gm)	22	19	17	15	14	13	12	11	141	112	40.5	29
Frankfurter, chicken—1 frankfurter (10/lb) (45 gm)	18	15	14	13	12	11	10	10	119	80	22.6	48
Frankfurter, chili-filled—1 frankfurter (8/pound) (57 gm)	19	16	14	13	12	10	10	9	134	95	35.3	20
Frankfurter, meat & poultry—1 frankfurter (8/12 oz pkg) (43 gm)	20	17	15	14	13	12	11	10	124	95	33.9	32
Frankfurters, canned—1 frankfurter (1.7 oz) (48 gm)	20	17	15	14	13	13	11	10	106	78	34.6	30

Luncheon Meats and Sausages (cont.)

	1200	1500	1800	2100	2400	2800	3200	3600	TOT CAL	FAT CAL	S/FAT CAL	CHOL MG
Frankfurters, raw, w/nonfat dry milk & cereal—1 frankfurter (1.6 oz) (45 gm)	21	18	16	14	13	12	11	10	123	88	36.5	29
Frankfurter, turkey—1 frankfurter (10/lb) (45 gm)	19	16	15	14	13	12	11	10	104	72	24.1	51
Ham and cheese loaf—1 sl (1 oz) (28 gm)	11	9	8	8	7	6	6	5	73	51	18.9	16
Ham, approx 11% fat, sliced—1 sl (6¼" x 4" x ¹⁄₁₆") (28 gm)	6	6	5	5	4	4	4	3	52	27	8.6	16
Ham & cheese spread—1 tbsp (15 gm)	7	6	5	5	4	4	3	3	37	25	11.6	9
Ham, chopped, not canned—1 square sl (4¼" x 4¼") (21 gm)	7	6	5	5	4	4	3	3	48	32	10.8	11
Ham, deviled or potted—1 tbsp (14 gm)	7	6	6	5	5	4	4	4	47	38	13.6	9
Ham, extra lean, approx 5% fat, sliced—1 sl (6¼" x 4" x ¹⁄₁₆") (28 gm)	4	3	3	3	3	3	2	2	37	13	4.1	13
Ham loaf, luncheon meat (inc ham sausage, hamettes)—1 sl (4" x 4" x ³⁄₃₂") (28 gm)	15	13	11	10	9	8	8	7	94	76	27.2	17
Ham, luncheon meat, chopped, minced, pressed, spiced (inc Spam, Treet)—1 sl (3⅓" x 3⅓" x ⅛") (23 gm)	8	7	6	6	5	5	4	4	56	39	13.2	13
Ham, luncheon meat, chopped, minced, pressed, spiced, low-fat—1 sl (3⅓" x 3⅓" x ⅛") (23 gm)	3	3	3	2	2	2	2	2	30	10	3.4	11
Ham salad spread—1 tbsp (16 gm)	4	4	3	3	3	2	2	2	35	22	7.3	6
Head cheese (inc jellied pork)—1 med sl (1 oz, 4" x 4" x ³⁄₃₂") (28 gm)	9	8	7	7	6	6	5	5	59	40	12.4	23

Food												
Honey loaf—1 sl (1 oz) (28 gm)	3	3	3	2	2	2	2	2	36	11	3.6	10
Honey roll sausage—1 oz (28 gm)	7	6	5	5	4	4	4	3	52	27	10.4	14
Italian sausage (inc hot links)—1 oz, ckd (28 gm)	14	12	10	9	9	8	7	7	90	65	22.9	22
Knockwurst (inc knoblauch)—1 link (4" long x 1⅛" dia) (68 gm)	34	29	26	23	21	19	17	16	209	170	62.4	39
Liver cheese—1 oz (28 gm)	18	16	14	13	12	11	10	10	86	66	22.9	49
Liverwurst (inc liver cheese, braunschweiger, liver sausage, liver loaf, liver pudding, liver bacon, sausage, goose bacon sausage, goose liver—1 sl (3⅛" dia x ¼") (28 gm)	19	17	15	14	13	12	11	10	101	81	27.5	44
Luncheon loaf (olive, pickle, or pimiento)—1 sl (1 oz, 4" x 4" x ³/₃₂") (28 gm)	9	8	7	6	6	5	5	4	70	47	17.3	10
Luncheon meat, jellied—1 oz (28 gm)	3	3	3	2	2	2	2	2	31	8	3.6	10
Luncheon meat, loaved—1 oz (28 gm)	15	13	12	11	10	9	8	7	87	67	28.5	18
Luxury loaf—2 sl (57 gm)	7	6	5	5	5	4	4	4	80	25	8.0	20
Mettwurst—1 oz (28 gm)	14	12	11	10	9	8	7	7	87	68	25.0	19
Mortadella—1 sl (15/8 oz pkg) (15 gm)	7	6	5	5	4	4	4	3	47	34	12.8	8
Mother's loaf—1 square sl (4½" x 4½") (21 gm)	8	7	6	5	5	4	4	4	59	42	15.0	9
New England brand sausage—1 oz (28 gm)	5	4	4	4	3	3	3	3	46	19	6.5	14
Olive loaf—2 sl (4" x 4" x ³/₃₂") (57 gm)	17	14	13	11	10	9	9	8	133	85	29.8	22
Pastrami (beef, smoked, spiced)—1 sl (28 gm)	17	15	13	12	11	10	9	8	99	76	28.3	27
Pâté, chicken liver, canned—1 tbsp (13 gm)	10	9	9	8	8	8	7	7	26	15	4.7	51
Pâté de foie gras, canned (goose liver pâté), smoked—1 tbsp (13 gm)	11	9	8	7	7	7	6	5	60	51	16.9	20
Peppered loaf—2 sl (4" x 4" x ³/₃₂") (57 gm)	9	8	7	6	6	6	5	5	84	32	11.7	26
Pepperoni—1 c, sliced (102 gm)	78	67	59	53	48	43	40	36	507	404	148.1	81

	1200	1500	1800	2100	2400	2800	3200	3600	TOT CAL	FAT CAL	S/FAT CAL	CHOL MG
Luncheon Meats and Sausages (cont.)												
Pickle & pimiento loaf—2 sl (4" x 4" x 3/32") (57 gm)	21	18	16	14	13	12	11	10	149	108	40.0	21
Picnic loaf—2 sl (4" x 4" x 3/32") (57 gm)	17	15	13	12	11	10	9	8	132	85	31.0	22
Polish sausage (kielbasa)—1 sl (6" x 3¾" x 1/16") (26 gm)	13	11	10	9	8	7	7	6	81	64	23.2	17
Pork & beef chopped together—2 sl (4" x 4" x 3/32") (57 gm)	31	27	23	21	19	17	16	14	200	164	59.1	31
Pork & beef sausage—2 patties, ckd (54 gm)	34	29	26	23	21	19	17	16	214	176	63.0	38
Pork, canned—1 square sl (4¼" x 4¼") (21 gm)	11	10	8	8	7	6	6	5	70	58	20.4	13
Pork sausage, brown & serve, cooked—2 patties (54 gm)	30	26	23	21	19	17	16	15	199	151	52.5	45
Pork sausage, canned, drained solids—1 link (approx 3" long) (12 gm)	7	6	6	5	5	4	4	4	46	35	12.8	11
Pork sausage, canned, solid & liquid—1 oz (28 gm)	18	16	14	12	11	10	9	9	116	97	34.8	19
Pork sausage, country style, fresh, cooked—2 patties (raw, 3⅞" dia x ¼") (cooked, 54 gm)	30	26	23	21	19	17	16	15	199	151	52.5	45
Roast beef spread—1 tbsp (14 gm)	4	4	3	3	3	2	2	2	33	22	7.5	5
Salami, beef (inc kosher salami)—1 sl (4" dia x ⅛") (10/8 oz pkg) (23 gm)	10	9	8	7	6	6	5	5	58	42	17.5	14
Salami, dry (inc Italian salami, Goteborg)—3 sl (3⅛" dia x 1/16") (30 gm)	18	16	14	13	12	10	9	9	125	93	32.9	24
Salami, smoked—1 oz (28 gm)	13	11	10	9	8	7	7	6	74	53	23.0	18

Salami, soft, cooked (inc beer bologna, beer sausage, beerwurst)—1 sl (4½" dia x 1⁄10") (28 gm)	12	10	9	8	8	7	6	6	70	51	20.4	18
Sandwich loaf, luncheon meat (inc old-fashioned loaf, Dutch loaf, pepperloaf, praski)—1 sl (1 oz) (28 gm)	7	6	5	5	4	4	4	3	54	30	10.9	13
Sandwich spread, lo-cal, 5 cal/tsp—1 tbsp (15 gm)	2	2	2	1	1	1	1	1	17	13	1.4	8
Sandwich spread, w/chopped pickle, regular, unspecified oils—1 tbsp (15 gm)	5	4	4	4	3	3	3	3	60	47	7.2	12
Scrapple, cooked—2 sl (2¾" x 2⅛" x ¼") (1⁄18 of loaf) (50 gm)	14	12	11	10	9	8	7	7	114	65	23.0	23
Smoked chopped beef—1 oz (28 gm)	4	4	3	3	3	3	2	2	35	12	4.6	13
Smoked link sausage, pork—1 link (4" long x 1⅛" dia) (68 gm)	38	33	29	26	24	21	19	18	265	194	69.3	46
Smoked link sausage, pork and beef—1 link (4" long x 1⅛" dia) (68 gm)	36	31	28	25	23	21	19	17	228	186	65.0	48
Souse—1 sl (1 oz, 3⅞" sq x ⅛") (28 gm)	7	7	6	5	5	5	4	4	49	33	10.2	19
Summer sausage—1 oz (28 gm)	17	15	13	12	11	10	9	8	95	76	30.7	21
Thin-sliced beef—1 oz (28 gm)	4	3	3	3	3	2	2	2	50	10	4.2	12
Thuringer (inc summer sausage)—1 sl (1 oz, 4⅜" dia x 1⁄10") (28 gm)	16	14	12	11	10	9	8	8	97	75	30.3	19
Turkey, breast, pressed, smoked, luncheon meat—1 sl (1 oz) (28 gm)	2	2	2	2	2	2	2	1	31	4	1.2	11
Turkey ham roll—1 sl (1 oz) (28 gm)	4	4	4	3	3	3	3	3	36	13	4.3	16
Turkey luncheon meat, turkey ham, cured turkey thigh meat—2 sl (2 oz) (57 gm)	9	8	7	7	6	6	5	5	73	26	8.7	32
Turkey luncheon meat, loaf, breast meat—2 sl (1.5 oz) (43 gm)	3	3	3	3	3	2	2	2	47	6	1.9	17
Turkey pastrami—1 sl (1 oz) (28 gm)	4	4	4	3	3	3	3	3	39	16	4.6	15

	1200	1500	1800	2100	2400	2800	3200	3600	TOT CAL	FAT CAL	S/FAT CAL	CHOL MG
Luncheon Meats and Sausages (cont.)												
Veal loaf—1 sl (1 oz) (28 gm)	15	13	12	10	10	9	8	7	86	66	28.2	18
Vienna sausage, canned—3 sausages (2" long x ⅞" dia) (48 gm)	22	19	16	15	13	12	11	10	134	109	40.1	25
Vienna sausage, chicken, canned—3 sausages (48 gm)	18	16	14	13	12	11	10	10	123	84	23.9	48
Imitation Meats												
Bacon bits, meatless—1 tbsp (7 gm)	1	1	1	1	1	1	1	½	31	16	2.6	0
Bacon strips, meatless (inc Morning Star Breakfast Strips, Stripple)—2 strips (16 gm)	4	3	3	2	2	2	2	1	70	52	8.1	0
Breakfast links, patties, or slices, meatless (inc Prosage, Morningstar)—1 link (25 gm)	2	1	1	1	1	1	1	1	48	25	3.9	0
—1 patty (38 gm)	3	2	2	2	2	1	1	1	72	38	5.9	0
Chicken, meatless—½ c (84 gm)	7	6	5	5	4	4	3	3	185	106	16.6	0
Chicken, meatless, breaded, fried (inc Loma Linda brand)—1 piece (36 gm)	3	2	2	2	1	1	1	1	61	39	5.8	0
Fish sticks, meatless—2 sticks (57 gm)	6	5	5	4	4	3	3	3	165	92	14.5	0
Frankfurter, meatless—1 frankfurter (70 gm)	4	4	3	3	3	2	2	2	140	63	9.8	0
Luncheon slices, meatless beef, chicken, salami, or turkey (inc vegetarian ham, Wham, Loma Linda, Worthington)—1 reg sl (1 oz) (28 gm)	3	2	2	2	2	1	1	1	78	40	6.3	0
Meatballs, meatless—2 meatballs (36 gm)	2	2	1	1	1	1	1	1	72	29	4.6	0

Food												
Sandwich spread, meat-substitute type —1 tbsp (16 gm)	1	1	1	1	1	½	½	½	24	13	2.0	0
Soyburger (inc vegetarian burger)—1 patty (70 gm)	4	3	3	3	2	2	2	2	140	57	8.9	0
Vegetarian bouillon, dry—1 cube (4 gm)	½	½	½	0	0	0	0	0	13	2	.8	0
Vegetarian fillets—1 fillet (85 gm)	10	8	7	6	6	5	4	4	247	138	21.6	0
Vegetarian meat loaf or patties (meat loaf made with meat substitute)—1 sl (56 gm)	3	3	2	2	2	2	1	1	112	45	7.1	0
Vegetarian pot pie (inc beeflike and chickenlike pot pies)—1 pie (227 gm)	41	35	30	27	24	22	20	18	524	304	85.3	20
Vegetarian stew—1 c (239 gm)	4	3	3	3	2	2	2	2	287	65	9.7	0
MEAL REPLACEMENTS and SUPPLEMENTS												
Bee pollen—1 tsp (3 gm)	0	0	0	0	0	0	0	0	9	1	.5	0
Diet beverage, liquid, canned (inc Metrecal, Sego, Slender)—1 can (10 fl oz) (313 gm)	3	2	2	2	2	2	2	2	219	39	5.1	6
Diet beverage, powder, reconstituted (inc Lookfit, Metrecal, Shape, Slender)—1 pkt (28 g) in 6 fl oz skim milk (212 gm)	3	2	2	2	2	2	2	2	170	8	4.8	6
Formulated diet meal, powdered, soy protein isolate, w/herbs (inc Optimum Plus)—1 oz (28 gm)	1	1	1	1	½	½	½	½	39	12	1.8	0
Gelatin, dessert, made w/water, plain—1 c (240 gm)	0	0	0	0	0	0	0	0	142	0	0	0
Gelatin, dessert, made w/water, w/fruit added—1 c (240 gm)	0	0	0	0	0	0	0	0	161	2	0	0
Gelatin, dessert powder—4 oz pkt (113 gm)	0	0	0	0	0	0	0	0	420	0	0	0
Gelatin, drinking, orange flavor, prepared w/water fr powder—1 pkt (136 gm) in 4 fl oz water	0	0	0	0	0	0	0	0	67	2	.3	0

Meat Replacements and Supplements *(cont.)*

	1200	1500	1800	2100	2400	2800	3200	3600	TOT CAL	FAT CAL	S/FAT CAL	CHOL MG
Gelatin drink, powder, flavored, w/lo-cal sweetener, reconstituted (inc Nutri System Orange Drink)—1 pkt (255 gm) in 8 fl oz water	0	0	0	0	0	0	0	0	51	0	0	0
Gelatin, dry—1 envelope (7 gm)	0	0	0	0	0	0	0	0	23	0	0	0
Gelatin, plain, drink—1 c (244 gm)	0	0	0	0	0	0	0	0	19	0	0	0
Gelatin, plain, prepared—1 pkt, prep (480 gm)	0	0	0	0	0	0	0	0	283	0	0	0
High-calorie beverage, canned or powdered, reconstituted (inc Nutrament)—1 c (250 gm)	4	3	3	3	2	2	2	2	235	59	8.8	0
High-protein bar, candylike, soy & milk base (inc Hoffman's, Tiger's Milk)—1 bar (50 gm)	4	3	3	3	2	2	2	2	223	64	9.3	0
High-protein bar, soy base (inc E.M.F. Shaklee's Protein Bar)—1 bar (2.5 oz) (71 gm)	8	7	6	5	4	4	4	3	324	121	17.5	0
High-protein wafers (inc hi-prot oatmeal wafer)—25 wafers (60 gm)	1	1	1	1	1	1	1	1	200	23	3.2	0
Instant breakfast, fluid, canned—1 can (10 fl oz) (313 gm)	31	27	23	21	19	17	16	15	307	87	54.5	42
Instant breakfast, powder, milk added (inc Carnation Instant Breakfast, Lucerne)—1 pkt (35.1 gm) in 8 fl oz milk (279 gm)	27	24	21	19	17	15	14	13	274	78	48.5	37
Meal replacement formula, Cambridge diet, reconstituted, all flavors—1 serving (1 scoop in 8.5 fl oz water) (286 gm)	6	5	5	4	4	4	3	3	113	9	5.9	19

Food												
Meal replacement or supplement, liquid, predigested protein (inc Pro-Linn, E.M.F.)—1 fl oz (37 gm)	0	0	0	0	0	0	0	0	70	0	0	0
Meal replacement or supplement, liquid, soy-base (inc Isocal liquid nutrition)—1 can (12 oz) (370 gm)	9	8	7	6	5	5	4	4	392	147	21.0	0
Meal replacement or supplement, liquid, soy-base, hi-prot (inc Ensure liquid nutrition)—1 c (252 gm)	7	6	5	5	4	4	3	3	350	111	16.1	0
Meal supplement or replacement, gelatin-base, powder, reconstituted (inc Physicians Weight Loss Chicken Nutrient)—1 c (252 gm)	0	0	0	0	0	0	0	0	50	0	0	0
Meal supplement or replacement, milk base, hi-prot, liquid (inc Sustacal)—1 c (256 gm)	3	3	2	2	2	2	2	1	241	51	7.6	0
Protein diet powder w/soy & casein (inc Nature Slim, Herbalife)—1 serving (20 gm)	½	½	½	½	½	0	0	0	78	6	.9	0
Protein supplement, powdered (inc Shaklee Instant Protein, Trim n' Easy, Slender Now, MLV Protein Additive)—3 tbsp (1 oz) (28 gm)	1	½	½	½	½	½	½	0	109	9	1.3	0
Protein supplement, tablet (inc milk tablet)—1 tablet (1 gm)	0	0	0	0	0	0	0	0	4	0	0	0
Textured vegetable protein, dry—1 c, coarse grain (68 gm)	½	½	0	½	0	0	0	0	222	6	.8	0
—1 c, fine grain (120 gm)	1	1	½	½	½	½	½	½	391	10	1.4	0

MEAT
Beef

Food												
Brisket, flat half, all grades, lean & fat, braised—4 oz (113 gm)	82	70	62	56	51	46	42	39	463	355	147.3	104
Brisket, flat half, all grades, lean only, braised—4 oz (113 gm)	44	39	35	32	30	27	25	23	297	162	63.7	103

	1200	1500	1800	2100	2400	2800	3200	3600	TOT CAL	FAT CAL	S/FAT CAL	CHOL MG
Beef (cont.)												
Brisket, point half, all grades, lean & fat, braised—4 oz (113 gm)	70	61	54	49	45	41	37	34	415	295	120.7	108
Brisket, point half, all grades, lean only, braised—4 oz (113 gm)	31	27	25	23	22	20	19	18	241	88	31.2	108
Brisket, whole, all grades, lean & fat, braised—4 oz (113 gm)	76	66	58	52	48	43	40	37	443	331	134.4	105
Brisket, whole, all grades, lean only, braised—4 oz (113 gm)	37	33	30	27	25	23	22	20	273	131	46.9	105
Chipped beef, dried, creamed—1 c (245 gm)	70	60	53	48	44	40	36	34	377	227	123.2	98
Chuck, arm pot roast, choice, lean & fat, braised—4 oz (113 gm)	67	58	51	47	43	39	36	33	401	270	111.0	112
Chuck, arm pot roast, choice, lean only, braised—4 oz (113 gm)	35	32	29	26	25	23	21	20	265	106	40.0	113
Chuck, arm pot roast, good, lean & fat, braised—4 oz (113 gm)	63	55	49	44	41	37	34	31	383	250	102.3	112
Chuck, arm pot roast, good, lean only, braised—4 oz (113 gm)	33	30	27	25	23	22	20	19	252	91	34.7	113
Chuck, blade roast, choice, lean & fat, braised—4 oz (113 gm)	77	66	59	53	49	44	40	37	440	317	132.0	116
Chuck, blade roast, choice, lean only, braised—4 oz (113 gm)	48	42	38	35	32	29	27	26	312	161	65.6	120
Chuck, blade roast, good, lean & fat, braised—4 oz (113 gm)	72	62	55	50	46	42	38	35	415	289	120.7	117
Chuck, blade roast, good, lean only, braised—4 oz (113 gm)	44	39	35	32	30	28	26	24	291	139	56.9	120
Corned beef, canned—1 oz (28 gm)	11	9	8	8	7	7	6	6	71	38	15.8	24

Food												
Corned beef, canned, ready-to-eat—1 sl (40 gm)	14	12	11	10	10	9	8	8	86	43	19.3	36
Corned beef, cooked, lean & fat—4 oz, boneless, ckd (113 gm)	64	55	49	45	41	37	34	32	381	285	104.7	111
Corned beef, cooked, lean only—3 thin sl (approx 4½" x 2½" x ⅛") (63 gm)	17	15	14	13	12	11	10	10	117	45	18.9	54
Corned beef hash, canned, w/potato—1 c (220 gm)	59	51	45	40	37	33	30	28	398	224	107.3	73
Corned beef, jellied loaf—1 oz (28 gm)	5	4	4	4	3	3	3	3	43	15	6.7	13
Dried beef—1 oz (28 gm)	4	3	3	3	3	2	2	2	47	10	4.1	12
Flank, choice, lean & fat, braised—4 oz (113 gm)	43	37	33	30	28	25	23	22	291	157	67.5	81
Flank, choice, lean & fat, broiled—4 oz (113 gm)	44	38	34	31	28	26	24	22	288	167	70.9	80
Flank, choice, lean only, braised—4 oz (113 gm)	39	34	31	28	26	23	22	20	277	142	60.5	80
Flank, choice, lean only, broiled—4 oz (113 gm)	41	36	32	29	27	24	23	21	276	152	65.1	80
Ground beef, meatballs, meat only—3 med meatballs (85 gm)	41	35	32	29	27	24	22	21	243	147	61.1	86
Ground beef or patty, breaded, cooked—1 med patty (4 oz, raw, 4 patties/lb) (yield after cooking, 101 gm)	35	30	27	24	22	20	19	17	341	206	59.0	56
Ground beef w/textured vegetable protein, cooked—4 oz, cooked (113 gm)	45	39	35	32	29	26	24	23	301	180	70.0	87
Ground, extra lean, broiled, medium—4 oz (113 gm)	44	38	34	31	29	26	24	23	289	167	65.4	95
Ground, extra lean, broiled, well done—4 oz (113 gm)	46	40	36	33	31	28	26	24	300	161	63.3	112
Ground, extra lean, pan-fried, medium—4 oz (113 gm)	43	38	34	31	29	26	24	22	288	168	65.7	92
Ground, extra lean, pan-fried, well done—4 oz (113 gm)	45	39	35	32	30	27	25	24	299	163	63.9	105

THE 2-IN-1 SYSTEM

	1200	1500	1800	2100	2400	2800	3200	3600	TOT CAL	FAT CAL	S/FAT CAL	CHOL MG
Beef (cont.)												
Ground, extra lean, raw—4 oz (113 gm)	43	37	33	30	28	25	23	22	265	174	69.3	78
Ground, lean, broiled, medium—4 oz (113 gm)	48	42	38	34	32	29	27	25	308	188	73.9	99
Ground, lean, broiled, well done—4 oz (113 gm)	49	43	39	36	33	30	28	26	317	180	70.7	115
Ground, lean, pan-fried, medium—4 oz (113 gm)	49	42	38	34	32	29	27	25	312	194	76.3	95
Ground, lean, pan-fried, well done—4 oz (113 gm)	48	42	38	35	32	29	27	25	313	180	70.8	108
Ground, lean, raw—4 oz (113 gm)	51	44	39	35	33	30	27	25	298	211	84.5	85
Ground, patties, frozen, broiled, medium—4 oz (113 gm)	52	45	40	37	34	31	28	27	320	200	78.7	107
Ground, patties, frozen, raw—4 oz (113 gm)	56	49	43	39	36	32	30	28	319	236	95.2	89
Ground, regular, broiled, medium—4 oz (113 gm)	53	46	41	37	34	31	29	27	328	211	82.9	101
Ground, regular, broiled, well done—4 oz (113 gm)	53	46	41	38	35	32	29	27	331	198	78.0	115
Ground, regular, pan-fried, medium—4 oz (113 gm)	56	48	43	39	36	33	30	28	347	230	90.3	100
Ground, regular, pan-fried, well done—4 oz (113 gm)	51	45	40	36	34	31	28	26	324	193	75.8	111
Ground, regular, raw—4 oz (113 gm)	64	55	49	44	40	37	33	31	351	270	109.6	96
Neckbones, cooked—1 oz, w/bone, ckd (yield after bone removed) (11 gm)	5	5	4	4	3	3	3	3	32	19	7.9	11
Oxtails, cooked—1 oz, w/bone, ckd (yield after bone removed) (16 gm)	6	6	5	5	4	4	4	3	40	20	8.3	17

Food											
Pickled—1 oz, boneless (28 gm)	11	10	9	8	8	7	6	6	70	48	16.0	27
Porterhouse steak, choice, lean & fat, broiled—4 oz (113 gm)	54	47	42	38	35	32	29	27	339	216	89.5	93
Porterhouse steak, choice, lean only, broiled—4 oz (113 gm)	34	30	27	25	23	21	20	18	247	110	44.1	91
Potpie, commercial, frozen, unheated—4 oz (113 gm)	17	14	13	11	10	9	9	8	217	101	30.5	20
Potpie, home recipe, baked—1 pie (9" dia) (630 gm)	122	104	92	83	75	68	62	57	1550	823	228.0	132
Rib, eye, small end (ribs 10–12), choice, lean & fat, broiled—4 oz (113 gm)	54	47	42	38	35	31	29	27	333	210	89.0	93
Rib, eye, small end (ribs 10–12), choice, lean only, broiled—4 oz (113 gm)	39	34	31	28	26	24	22	20	267	131	55.7	91
Rib, whole (ribs 6–12), choice, lean & fat, broiled—4 oz (113 gm)	74	64	56	51	46	42	38	35	417	313	132.2	97
Rib, whole (ribs 6–12), choice, lean & fat, roasted—4 oz (113 gm)	77	66	59	53	48	44	40	37	437	331	140.4	96
Rib, whole (ribs 6–12), choice, lean only, broiled—4 oz (113 gm)	40	35	32	29	27	24	23	21	264	138	58.1	92
Rib, whole (ribs 6–12), choice, lean only, roasted—4 oz (113 gm)	42	37	33	30	28	25	23	22	279	146	62.4	91
Rib, whole (ribs 6–12), good, lean & fat, roasted—4 oz (113 gm)	71	61	54	49	45	40	37	34	408	299	126.8	96
Rib, whole (ribs 6–12), good, lean only, broiled—4 oz (113 gm)	36	32	29	26	24	22	21	19	241	115	48.9	92
Rib, whole (ribs 6–12), good, lean only, roasted—4 oz (113 gm)	37	33	30	27	25	23	21	20	255	124	52.5	91
Rib, whole (ribs 6–12), good, lean & fat, broiled—4 oz (113 gm)	67	58	52	47	43	39	35	33	385	280	118.4	96
Rib, whole (ribs 6–12), prime, lean & fat, broiled—4 oz (113 gm)	83	71	63	57	52	46	42	39	463	359	152.4	97
Rib, whole (ribs 6–12), prime, lean & fat, roasted—4 oz (113 gm)	86	74	65	59	53	48	44	40	481	377	159.6	97

	1200	1500	1800	2100	2400	2800	3200	3600	TOT CAL	FAT CAL	S/FAT CAL	CHOL MG
Beef (cont.)												
Rib, whole (ribs 6–12), prime, lean only, broiled—4 oz (113 gm)	50	44	39	36	33	30	27	25	317	191	81.3	92
Rib, whole (ribs 6–12), prime, lean only, roasted—4 oz (113 gm)	52	45	40	36	33	30	28	26	331	198	84.6	91
Roast, canned (inc beef, steak, canned)—4 oz, boneless (113 gm)	39	34	31	28	26	24	22	20	254	133	55.2	91
Round, bottom round, choice, lean & fat, braised—4 oz (113 gm)	43	38	34	31	29	27	25	23	299	155	59.0	108
Round, bottom round, choice, lean only, braised—4 oz (113 gm)	33	29	27	25	23	21	20	19	255	102	36.1	108
Round, bottom round, good, lean & fat, braised—4 oz (113 gm)	41	36	33	30	28	25	24	22	287	142	54.2	108
Round, bottom round, good, lean only, braised—4 oz (113 gm)	31	28	25	23	22	20	19	18	243	89	31.8	108
Round, bottom round, prime, lean & fat, braised—4 oz (113 gm)	50	44	39	36	33	30	28	26	337	194	74.6	108
Round, bottom round, prime, lean only, braised—4 oz (113 gm)	37	33	30	27	26	24	22	21	283	130	45.8	108
Round, eye of round, choice, lean & fat, roasted—4 oz (113 gm)	39	34	31	28	26	24	22	20	276	146	59.6	83
Round, eye of round, choice, lean only, roasted—4 oz (113 gm)	24	21	19	18	17	16	14	14	208	68	26.0	79
Round, eye of round, good, lean and fat, roasted—4 oz (113 gm)	38	33	30	27	25	23	21	20	268	138	56.4	83
Round, eye of round, good, lean only, roasted—4 oz (113 gm)	23	20	19	17	16	15	14	13	201	61	23.4	79
Round, eye of round, prime, lean & fat, roasted—4 oz (113 gm)	40	35	31	28	26	24	22	20	284	152	61.4	81

Food												
Round, eye of round, prime, lean only, roasted—4 oz (113 gm)	27	24	21	20	18	17	16	15	224	84	32.2	79
Round, full cut, choice, lean & fat, broiled—4 oz (113 gm)	48	42	37	34	31	28	26	24	311	186	74.5	95
Round, full cut, choice, lean only, broiled—4 oz (113 gm)	28	25	23	21	19	18	17	16	220	82	29.6	93
Round, full cut, good, lean & fat, broiled—4 oz (113 gm)	45	40	36	32	30	27	25	23	296	172	68.9	95
Round, full cut, good, lean only, broiled—4 oz (113 gm)	26	23	21	20	18	17	16	15	209	71	25.6	93
Round, tip round, choice, lean & fat, roasted—4 oz (113 gm)	43	37	34	31	28	26	24	22	288	160	63.7	93
Round, tip round, choice, lean only, roasted—4 oz (113 gm)	27	24	22	21	19	18	17	16	219	79	28.9	92
Round, tip round, good, lean & fat, roasted—4 oz (113 gm)	40	35	31	29	27	24	22	21	273	144	57.3	93
Round, tip round, good, lean only, roasted—4 oz (113 gm)	25	23	21	19	18	17	16	15	208	68	25.1	92
Round, tip round, prime, lean & fat, roasted—4 oz (113 gm)	49	43	38	35	32	29	27	25	323	196	78.0	95
Round, tip round, prime, lean only, roasted—4 oz (113 gm)	31	28	25	23	21	20	18	17	241	103	37.6	92
Round, top round, choice, lean & fat, pan-fried—4 oz (113 gm)	30	27	24	23	21	19	18	17	241	92	34.2	96
Round, top round, choice, lean & fat, broiled—4 oz (113 gm)	46	41	37	34	31	28	26	24	328	174	66.5	109
Round, top round, choice, lean only, broiled—4 oz (113 gm)	25	23	21	19	18	17	16	15	220	66	23.0	96
Round, top round, choice, lean only, pan-fried—4 oz (113 gm)	30	27	25	23	21	20	19	18	257	88	28.4	111
Round, top round, good, lean & fat, broiled—4 oz (113 gm)	29	26	24	22	21	19	18	17	235	85	32.3	96
Round, top round, good, lean only, broiled—4 oz (113 gm)	23	21	20	18	17	16	15	14	208	55	19.2	96

	1200	1500	1800	2100	2400	2800	3200	3600	TOT CAL	FAT CAL	S/FAT CAL	CHOL MG
Beef (cont.)												
Round, top round, prime, lean & fat, broiled—4 oz (113 gm)	34	31	28	25	24	22	20	19	268	119	44.0	96
Round, top round, prime, lean only, broiled—4 oz (113 gm)	29	26	24	22	20	19	18	17	244	90	31.7	96
Sandwich steak, flaked, formed, thinly sliced (inc Steak-umms)—1 sandwich steak (41 gm)	16	14	12	11	10	9	9	8	105	61	23.8	33
Shank crosscuts, choice, lean & fat, simmered—4 oz (113 gm)	35	31	28	26	24	22	20	19	277	124	48.5	89
Shank crosscuts, choice, lean only, simmered—4 oz (113 gm)	24	22	20	18	17	16	15	14	228	65	23.3	88
Short ribs, barbecued, w/sauce, lean & fat—1 med rib (yield after cooking, bone removed) (66 gm)	29	25	22	20	18	17	15	14	184	129	52.5	38
Short ribs, barbecued, w/sauce, lean only—1 med rib (yield after cooking, bone and fat removed) (51 gm)	12	10	9	9	8	7	7	6	91	40	15.7	31
Short ribs, choice, lean & fat, braised—4 oz (113 gm)	97	83	74	66	60	54	49	45	533	428	181.5	107
Short ribs, choice, lean only, braised—4 oz (113 gm)	51	45	40	37	34	31	28	26	335	185	78.9	105
Short ribs, cooked, lean & fat (inc beef rib tips)—1 med rib (yield after cooking, bone removed) (66 gm)	50	43	38	34	31	28	25	23	283	220	91.6	58
Short ribs, cooked, lean only (inc beef rib tips)—1 med rib (yield after cooking, bone & fat removed) (51 gm)	13	12	11	10	9	9	8	8	101	35	14.6	43

Food												
Stew meat, cooked, lean & fat—4 oz, boneless, ckd (113 gm)	53	46	41	38	35	32	29	27	325	195	81.0	109
Stew meat, cooked, lean only—4 oz, boneless, ckd (yield after fat removed) (100 gm)	27	24	22	20	19	18	16	16	192	63	26.0	96
Tallow—1 tbsp (13 gm)	28	23	20	18	17	15	13	12	115	115	57.3	14
T-bone steak, choice, lean & fat, broiled—4 oz (113 gm)	61	53	47	42	39	35	32	30	368	251	104.3	95
T-bone steak, choice, lean only, broiled—4 oz (113 gm)	33	29	26	24	22	21	19	18	243	106	42.3	91
Tenderloin, choice, lean & fat, broiled—4 oz (113 gm)	48	42	37	34	31	29	26	25	307	181	73.9	97
Tenderloin, choice, lean & fat, roasted—4 oz (113 gm)	57	50	44	40	37	33	31	29	349	230	94.4	99
Tenderloin, choice, lean only, broiled—4 oz (113 gm)	32	28	26	24	22	20	19	18	235	97	38.2	96
Tenderloin, choice, lean only, roasted—4 oz (113 gm)	36	32	29	26	24	22	21	19	252	119	46.4	97
Tenderloin, good, lean & fat, broiled—4 oz (113 gm)	44	39	35	32	29	27	25	23	288	161	65.4	97
Tenderloin, good, lean & fat, roasted—4 oz (113 gm)	53	46	41	37	34	31	29	27	327	208	84.8	99
Tenderloin, good, lean only, broiled—4 oz (113 gm)	30	26	24	22	21	19	18	17	223	85	33.2	96
Tenderloin, good, lean only, roasted—4 oz (113 gm)	33	29	26	24	23	21	19	18	236	103	40.1	97
Tenderloin, prime, lean & fat, broiled—4 oz (113 gm)	59	51	45	41	38	34	31	29	360	239	97.8	97
Tenderloin, prime, lean & fat, roasted—4 oz (113 gm)	69	59	53	48	44	39	36	33	407	293	119.7	100
Tenderloin, prime, lean only, broiled—4 oz (113 gm)	37	32	29	27	25	23	21	20	263	126	49.3	96
Tenderloin, prime, lean only, roasted—4 oz (113 gm)	42	37	33	30	28	26	24	22	289	156	60.8	97

	1200	1500	1800	2100	2400	2800	3200	3600	TOT CAL	FAT CAL	S/FAT CAL	CHOL MG
Beef *(cont.)*												
Top loin, choice, lean & fat, broiled—4 oz (113 gm)	51	44	39	36	33	30	27	25	324	199	82.5	91
Top loin, choice, lean only, broiled—4 oz (113 gm)	31	27	25	23	21	19	18	17	235	96	38.4	87
Top loin, good, lean & fat, broiled—4 oz (113 gm)	45	39	35	32	30	27	25	23	297	172	70.8	89
Top loin, good, lean only, broiled—4 oz (113 gm)	27	24	22	20	19	18	16	15	216	77	30.7	87
Top loin, prime, lean & fat, broiled—4 oz (113 gm)	63	54	48	43	40	36	33	30	384	264	109.3	91
Top loin, prime, lean only, broiled—4 oz (113 gm)	38	33	30	27	25	23	21	20	277	139	55.5	87
Wedge-bone sirloin, choice, lean & fat, broiled—4 oz (113 gm)	51	44	40	36	33	30	28	26	320	188	78.2	103
Wedge-bone sirloin, choice, lean & fat, pan-fried—4 oz (113 gm)	63	55	49	44	41	37	34	31	384	251	102.6	112
Wedge-bone sirloin, choice, lean only, broiled—4 oz (113 gm)	32	29	26	24	23	21	19	18	240	92	37.7	101
Wedge-bone sirloin, choice, lean only, pan-fried—4 oz (113 gm)	37	32	30	27	25	23	22	21	269	112	42.6	113
Wedge-bone sirloin, good, lean & fat, broiled—4 oz (113 gm)	49	43	38	35	32	29	27	25	309	178	73.9	103
Wedge-bone sirloin, good, lean only, broiled—4 oz (113 gm)	30	27	25	23	21	20	18	17	227	79	32.5	101
Wedge-bone sirloin, prime, lean & fat, broiled—4 oz (113 gm)	59	51	46	41	38	34	32	29	361	233	96.8	103
Wedge-bone sirloin, prime, lean only, broiled—4 oz (113 gm)	38	33	30	28	26	23	22	20	268	121	49.3	101

Game

Food												
Caribou—4 oz, boneless, ckd (113 gm)	25	23	21	19	18	17	16	15	212	61	21.8	98
Deer bologna (inc venison sausage, caribou sausage)—1 oz (28 gm)	16	13	12	11	10	9	8	7	88	71	29.4	16
Deer chop, (inc venison chop)—4 oz, w/bone, ckd (yield after bone removed) (92 gm)	23	20	18	17	16	15	14	13	198	94	28.1	67
Deer ribs—4 oz, w/bone, ckd (yield after bone removed) (92 gm)	20	18	16	15	14	13	12	12	166	49	17.7	76
Goat, baked—4 oz, boneless, ckd (113 gm)	34	31	29	27	26	24	23	21	179	57	23.8	152
Goat, boiled—4 oz, boneless, ckd (113 gm)	44	40	37	35	33	31	29	28	237	79	33.0	191
Goat, fried—4 oz, boneless, ckd (113 gm)	32	30	28	26	25	23	22	21	204	46	19.0	154
Moose, cooked—4 oz, boneless, ckd (113 gm)	24	21	20	18	17	16	15	14	203	58	20.9	93
Rabbit, domestic, breaded, fried—4 oz, boneless, ckd (113 gm)	34	30	28	26	24	22	21	19	265	105	38.8	109
Rabbit, domestic, cooked—1 body piece (use for breast) (105 gm)	33	30	27	25	24	22	20	19	225	95	36.5	111
Rabbit, wild, cooked—4 oz, w/bone, ckd (yield after bone removed) (92 gm)	29	26	24	22	21	19	18	17	197	83	32.0	97
Venison, cured—4 oz, boneless, cooked (113 gm)	24	21	20	18	17	16	15	14	201	59	21.5	91
Venison, roasted (inc roast antelope)—4 oz, boneless, ckd (113 gm)	24	22	20	19	18	16	15	14	204	60	21.9	93
Venison steak, cooked (inc fried venison, ground venison)—4 oz, boneless, ckd (113 gm)	24	22	20	19	18	16	15	14	204	60	21.9	93
Venison, stewed—4 oz, boneless, ckd (113 gm)	24	22	20	19	18	16	15	14	204	60	21.9	93

	1200	1500	1800	2100	2400	2800	3200	3600	TOT CAL	FAT CAL	S/FAT CAL	CHOL MG
Game (cont.)												
Wild pig, smoked—4 oz, boneless, ckd (113 gm)	37	32	29	27	25	23	21	20	262	132	45.5	105
Lamb												
Ground or patty, cooked—1 patty (4 oz, raw) (yield after cooking) (77 gm)	52	45	40	36	33	30	27	25	275	203	92.2	74
Hocks, cooked—4 oz, w/bone, ckd (yield after bone removed) (76 gm)	36	31	28	25	23	21	20	18	211	128	56.6	69
Loin chop, cooked, lean and fat (inc lamb steak)—1 sm (4 oz, w/bone, raw) (yield after cooking, bone removed) (71 gm)	48	42	37	33	31	28	25	23	253	187	85.0	68
—1 med (5 oz, w/bone, raw) (yield after cooking, bone removed) (89 gm)	61	52	46	42	38	35	32	29	318	234	106.5	86
—1 lg (6 oz, w/bone, raw) (yield after cooking, bone removed) (107 gm)	73	63	56	50	46	42	38	35	382	281	128.1	103
Loin chop, cooked, lean only (inc lamb steak)—1 sm (4 oz, w/bone, raw) (yield after cooking, bone & fat removed) (50 gm)	13	12	11	10	10	9	8	8	93	34	13.8	47
—1 med (5 oz, w/bone, raw) (yield after cooking, bone & fat removed) (62 gm)	17	15	14	13	12	11	10	10	116	42	17.1	58
—1 lg (6 oz, w/bone, raw) (yield after cooking, bone & fat removed) (75 gm)	20	18	16	15	14	13	12	12	140	50	20.7	70

Food												
Ribs, cooked, no fat (inc mutton ribs)—1 rib (yield after cooking, bone removed) (25 gm)	8	7	6	6	5	5	5	4	52	23	10.3	22
Ribs, cooked, lean & fat, (inc mutton ribs)—1 rib (yield after cooking, bone removed) (46 gm)	36	31	28	25	23	20	18	17	187	147	67.5	41
Roast, cooked, lean & fat, (inc mutton roast, lamb leg, mutton leg)—4 oz, boneless, ckd (113 gm)	54	47	42	38	35	32	29	27	316	193	84.8	104
Roast, cooked, lean only, (inc mutton roast, lamb leg, mutton leg)—4 oz, boneless, ckd (yield after fat removed) (84 gm)	22	19	18	16	15	14	13	13	156	53	22.2	75
Shoulder chop, cooked, lean & fat—1 sm (5.5 oz, w/bone, raw) (yield after cooking, bone removed) (100 gm)	65	56	50	45	41	37	34	32	336	243	114.0	95
—1 med (7 oz, w/bone, raw) (yield after cooking, bone removed) (127 gm)	83	72	64	58	53	48	44	40	427	309	144.7	121
—1 lg (8 oz, w/bone, raw) (yield after cooking, bone removed) (145 gm)	95	82	73	66	60	54	50	46	487	353	165.3	138
Shoulder chop, cooked, lean only —1 sm (5.5 oz, w/bone, raw) (yield after cooking, bone and fat removed) (72 gm)	23	20	18	17	16	14	13	12	147	64	27.6	66
—1 med (7 oz, w/bone, raw) (yield after cooking, bone & fat removed) (91 gm)	28	25	23	21	20	18	17	16	185	81	34.9	83
—1 lg (8 oz, w/bone, raw) (yield after cooking, bone and fat removed) (104 gm)	33	29	26	24	22	21	19	18	212	93	39.9	95
Shoulder, cooked, lean & fat—4 oz, boneless, ckd (113 gm)	74	64	57	51	47	42	39	36	381	276	129.1	108

	1200	1500	1800	2100	2400	2800	3200	3600	TOT CAL	FAT CAL	S/FAT CAL	CHOL MG
Lamb (cont.)												
Shoulder, cooked, lean only—4 oz, w/bone, cooked (yield after bone & fat removed) (63 gm)	20	18	16	15	14	13	12	11	128	56	24.2	58
Pork												
Chop, battered, fried, lean & fat—1 sm or thin cut (3 oz, w/bone, raw) (yield after cooking, bone removed) (60 gm)	32	27	24	22	20	18	17	16	221	156	54.0	50
—1 med (5.5 oz, w/bone, raw) (yield after cooking, bone removed) (110 gm)	58	50	45	41	37	34	31	29	404	286	99.0	92
—1 lg (8 oz, w/bone, raw) (yield after cooking, bone removed) (159 gm)	84	73	65	59	54	49	45	41	584	414	143.1	134
Chop, battered, fried, lean only—1 sm or thin cut (3 oz, w/bone, raw) (yield after cooking, bone & fat removed) (45 gm)	16	14	12	11	11	10	9	8	126	70	22.0	38
—1 med (5.5 oz, w/bone, raw) (yield after cooking, bone & fat removed) (84 gm)	29	26	23	21	20	18	17	15	235	130	41.2	70
—1 lg (8 oz, w/bone, raw) (yield after cooking, bone & fat removed) (122 gm)	42	37	34	31	28	26	24	22	341	189	59.8	102
Chop, breaded, broiled or baked, lean & fat (inc Shake-n-Bake)—1 sm or thin cut (3 oz, with bone, raw) (yield after cooking, bone removed) (55 gm)	27	24	21	19	18	16	15	14	194	129	45.6	46

Food												
—1 med (5.5 oz, w/bone, raw) (yield after cooking, bone removed) (100 gm)	50	43	38	35	32	29	27	25	352	234	82.8	84
—1 lg (8 oz, w/bone, raw) (yield after cooking, bone removed) (146 gm)	73	63	56	51	47	42	39	36	514	342	121.0	123
Chop, breaded, broiled or baked, lean only (inc Shake-n-Bake)—1 sm or thin cut (3 oz, w/bone, raw) (yield after cooking, bone & fat removed) (43 gm)	15	13	12	11	10	9	8	8	117	59	19.8	37
—1 med (5.5 oz, w/bone, raw) (yield after cooking, bone & fat removed) (80 gm)	27	24	21	20	18	17	15	14	218	110	36.8	68
—1 lg (8 oz, w/bone, raw) (yield after cooking, bone & fat removed) (116 gm)	39	34	31	28	26	24	22	21	316	159	53.3	99
Chop, breaded, fried, lean & fat—1 sm or thin cut (3 oz, w/bone, raw) (yield after cooking, bone removed) (60 gm)	33	29	26	24	22	20	18	17	230	158	52.7	64
—1 med (5.5 oz, w/bone, raw) (yield after cooking, bone removed) (110 gm)	61	53	48	43	40	36	33	31	422	290	96.7	117
—1 lg (8 oz, w/bone, raw) (yield after cooking, bone removed) (159 gm)	88	77	69	62	57	52	48	45	610	419	139.7	168
Chop, breaded, fried, lean only—1 sm or thin cut (3 oz, w/bone, raw) (yield after cooking, bone & fat removed) (45 gm)	16	15	13	12	11	10	10	9	131	68	20.2	48
—1 med (5.5 oz, w/bone, raw) (yield after cooking, bone & fat removed) (84 gm)	31	27	25	23	21	19	18	17	244	126	37.6	90

Pork (cont.)

	1200	1500	1800	2100	2400	2800	3200	3600	TOT CAL	FAT CAL	S/FAT CAL	CHOL MG
—1 lg (8 oz, w/bone, raw) (yield after cooking, bone & fat removed) (122 gm)	44	39	36	33	31	28	26	25	354	183	54.6	130
Chop, broiled or baked, lean & fat (inc floured)—1 sm or thin cut (3 oz, w/bone, raw) (yield after cooking, bone removed) (46 gm)	25	21	19	17	16	14	13	12	158	112	40.5	43
—1 med (5.5 oz, w/bone, raw) (yield after cooking, bone removed) (85 gm)	45	39	35	32	29	27	24	23	292	207	74.7	79
—1 lg (8 oz, w/bone, raw) (yield after cooking, bone removed) (123 gm)	66	57	51	46	42	39	35	33	423	300	108.2	115
Chop, broiled or baked, lean only (inc floured)—1 sm or thin cut (3 oz, w/bone, raw) (yield after cooking, bone & fat removed (37 gm)	13	12	11	10	9	8	8	7	95	51	17.4	35
—1 med (5.5 oz, w/bone, raw) (yield after cooking, bone & fat removed) (67 gm)	24	21	19	17	16	15	14	13	171	92	31.6	63
—1 lg (8 oz, w/bone, raw) (yield after cooking, bone & fat removed) (98 gm)	35	31	28	26	24	22	20	19	250	134	46.2	93
Chop, fried, lean & fat (inc floured)—1 sm or thin cut (3 oz, w/bone, raw) (yield after cooking, bone removed) (50 gm)	30	26	23	21	19	17	16	15	190	142	51.3	46

Food											
—1 med (5.5 oz, w/bone, raw) (yield after cooking, bone removed) (92 gm)	55	48	42	38	35	32	29	27	350	261	94.3	86
—1 lg (8 oz, w/bone, raw) (yield after cooking, bone removed) (134 gm)	80	70	62	56	51	46	42	39	509	381	137.4	125
Chop, fried, lean only (inc floured)—1 sm or thin cut (3 oz, w/bone, raw) (yield after cooking, bone & fat removed) (38 gm)	14	12	11	10	9	9	8	7	99	53	18.2	36
—1 med (5.5 oz, w/bone, raw) (yield after cooking, bone & fat removed) (70 gm)	25	22	20	18	17	16	14	13	182	98	33.5	65
—1 lg (8 oz, w/bone, raw) (yield after cooking, bone & fat removed) (102 gm)	37	32	29	27	25	23	21	20	265	142	48.9	95
Chop, smoked or cured, cooked, lean & fat—1 sm or thin cut (3 oz, w/bone, raw) (yield after cooking, bone removed) (46 gm)	18	16	14	13	12	10	10	9	129	88	31.8	27
—1 med (5.5 oz, w/bone, raw) (yield after cooking, bone removed) (84 gm)	33	29	26	23	21	19	17	16	235	161	58.0	49
—1 lg (8 oz, w/bone, raw) (yield after cooking, bone removed) (123 gm)	49	42	37	34	31	28	26	24	344	236	84.9	71
Chop, smoked or cured, cooked, lean—1 sm or thin cut (3 oz, w/bone, raw) (yield after cooking, bone & fat removed) (37 gm)	6	6	5	5	4	4	4	3	63	23	7.9	18
—1 med (5.5 oz, w/bone, raw) (yield after cooking, bone & fat removed) (67 gm)	11	10	9	8	8	7	7	6	114	42	14.2	32
—1 lg (8 oz, w/bone, raw) (yield after cooking, bone & fat removed) (98 gm)	17	15	13	12	11	10	10	9	167	62	20.8	47

Pork (cont.)

	1200	1500	1800	2100	2400	2800	3200	3600	TOT CAL	FAT CAL	S/FAT CAL	CHOL MG
Chop, stewed, lean & fat—1 sm or thin cut (3 oz, w/bone, raw) (yield after cooking, bone removed) (42 gm)	24	21	18	17	15	14	13	12	155	105	38.1	43
—1 med (5.5 oz, w/bone, raw) (yield after cooking, bone removed) (77 gm)	43	38	34	31	28	25	23	22	283	193	69.9	79
—1 lg (8 oz, w/bone, raw) (yield after cooking, bone removed) (112 gm)	63	55	49	44	41	37	34	32	412	281	101.7	114
Chop, stewed, lean only—1 sm or thin cut (3 oz, w/bone, raw) (yield after cooking, bone & fat removed) (36 gm)	13	12	11	10	9	8	8	7	98	47	16.3	38
—1 med (5.5 oz, w/bone, raw) (yield after cooking, bone & fat removed) (66 gm)	24	21	19	18	16	15	14	13	180	87	29.9	69
—1 lg (8 oz, w/bone, raw) (yield after cooking, bone & fat removed) (96 gm)	35	31	28	26	24	22	21	19	262	126	43.5	101
Cracklings, cooked—1 c (91 gm)	75	65	57	51	47	42	38	35	524	403	142.7	77
Cured, blade roll, boneless, lean & fat, roasted—4 oz, yield/3.6 oz unheated (113 gm)	50	43	38	35	32	29	26	24	325	239	85.4	76
Cured, blade roll, boneless, lean & fat, unheated—4 oz (113 gm)	45	39	34	31	28	26	23	22	304	223	80.8	60
Cured, ham, boneless, extra lean (approximately 5% fat), **roasted**—4 oz, yield/3.4 oz unheated (113 gm)	18	16	14	13	12	11	11	10	164	56	18.5	60

Food											
Cured, ham, boneless, extra lean (approximately 5% fat), unheated—1 sl (6¼" x 4") (28 gm)	4	3	3	3	3	3	2	2	37	13	4.1	13
Cured, ham, boneless, extra lean & reg, unheated—1 sl (6¼" x 4") (28 gm)	5	5	4	4	4	3	3	3	46	22	6.9	15
Cured, ham, boneless, reg (approximately 11% fat), unheated—1 sl (6¼" x 4") (28 gm)	6	6	5	5	4	4	4	3	52	27	8.6	16
Cured, ham, canned, extra lean (approximately 4% fat), roasted—4 oz, yield 3.4 oz unheated (113 gm)	12	11	10	9	8	8	7	7	155	49	16.3	33
Cured, ham, canned, reg (approximately 13% fat), roasted—4 oz, yield/3.6 oz unheated (113 gm)	34	29	26	24	22	20	18	17	256	155	51.3	69
Cured, ham, center slice, country style, lean only, raw—4 oz (113 gm)	25	22	20	19	17	16	15	14	220	85	28.4	79
Cured, ham, center slice, lean & fat, unheated—4 oz (113 gm)	30	26	24	21	20	18	17	15	229	131	46.5	61
Cured, ham patties, grilled—1 patty, yield fr unheated (60 gm)	33	28	25	23	21	19	17	16	203	166	59.4	43
Cured, shoulder, arm picnic, lean & fat, roasted—4 oz (113 gm)	45	39	34	31	28	26	23	22	317	217	78.2	65
Cured, shoulder, arm picnic, lean only, roasted—4 oz (113 gm)	19	17	15	14	13	12	11	11	193	72	24.1	55
Dehydrated, Oriental style—1 c (22 gm)	23	20	17	15	14	12	11	10	135	124	46.4	15
Ears, frozen, simmered—1 ear, yield from raw (111 gm)	32	29	26	24	22	21	19	18	183	107	38.3	99
Ground or patty—4 oz, ckd (113 gm)	58	51	45	41	38	34	32	29	366	259	93.5	108
Ground or patty, breaded—4 oz, ckd (113 gm)	42	37	33	29	27	24	22	21	412	268	73.0	64
Leg (ham), whole, lean & fat, roasted—4 oz (113 gm)	50	44	39	36	33	30	28	26	333	211	76.5	105
Leg (ham), whole, lean only, roasted—4 oz (113 gm)	34	30	27	25	24	22	20	19	249	113	38.8	107

Pork (cont.)

	1200	1500	1800	2100	2400	2800	3200	3600	TOT CAL	FAT CAL	S/FAT CAL	CHOL MG
Loin, center rib, lean & fat, braised—1 chop (67 gm)	36	31	28	25	23	21	19	18	246	164	59.2	64
Loin, center rib, lean & fat, broiled—1 chop (77 gm)	41	35	31	28	26	24	22	20	264	183	66.0	72
Loin, center rib, lean & fat, pan-fried—1 chop (88 gm)	53	46	41	37	34	30	28	26	343	261	94.4	74
Loin, center rib, lean and fat, roasted—1 chop (79 gm)	37	32	29	26	24	22	20	18	252	167	60.6	64
Loin, center rib, lean only, braised—1 chop (53 gm)	19	16	15	14	13	12	11	10	147	68	23.8	51
Loin, center rib, lean only, broiled—1 chop (63 gm)	22	20	18	16	15	14	13	12	162	85	29.2	59
Loin, center rib, lean only, pan-fried—1 chop (62 gm)	21	18	16	15	14	13	12	11	160	86	29.3	50
Loin, center rib, lean only, roasted—1 chop (66 gm)	21	18	16	15	14	13	12	11	162	82	28.3	52
Loin, tenderloin, lean only, roasted—4 oz (113 gm)	24	22	20	19	18	17	16	15	188	49	16.9	105
Neck bones, cooked (inc pork backbone)—1 c, w/bone (yield after bone removed) (56 gm)	14	13	12	11	10	9	9	8	116	48	16.5	45
Pickled, cut unspecified—4 oz, boneless (113 gm)	37	33	29	27	25	23	22	20	264	133	45.8	105
Pig's feet, cooked—1 foot, w/bone (yield after bone removed) (87 gm)	28	25	23	21	20	18	17	16	168	97	33.3	86
Pig's feet, pickled—1 foot, w/bone (yield after bone removed) (87 gm)	32	28	25	23	21	20	18	17	177	126	43.6	80

Food											
Pig's hocks, cooked (inc ham hocks)—1 ham hock, w/bone, ckd (yield after bone removed) (51 gm)	27	24	21	19	18	16	15	14	175	116	42.3	55
Roast, loin, cooked, lean & fat—4 oz, boneless, cooked (113 gm)	55	48	43	39	36	32	30	28	359	246	89.0	101
Roast, loin, cooked, lean only—4 oz, boneless, ckd (yield after fat removed) (100 gm)	33	29	26	24	22	21	19	18	239	124	42.8	89
Roast, shoulder, cooked, lean & fat—4 oz, boneless, ckd (113 gm)	58	51	45	41	38	34	32	29	367	260	93.9	108
Roast, shoulder, cooked, lean only—4 oz, boneless, ckd (yield after fat removed) (96 gm)	34	30	27	25	23	21	20	19	233	129	44.4	93
Roast, smoked or cured, cooked, lean & fat—4 oz, boneless, ckd (113 gm)	25	22	20	18	17	15	14	13	202	92	31.8	67
Roast, smoked or cured, cooked, lean only—4 oz, boneless, ckd (yield after fat removed) (92 gm)	14	13	12	11	10	9	9	8	133	46	15.0	49
Roll, cured, fried—1 sl (1 oz) (28 gm)	16	14	12	11	10	9	8	8	99	82	28.8	20
Skin, boiled—3 sl (33 gm)	41	35	30	27	25	22	20	18	231	222	80.8	30
Skin, rinds, deep-fried (inc snack type)—1 c (32 gm)	19	16	14	13	12	11	10	9	172	88	31.4	31
Spareribs, cooked (inc flank end, ribs)—1 med cut (105 gm)	69	60	53	48	44	40	37	35	414	285	110.5	126
Steak or cutlet, battered, fried, lean & fat—4 oz, boneless, ckd (113 gm)	59	51	46	41	38	35	32	29	406	285	97.0	103
Steak or cutlet, battered, fried, lean only—4 oz, boneless, ckd (yield after fat removed) (96 gm)	36	32	29	26	24	22	21	19	275	159	50.7	89
Steak or cutlet, breaded, broiled or baked, lean & fat—4 oz, boneless, ckd (113 gm)	57	49	44	40	37	33	31	28	394	262	92.5	102
Steak or cutlet, breaded, broiled or baked, lean only—4 oz, boneless, ckd, (yield after fat removed) (96 gm)	34	30	27	25	23	21	19	18	263	137	46.0	87

	1200	1500	1800	2100	2400	2800	3200	3600	TOT CAL	FAT CAL	S/FAT CAL	CHOL MG
Pork (cont.)												
Steak or cutlet, broiled or baked, lean & fat—4 oz, boneless, ckd (113 gm)	61	53	47	43	40	36	33	31	384	272	98.1	113
Steak or cutlet, broiled or baked, lean only—4 oz, boneless, ckd (yield after fat removed) (96 gm)	36	32	29	26	25	23	21	20	247	137	47.4	97
Steak or cutlet, fried, lean & fat (inc floured)—4 oz, boneless, ckd (113 gm)	68	59	52	47	43	39	36	34	422	309	111.5	116
Steak or cutlet, fried, lean only (inc floured)—4 oz, boneless, ckd (yield after fat removed) (96 gm)	37	32	29	27	25	23	21	20	254	137	46.9	102
Tail, simmered—4 oz (113 gm)	79	69	62	56	51	47	43	40	448	365	126.9	147
Tenderloin, braised—4 oz, boneless, ckd (113 gm)	41	37	33	31	28	26	24	23	309	149	51.4	119
Tenderloin, breaded, fried—4 oz, boneless, ckd (113 gm)	26	24	22	20	19	18	17	16	246	98	20.8	108
Pork Products												
Bacon, formed, lean meat added, cooked (inc breakfast strips, Sizzlean)—1 strip (11 gm)	7	6	6	5	5	4	4	4	50	36	12.6	12
Bacon or side pork, fresh, cooked—1 sl (¼" thick) (26 gm)	17	15	13	12	11	10	9	9	103	71	27.5	31
Bacon, smoked or cured, cooked—2 med sl (yield after cooking) (16 gm)	13	11	10	9	8	7	7	6	92	71	25.1	14

Food												
Bacon, smoked or cured, cooked, lean only—2 med sl (yield after cooking, fat removed) (12 gm)	2	2	2	2	2		1	1	22	9	3.1	7
Bacon, smoked or cured, lower sodium—2 med sl (yield after cooking) (16 gm)	13	11	10	9	8	7	7	6	92	71	25.1	14
Canadian bacon, cooked—1 sl (23 gm)	5	4	4	3	3	3		3	43	17	5.9	13
Ham, breaded, fried (inc smoked, cured, canned, chicken fried ham)—1 med sl (approx 4½" x 2½" x ¼") (42 gm)	11	10	9	8	8	7	6	6	97	52	14.7	30
Ham croquette—1 croquette (1" dia x 3") (65 gm)	23	20	18	16	15	14	12	12	163	88	35.0	46
Ham, fried (inc smoked, cured, canned)—1 med sl (approx 4½" x 2½" x ¼") (42 gm)	10	9	8	7	7	6	6	5	94	57	15.8	22
Ham, prosciutto—1 oz, boneless (28 gm)	6	6	5	5	4	4	4	4	55	21	7.0	20
Ham, smoked or cured, ground patty—4 oz, ckd (113 gm)	63	54	48	43	40	36	33	30	388	315	113.1	82
Salt pork or fat back, cooked (inc hog jowl, boiled)—1 sl (2¼" x 1¾" x ¼") (26 gm)	32	27	24	21	19	17	15	14	186	178	63.4	22

Veal

Food												
Chop, cooked, lean & fat—1 sm chop (4.75 oz, w/bone, raw) (yield after cooking, bone removed) (78 gm)	30	26	24	22	20	19	17	16	181	94	39.8	78
—1 med (6.5 oz, w/bone, raw) (yield after cooking, bone removed) (107 gm)	41	36	33	30	28	25	24	22	249	128	54.7	107
—1 lg (8.25 oz, w/bone, raw) (yield after cooking, bone removed) (136 gm)	52	46	42	38	35	32	30	28	316	163	69.5	137

THE 2-IN-1 SYSTEM

	1200	1500	1800	2100	2400	2800	3200	3600	TOT CAL	FAT CAL	S/FAT CAL	CHOL MG
Veal (cont.)												
Chop, cooked, lean only—1 sm chop (4.75 oz, w/bone, raw) (yield after cooking, bone & fat removed) (62 gm)	24	22	20	19	18	17	16	15	129	43	18.1	105
—1 med (6.5 oz, w/bone, raw) (yield after cooking, bone & fat removed) (85 gm)	33	30	28	26	25	23	22	21	177	59	24.8	144
—1 lg (8.25 oz, w/bone, raw) (yield after cooking, bone & fat removed) (108 gm)	43	39	36	33	31	29	28	26	225	75	31.5	183
Chop, fried, lean & fat (inc breaded)—1 sm (4.75 oz, w/bone, raw) (yield after cooking, bone removed) (78 gm)	32	28	25	23	21	20	18	17	205	122	46.3	75
—1 sm, breaded (4.75 oz, w/bone, raw) (yield after cooking, bone removed) (93 gm)	38	34	30	28	26	23	22	20	245	146	55.2	89
—1 med (6.5 oz, w/bone, raw) (yield after cooking, bone removed) (107 gm)	44	39	35	32	29	27	25	23	282	168	63.5	102
—1 med, breaded (6.5 oz, w/bone, raw) (yield after cooking, bone removed) (127 gm)	52	46	41	38	35	32	29	27	335	199	75.3	121
—1 lg (8.25 oz, w/bone, raw) (yield after cooking, bone removed) (136 gm)	56	49	44	40	37	34	32	29	358	213	80.7	130
—1 lg, breaded (8.25 oz, w/bone, raw) (yield after cooking, bone removed) (161 gm)	66	58	52	48	44	40	37	35	424	252	95.5	154

Food												
Chop, fried, lean only (inc breaded)—1 sm (4.75 oz, w/bone, raw) (yield after cooking, bone & fat removed) (62 gm)	24	22	20	19	18	17	16	15	129	43	18.1	105
—1 sm, breaded (4.75 oz, w/bone, raw) (yield after cooking, bone & fat removed) (74 gm)	29	26	24	23	22	20	19	18	154	52	21.6	125
—1 med (6.5 oz, w/bone, raw) (yield after cooking, bone & fat removed) (85 gm)	33	30	28	26	25	23	22	21	177	59	24.8	144
—1 med, breaded (6.5 oz, w/bone, raw) (yield after cooking, bone & fat removed) (101 gm)	40	36	33	31	29	27	26	25	211	70	29.5	171
—1 lg (8.25 oz, w/bone, raw) (yield after cooking, bone & fat removed) (108 gm)	43	39	36	33	31	29	28	26	225	75	31.5	183
—1 lg, breaded (8.25 oz, w/bone, raw) (yield after cooking, bone and fat removed) (128 gm)	50	46	42	39	37	35	33	31	267	89	37.3	216
Chuck, med fat, braised—8.4 oz (240 gm)	97	85	77	70	65	60	55	52	564	276	132.7	242
Cutlet or steak, cooked, lean & fat—4 oz, boneless, ckd (113 gm)	39	34	31	29	27	25	23	22	243	113	47.4	114
Cutlet or steak, cooked, lean only—4 oz, boneless, ckd (yield after fat removed) (104 gm)	30	27	25	24	23	21	20	19	187	42	17.4	142
Cutlet or steak, fried, lean & fat (inc breaded or floured)—4 oz, boneless, ckd (113 gm)	42	37	34	31	29	26	24	23	279	155	57.3	108
Cutlet or steak, fried, lean only (inc breaded or floured)—4 oz, boneless, ckd (yield after fat removed) (104 gm)	30	27	25	24	23	21	20	19	187	42	17.4	142
Ground or patty, cooked (inc fried)—1 med patty (4 oz, raw, 4/lb) (yield after cooking) (67 gm)	26	23	20	19	17	16	15	14	156	80	34.2	67

341

	1200	1500	1800	2100	2400	2800	3200	3600	TOT CAL	FAT CAL	S/FAT CAL	CHOL MG
Veal (cont.)												
Loin, med fat, broiled—9.5 oz (269 gm)	111	98	88	81	75	68	63	59	629	324	155.7	272
Mock chicken legs, cooked (inc city chicken, pork & veal on a stick)—4 oz, ckd (113 gm)	46	40	36	33	30	27	25	23	328	209	73.1	89
Patty, breaded (inc fried)—1 med patty (4 oz, raw, 4/lb) (yield after cooking) (79 gm)	32	29	26	24	22	20	18	17	223	127	45.5	79
Roasted, lean only—4 oz, boneless, ckd (yield after fat removed) (104 gm)	32	29	27	25	24	22	21	20	164	52	21.9	140
Round w/rump, med fat, broiled—8.7 oz (247 gm)	91	81	73	67	62	57	53	50	534	247	118.4	249
Organ Meats												
Beef brain, pan-fried—4 oz (113 gm)	369	346	327	312	299	284	271	261	223	162	38.2	2261
Beef, brains and eggs, cooked—½ c (121 gm)	249	232	219	209	200	190	181	174	165	107	35.3	1494
Beef brain, simmered—4 oz (113 gm)	376	352	334	318	305	290	277	267	181	127	29.8	2327
Beef heart, simmered—4 oz (113 gm)	42	39	36	34	32	31	29	28	199	58	17.2	219
Beef kidneys, simmered—4 oz (113 gm)	73	69	65	62	59	56	53	51	163	35	11.2	439
Beef liver, battered, fried—4 oz, ckd (113 gm)	77	71	67	63	60	57	54	52	262	87	28.3	415
Beef liver, braised—1 med sl (6½" x 2⅜" x ⅜") (85 gm)	66	62	58	55	53	50	48	46	177	43	17.0	378
Beef liver, breaded, fried—1 med sl (6½" x 2⅜" x ⅜") (101 gm)	74	68	64	61	58	55	52	50	244	91	24.8	403
Beef liver, fried or broiled, no coating—1 med sl (6½" x 2⅜" x ⅜") (85 gm)	74	68	64	61	58	55	52	50	184	61	21.6	410

Food												
Beef lungs, braised—4 oz (113 gm)	55	51	48	46	44	41	39	38	136	37	13.0	315
Beef pancreas, braised—4 oz (113 gm)	73	66	61	57	53	50	47	44	307	175	60.1	297
Beef spleen, braised—4 oz (113 gm)	68	63	59	56	54	51	49	47	164	43	14.2	393
Beef thymus, braised—4 oz (113 gm)	91	82	75	69	65	60	57	54	361	254	87.8	333
Beef tongue, simmered—4 oz (113 gm)	59	52	46	42	39	35	33	30	321	211	91.1	121
Calf tongue, braised—1 sl (3" x 2" x ⅛") (20 gm)	6	5	5	4	4	4	3	3	32	11	5.4	20
Calf heart, braised—1 c chopped or diced (145 gm)	91	83	76	71	67	63	60	57	302	119	65.3	397
Calf liver, braised—4 oz, ckd (113 gm)	94	88	83	79	75	72	68	66	198	49	14.1	562
Calf liver, breaded, fried—4 oz, ckd (113 gm)	68	63	59	55	52	49	47	45	306	126	34.4	341
Calf liver, fried—4 oz sl (113 gm)	91	84	79	75	71	67	64	61	296	134	30.6	496
Calf liver, fried or broiled, no coating—1 med sl (6½" x 2⅜" x ⅜") (85 gm)	56	52	48	45	43	40	38	37	222	101	27.6	281
Heart, boiled or simmered (inc pork, veal, lamb, venison)—4 oz, ckd (113 gm)	56	52	49	46	44	42	40	38	212	58	18.2	309
Heart, braised (inc beef, veal, venison)—4 oz, ckd (113 gm)	56	52	49	46	44	42	40	38	212	58	18.2	309
Heart, fried (inc pork, veal, lamb, venison)—4 oz, ckd (113 gm)	51	47	43	41	39	37	35	33	292	123	25.8	252
Hog lights (lungs), cooked—1 oz, ckd (28 gm)	18	17	16	15	14	14	13	13	28	8	2.7	108
Hog maws (stomach), cooked—½ c (70 gm)	48	44	41	38	36	34	32	30	173	95	33.8	213
Kidney, braised (inc beef, veal, lamb, venison)—½ kidney, ckd (85 gm)	119	111	104	99	94	89	85	82	213	91	30.0	679
Kidney, breaded, fried floured, fried (inc beef, pork, veal, lamb, venison)—½ kidney, ckd (112 gm)	97	90	84	80	76	72	69	66	237	100	25.7	546
Kidney, broiled or simmered (inc pork, veal, lamb, venison)—½ kidney, ckd (85 gm)	119	111	104	99	94	89	85	82	213	91	30.0	679

	1200	1500	1800	2100	2400	2800	3200	3600	TOT CAL	FAT CAL	S/FAT CAL	CHOL MG
Organ Meats (cont.)												
Lamb heart, braised—1 c (145 gm)	114	102	93	86	81	75	70	66	377	188	117.4	397
Lamb liver, cooked (inc mutton)—1 oz, ckd (28 gm)	32	30	28	27	26	24	23	22	73	31	8.6	183
Lamb sweetbreads (thymus), braised—4 oz (113 gm)	91	85	80	76	73	69	66	63	199	62	20.4	528
Lamb tongue, braised—1 sl (3" x 2" x ⅙") (20 gm)	11	10	9	8	7	7	6	6	51	32	18.0	20
Pork brains, braised—4 oz (113 gm)	460	432	409	391	375	357	342	328	156	97	21.8	2891
Pork chitterlings (intestines), simmered—4 oz (113 gm)	71	62	56	51	47	43	40	37	344	293	103.1	163
Pork heart, braised—1 heart (129 gm)	51	48	45	42	40	38	36	35	191	59	15.5	285
Pork kidneys, braised—4 oz (113 gm)	92	86	81	77	73	70	67	64	171	48	15.4	544
Pork liver, braised—4 oz (113 gm)	69	64	60	57	55	52	50	48	186	45	14.3	400
Pork liver, breaded, fried—4 oz (113 gm)	81	75	70	66	63	60	57	55	270	89	25.5	444
Pork tongue, braised—4 oz (113 gm)	55	49	44	41	38	35	33	31	307	190	65.7	165
Sheep tongue, braised—1 sl (3" x 2" x ⅙") (20 gm)	14	12	11	10	9	8	7	7	65	46	25.2	20
Sweetbreads—4 oz, ckd (113 gm)	86	81	76	73	70	66	63	61	189	32	10.1	525
Tongue, braised (inc beef, veal)—1 oz, ckd (28 gm)	12	11	10	9	8	7	7	6	68	42	18.0	28
Tongue, deviled (inc beef, pork, lamb veal)—1 c (225 gm)	129	113	101	92	84	77	70	66	653	466	204.5	248
Tongue, smoked, cured, or pickled, cooked (inc beef, pork, lamb, veal)—1 oz, ckd (28 gm)	13	11	10	9	8	8	7	7	75	51	22.0	22
Tripe, battered, fried—1 oz, raw (28 gm)	7	6	6	5	5	4	4	4	43	19	6.3	25
Tripe, beef—1 oz, ckd (28 gm)	5	5	5	4	4	4	4	4	28	5	2.8	27

MILK and MILK PRODUCTS
Cream (Sweet and Sour)

Food												
Cream, half-and-half—1 ind container (0.5 fl oz) (15 gm)	5	5	4	4	3	3	3	2	20	16	9.7	6
Cream, heavy—1 tbsp (15 gm)	17	15	13	12	11	10	9	8	52	50	31.1	21
Cream, heavy, whipped, sweetened—1 c (119 gm)	126	109	96	86	79	71	65	60	413	371	231.1	153
Cream, heavy, whipped, unsweetened—1 c (119 gm)	135	116	102	92	84	76	69	64	410	396	246.7	163
Cream, light (inc coffee cream, table cream)—1 tbsp (15 gm)	9	8	7	6	5	5	4	4	29	26	16.2	10
Cream, light, powdered (inc coffee lightener, Instant Cream Powder, lo-cal)—1 pkt (2 gm)	4	3	3	2	2	2	2	2	13	10	6.5	4
Cream, light, whipped, unsweetened (inc coffee cream, table cream)—1 c (120 gm)	113	97	86	77	71	64	58	53	351	334	208.8	133
Cream, medium, 25% fat—1 tbsp (15 gm)	11	10	9	8	7	6	6	5	37	34	21.0	13
Cream, whipped, pressurized container (inc Reddiwhip, Fashion Whip, Quip)—1 c (60 gm)	40	35	31	27	25	23	21	19	154	120	74.7	46
Dip, sour cream base (inc buttermilk type, onion dip)—2 tbsp (31 gm)	17	15	13	11	10	9	8	8	68	55	33.9	13
Sour cream—2 tbsp (29 gm)	17	15	13	11	10	8	8	8	62	55	34.1	13
Sour cream, sour dressing, nonbutterfat (inc King Sour, Zest)—2 tbsp (29 gm)	16	13	11	10	9	8	7	7	52	43	34.6	2
Sour cream, half-and-half—2 tbsp (30 gm)	11	9	8	7	7	6	6	5	40	32	20.2	12
Sour cream, imitation, nondairy (inc Imo, Zero)—2 tbsp (29 gm)	21	17	15	13	12	11	9	9	60	51	46.4	0

345

Ice Cream (Frozen Desserts)

	1200	1500	1800	2100	2400	2800	3200	3600	TOT CAL	FAT CAL	S/FAT CAL	CHOL MG
Baked Alaska—1 piece (⅛ of whole) (103 gm)	29	25	23	21	19	18	16	15	256	90	40.2	70
Ice cream bar, cake-covered—1 bar (59 gm)	16	14	12	11	11	10	9	8	162	54	23.1	36
Ice cream bar or stick, chocolate-covered—1 bar (3 fl oz) (56 gm)	41	35	31	27	25	22	20	18	172	112	86.3	20
Ice cream bar or stick, not chocolate or cake-covered—1 bar (3 fl oz) (56 gm)	19	16	14	13	12	11	10	9	113	54	33.8	25
Ice cream, chocolate—1 dixie cup (3.5 fl oz) (60 gm)	17	14	13	11	10	9	8	8	123	52	32.4	15
—1 med dip or sl (67 gm)	19	16	14	13	11	10	9	9	137	58	36.2	17
—1 c (133 gm)	37	32	28	25	23	20	18	17	272	116	71.9	33
Ice cream cone, chocolate-covered or dipped, chocolate ice cream—1 cone (77 gm)	24	20	18	16	14	13	12	11	187	79	47.9	16
Ice cream cone, chocolate-covered or dipped, flavors other than chocolate—1 cone & single dip (78 gm)	28	24	21	19	17	15	14	13	188	86	52.4	29
—1 cone & double dip (143 gm)	51	44	39	35	32	28	26	24	344	158	96.0	54
Ice cream cone, chocolate-covered, w/nuts, chocolate ice cream—1 cone (78 gm)	23	20	17	16	14	13	11	10	219	113	47.5	15
Ice cream cone, chocolate- covered, w/nuts, flavors other than chocolate (inc Nutty Buddy)—1 cone & single dip (78 gm)	27	23	20	18	16	15	13	12	217	119	50.9	26
—1 cone & double dip (143 gm)	49	42	37	33	30	27	25	23	398	217	93.4	49
Ice cream cone, no topping, chocolate ice cream—1 cone (78 gm)	20	17	15	14	12	11	10	9	170	64	39.4	18

Food												
Ice cream cone, no topping, flavors other than chocolate—1 cone & single dip (78 gm)	24	21	19	17	15	14	13	12	168	71	43.6	32
Ice cream cone w/nuts, chocolate ice cream—1 cone (60 gm)	15	13	11	10	9	8	8	7	157	78	30.2	12
Ice cream cone w/nuts, flavors other than chocolate—1 cone (60 gm)	18	16	14	12	11	10	9	9	156	83	33.1	22
Ice cream cookie sandwich (inc Chipwich)—1 sandwich (59 gm)	16	14	12	11	10	9	8	8	144	50	28.9	20
Ice cream, flavors other than chocolate—1 dixie cup (3.5 fl oz) (60 gm)	20	17	15	14	13	11	11	10	121	58	36.2	27
—1 med dip or sl (67 gm)	23	19	17	16	14	13	12	11	136	65	40.4	30
—1 cup (133 gm)	45	39	34	31	28	25	23	21	269	129	80.2	59
Ice cream, French vanilla, soft serve—1 c (173 gm)	78	68	61	55	51	46	43	40	377	203	121.6	153
Ice cream, fried—1 c (133 gm)	43	37	33	29	27	24	22	20	359	164	81.4	45
Ice cream, imitation, chocolate, w/vegetable fat (inc Mellorine)—1 c (133 gm)	47	40	35	31	27	24	22	20	268	126	107.0	0
Ice cream, imitation, flavors other than chocolate, w/vegetable fat (inc Mellorine)—1 c (133 gm)	47	39	34	30	27	24	21	19	265	122	105.6	0
Ice cream pie (no crust)—1 c (142 gm)	63	54	48	43	39	35	32	30	314	181	112.8	81
Ice cream sandwich (5" x 1¾" x ¾") (59 gm)	16	14	12	11	10	9	8	8	144	50	28.9	20
Ice cream soda, chocolate—10 fl oz (248 gm)	30	26	23	20	18	17	15	14	229	91	56.7	30
Ice cream soda, flavors other than chocolate—10 fl oz (248 gm)	21	18	16	15	13	12	11	10	210	61	38.2	28
Ice cream, soft serve, chocolate—1 c (173 gm)	48	41	36	32	30	26	24	22	354	151	93.5	43
Ice cream, soft serve, flavors other than chocolate—1 cup (173 gm)	58	50	44	40	37	33	30	28	350	168	104.4	77
Ice cream sundae, chocolate or fudge topping, w/whipped cream—1 sundae (165 gm)	75	65	57	51	47	42	38	35	413	227	141.2	82

Ice Cream (Frozen Desserts) (cont.)

	1200	1500	1800	2100	2400	2800	3200	3600	TOT CAL	FAT CAL	S/FAT CAL	CHOL MG
Ice cream sundae, fruit topping, w/whipped cream (inc banana split, fruit parfait)—1 sundae (165 gm)	64	55	49	44	40	36	33	30	326	186	115.9	80
—1 banana split (425 gm)	165	142	125	113	103	93	85	78	841	479	298.5	206
Ice cream sundae, fudge topping, w/cake, w/whipped cream—1 sundae (175 gm)	74	63	56	51	46	42	38	35	472	236	133.1	94
Ice cream sundae, not fruit or chocolate topping, w/whipped cream (inc peanut butter parfait, butterscotch topping)—1 sundae (165 gm)	43	37	32	29	26	23	21	19	536	288	86.9	30
Ice cream sundae, prepackaged type, flavors other than chocolate—1 sundae (3.5 fl oz) (65 gm)	11	9	8	7	7	6	5	5	120	35	21.9	8
Ice cream, vanilla, regular, 10% fat—1 c (133 gm)	45	39	34	31	28	25	23	21	269	129	80.3	59
Ice cream, vanilla, rich, 16% fat—1 c (148 gm)	73	62	55	50	45	41	37	34	349	213	132.7	88
Ice cream, w/sherbet—1 c (163 gm)	28	24	22	19	18	16	15	14	270	82	50.8	37
Ice milk bar or stick, chocolate-coated—1 bar (3 fl oz) (56 gm)	29	24	21	19	17	15	13	12	142	78	62.8	6
Ice milk, chocolate—1 c (131 gm)	12	10	9	8	8	7	6	6	189	38	23.4	12
Ice milk cone, chocolate—1 cone & single dip (78 gm)	7	6	5	5	4	4	3	3	127	22	13.1	7
—1 cone & double dip (143 gm)	13	11	9	8	8	7	6	6	232	41	24.1	12
Ice milk cone, flavors other than chocolate—1 cone & single dip (78 gm)	6	5	4	4	3	3	3	3	115	18	10.8	5

Food	1	2	3	4	5	6	7	8	9	10	11	12
—1 cone & double dip (143 gm)	10	9	8	7	6	6	5	5	210	34	19.9	10
	6	5	5	4	4	3	3	3	92	18	11.4	7
Ice milk creamsicle or dreamsicle—1 sicle (3 fl oz) (66 gm)												
Ice milk, flavors other than chocolate—1 c (131 gm)	17	14	13	11	10	9	9	8	184	51	31.6	18
Ice milk, fudgesicle—1 sicle (2.5 fl oz) (73 gm)	1	1	1	½	½	½	½	½	91	2	1.2	1
Ice milk sandwich (inc Dairy Queen)—1 sandwich (60 gm)	7	6	5	4	4	4	3	3	118	24	12.8	6
Ice milk, soft serve, chocolate (inc frozen custard, Tastee Freeze, Dairy Queen, Dairy Queen Blizzard)—1 c (175 gm)	13	12	10	9	8	7	7	6	243	41	25.4	14
Ice milk, soft serve, flavors other than chocolate (inc frozen custard, Tastee Freeze, Dairy Queen, Dairy Queen Blizzard)—1 c (175 gm)	14	12	10	9	8	8	7	6	223	42	25.9	13
Ice milk sundae, chocolate or fudge topping, w/whipped cream (inc McDonald's sundaes)—1 McDonald's sundae (106 gm)	36	31	27	24	22	20	18	17	228	111	69.0	34
—1 sundae (165 gm)	56	48	42	38	34	31	28	26	355	173	107.5	53
Ice milk sundae, fruit topping, w/whipped cream—1 sundae (165 gm)	44	37	33	30	27	24	22	21	266	130	80.8	50
Ice milk sundae, not fruit or chocolate topping, w/whipped cream (inc McDonald's sundaes)—1 McDonald's sundae (106 gm)	33	28	25	22	20	18	16	15	243	99	63.7	28
—1 sundae (165 gm)	51	43	38	34	31	28	25	23	379	153	99.1	43
Ice milk, w/sherbet or ice cream—1 bar (2.5 fl oz) (60 gm)	6	5	4	4	3	3	3	3	84	17	10.4	6
Milk dessert bar, frozen, dietary, made fr low-fat milk (inc Weight Watchers Treat Bars, all flavors)—1 bar (2.75 fl oz) (81 gm)	1	1	1	1	1	1	1	½	90	9	2.1	1

Ice Cream (Frozen Desserts) (cont.)

	1200	1500	1800	2100	2400	2800	3200	3600	TOT CAL	FAT CAL	S/FAT CAL	CHOL MG
Milk dessert, frozen, dietary, made w/low-fat milk (inc Weight Watchers)—1 c (131 gm)	4	3	3	3	2	2	2	2	147	9	6.0	7
Milk dessert sandwich bar, frozen, dietary, made fr low-fat milk (inc Weight Watchers Sandwich Bars)—1 sandwich bar (2.75 fl oz plus 2 wafers) (77 gm)	3	3	3	2	2	2	2	2	115	21	6.2	4
Sherbet, all flavors—1 c (193 gm)	12	10	9	8	7	7	6	6	270	34	21.4	14
Tofu, frozen dessert, chocolate (inc Tofutti)—½ c (97 gm)	13	11	9	8	7	6	6	5	214	128	28.6	0
Tofu, frozen dessert, flavors other than chocolate (inc Tofutti)—½ c (97 gm)	9	8	7	6	5	5	4	4	254	159	20.4	0
Milk, including Evaporated and Condensed												
1% fat—1 c (245 gm)	7	6	6	5	5	4	4	4	104	21	13.3	10
2% fat (inc hi-prot milk, fortified, milk, 1.5% fat milk)—1 c (245 gm)	14	12	11	10	9	8	7	7	125	42	26.3	18
3.7% fat, whole, pasteurized and raw, fluid—1 c (244 gm)	28	24	21	19	17	16	14	13	157	80	50.0	35
Acidophilus (1% fat)—1 c (242 gm)	7	6	6	5	5	4	4	4	103	21	13.2	10
Acidophilus, 2% fat (inc 1.5% fat acidophilus milk)—1 c (245 gm)	14	12	11	10	9	8	7	7	125	42	26.3	18
Buttermilk, dry, reconstituted—1 c (245 gm)	6	5	5	4	4	4	3	3	93	13	7.8	17
Buttermilk (inc Kefir milk)—1 c (245 gm)	7	6	5	5	4	4	4	3	99	19	12.1	9

Food												
Condensed, sweetened, diluted—1 tbsp (17 gm)	3	2	2	2	2	1	1	1	31	8	4.7	3
Condensed, sweetened, undiluted—1 tbsp (19 gm)	5	4	4	3	3	3	3	2	61	15	9.4	6
Dry, reconstituted, low-fat—1 c (245 gm)	2	2	1	1	1	1	1	1	84	4	2.3	5
Dry, reconstituted, nonfat—1 c (245 gm)	1	1	1	1	1	1	1	1	81	1	.9	4
Dry, reconstituted, whole—1 c (244 gm)	25	21	19	17	15	14	13	12	149	72	45.2	29
Evaporated, 2% fat, undiluted—1 tbsp (16 gm)	1	1	1	1	1	½	½	15	3	1.7	1	
Evaporated, skim, diluted—1 tbsp (15 gm)	0	0	0	0	0	0	0	0	6	0	.1	0
Evaporated, skim, undiluted—1 tbsp (16 gm)	0	0	0	0	0	0	0	0	12	0	.2	1
Evaporated, whole, diluted—1 tbsp (15 gm)	2	1	1	1	1	1	1	1	10	5	3.2	2
Evaporated, whole, undiluted—1 tbsp (16 gm)	4	3	3	3	2	2	2	2	22	11	6.6	5
Goat's, whole—1 c (244 gm)	30	26	23	20	19	17	15	14	168	91	58.6	28
Human—1 c (246 gm)	25	22	19	17	16	14	13	12	171	97	44.5	34
Skim or nonfat (5% or less butterfat)—1 c (245 gm)	2	2	1	1	1	1	1	1	86	4	2.6	4
Whey, acid, dried—1 tbsp (3 gm)	0	0	0	0	½	0	0	0	10	0	0	0
Whey, acid, fluid—1 c (246 gm)	1	1	1	1	0	½	½	½	59	2	1.3	1
Whey, sweet, dried—1 tbsp (8 gm)	0	0	0	0	0	0	0	0	26	1	.4	0
Whey, sweet, fluid—1 c (246 gm)	3	3	2	2	2	2	2	2	66	8	5.1	5

Milk-Flavored Drinks

Food												
Carob flavor beverage mix prep fr powder w/milk—1 c milk & 3 tsp powder (256 gm)	25	22	19	17	16	14	13	12	195	74	45.7	33

	1200	1500	1800	2100	2400	2800	3200	3600	TOT CAL	FAT CAL	S/FAT CAL	CHOL MG
Milk-Flavored Drinks (cont.)												
Chocolate flavor beverage mix prep fr powder w/milk—1 c milk & 2-3 heaping tsp (266 gm)	27	23	21	18	17	15	14	13	226	79	49.3	33
Chocolate-flavored drink, whey- & milk-based (inc Yoo-hoo)—1 bottle (9 fl oz) (274 gm)	4	3	3	3	2	2	2	2	145	10	7.9	3
Chocolate milk, low-fat milk base (inc chocolate milk drink)—1 c (250 gm)	15	13	11	10	9	8	8	7	179	45	27.9	17
Chocolate milk, skim milk base—1 c (250 gm)	4	3	3	2	2	2	2	2	144	11	6.5	4
Chocolate milk, whole milk base—1 c (250 gm)	26	22	20	18	16	14	13	12	208	76	47.3	31
Cocoa & sugar mixture, low-fat milk added (inc Nestle's Quik, Hershey's Instant)—1 c (250 gm)	14	12	11	10	9	8	7	7	186	43	26.5	17
Cocoa & sugar mixture, skim milk added (inc Nestle's Quik, Hershey's Instant)—1 c (250 gm)	2	2	2	2	2	1	1	1	152	7	4.2	4
Cocoa & sugar mixture, whole milk added (inc Nestle's Quik, Hershey's Instant)—1 c (250 gm)	25	21	19	17	15	14	13	12	213	73	44.7	31
Cocoa, hot chocolate, fr dry mix, made w/whole milk—1 c (250 gm)	28	24	21	19	17	15	14	13	218	81	50.5	33
Cocoa, sugar & dry milk mixture, water added (inc Swiss Miss Hot Chocolate, Hershey's Hot Chocolate, Nestle's Hot Chocolate)—1 oz pkt w/6 fl oz water (206 gm)	3	2	2	2	2	1	1	1	100	10	6.0	1

Food												
Cocoa, whey & lo-cal sweetener, fortified, water added (inc Ovaltine Sugar-Free Hot Cocoa)—1 c (249 gm)	3	2	2	2	2	1	1	1	53	12	6.4	0
Cocoa, whey & lo-cal sweetener mixture, low-fat milk added (inc Sugar-Free Nestle's Quik, Swiss Miss Sugar-Free Chocolate Milk Maker)—1 c (250 gm)	16	14	12	11	10	9	8	8	136	50	29.8	18
Cocoa w/nonfat dry milk & lo-cal sweetner, water added (inc Swiss Miss Sugar-Free Hot Cocoa Mix, Alba Sugar-Free Hot Cocoa Mix—1 pkt dry mix w/6 fl oz water (197 gm)	1	1	1	1	1	½	½	½	51	4	2.2	0
Dairy drink mix, chocolate, reduced calorie, w/aspartame, powder—0.75 oz pkt (21 gm)	2	2	1	1	1	1	1	1	63	5	3.6	2
Dairy drink mix, chocolate, reduced calorie, w/aspartame, prep w/water—½ c water & 3 cube & 1 pkt (204 gm)	2	2	1	1	1	1	1	1	64	5	3.6	2
Eggnog, made w/2% low-fat milk—1 c (254 gm)	59	54	50	47	44	42	39	37	241	82	34.6	277
Eggnog, made w/whole milk—1 c (254 gm)	68	60	54	49	45	41	38	35	342	171	101.6	149
Flavored milk drink, whey- & milk-based, flavors other than chocolate (inc Yoo-hoo)—1 bottle (9 fl oz) (274 gm)	4	3	3	3	2	2	2	2	145	10	7.9	3
Malted milk flavor mix, chocolate, added nutrients, prepared w/milk—1 c milk & 4-5 heaping tsp (265 gm)	27	23	21	19	17	15	14	13	225	80	49.4	33
Malted milk flavor mix, natural, added nutrients, prepared w/milk—1 c milk & 4-5 heaping tsp (265 gm)	27	23	20	18	17	15	14	13	230	78	48.5	33
Malted milk, fortified, chocolate (inc Ovaltine)—1 c (235 gm)	24	21	18	16	15	14	12	11	200	70	43.5	30

Milk-Flavored Drinks (cont.)

	1200	1500	1800	2100	2400	2800	3200	3600	TOT CAL	FAT CAL	S/FAT CAL	CHOL MG
Malted milk, fortified, natural flavor (inc Ovaltine)—1 c (235 gm)	23	20	18	16	14	13	12	11	201	67	41.2	30
Malted milk, unfortified, chocolate—1 c (235 gm)	24	21	18	17	15	14	12	11	207	73	44.1	30
Malted milk, unfortified, chocolate, made w/skim milk—1 c (235 gm)	3	3	3	2	2	2	2	2	149	11	5.9	5
Malted milk, unfortified, natural flavor—1 c (235 gm)	26	23	20	18	16	15	14	12	209	79	47.6	33
Milk-based fruit drink (inc orange julius, strawberry julius, piña colada julius)—1 c (236 gm)	1	1	1	1	1	1	1	1	172	2	1.0	4
Milk beverage, beads, chocolate, whole milk added (inc PDQ)—1 c (250 gm)	25	21	19	17	15	14	13	12	213	73	44.7	31
Milk beverage, made w/whole milk, flavors other than chcocolate (inc strawberry Nestle's Quik)—1 c (250 gm)	24	21	18	16	15	14	12	11	221	69	43.1	31
Milk beverage, w/nonfat dry milk & lo-cal sweetener, water added, chocolate (inc Alba)—1 c (276 gm)	4	3	3	2	2	2	2	2	86	7	6.5	4
Milk beverage w/nonfat dry milk & lo-cal sweetener, water added, flavors other than chocolate (inc Alba)—1 c (276 gm)	3	3	2	2	2	2	2	2	105	7	6.0	4
Milk shake, fountain type, chocolate—1 milk shake (10 fl oz) (283 gm)	39	34	30	27	24	22	20	18	362	120	74.2	41
Milk shake, fountain type, flavors other than chocolate—1 milk shake (10 fl oz) (283 gm)	43	37	33	29	27	24	22	20	357	123	76.7	56

Food												
Milk shake w/malt (inc malted milk w/ice cream)—1 milk shake (10 fl oz) (283 gm)	44	38	33	30	27	25	23	21	405	129	78.6	56
Strawberry flavor beverage mix, powder, prepared w/milk—1 c milk & 2–3 heaping tsp (266 gm)	25	22	19	17	16	14	13	12	234	74	45.7	33
Thick shake, carry-out type, chocolate (inc thick shake mix, milk added; Wendy's Frosty)—1 container (12.5 fl oz, net wt 11 oz) (313 gm)	26	23	20	18	16	15	13	12	371	76	47.4	33
Thick shake, carry-out type, flavors other than chocolate (inc thick shake mix, milk added)—1 container (12.5 fl oz, net wt 11 oz) (313 gm)	29	25	22	20	18	17	15	14	350	85	53.1	37

Milk and Cream Imitations

Food												
Cream substitute, frozen (inc Coffee Rich, Coffee Tone, Freezer Pak, Poly Perx, Poly Rich)—1 tbsp (15 gm)	1	1	1	1	1	1	1	½	20	13	2.6	0
Cream substitute, liquid (inc Coffee Whitner, Dairy Rich Moca Mix, mocha mix)—1 ind container (15 gm)	1	1	1	1	1	1	1	½	20	13	2.6	0
Cream substitute, nondairy, powdered —1 tsp (2 gm)	3	2	2	2	2	2	2	1	11	6	5.8	0
Cream substitute, powdered (inc Coffee Mate, Coffee Tone, Cremora, Instant Coffee Creamer, Instant Creamer, Please, Pream)—1 pkt (3 gm)	4	3	3	3	2	2	2	2	16	10	8.8	0
Soy milk, fluid—½ c (120 gm)	1	1	1	1	1	1	½	½	39	21	2.3	0
Whipped cream substitute, nondairy, dietetic, made fr powdered mix (inc Feather Weight Low Calorie Whipped Topping)—1 c (80 gm)	18	15	13	11	10	9	8	7	42	42	39.7	0

	CHOL MG	S/FAT CAL	FAT CAL	TOT CAL	3600	3200	2800	2400	2100	1800	1500	1200
Milk and Cream Imitations (cont.)												
Whipped cream substitute, nondairy, made fr powdered mix (inc Dream Whip, Lucky Whip, Smooth Whip)—1 c (80 gm)	8	76.9	89	151	15	17	18	21	23	26	30	35
Whipped topping, nondairy, frozen (inc Cool Whip, Handiwhip Whip Topping, Pet Whip)—1 c (75 gm)	0	147.0	171	239	27	30	33	38	42	48	55	65
Whipped topping, nondairy, pressurized can (inc Lucky Whip, Reddiwhip in blue can)—1 c (70 gm)	0	119.1	140	184	22	24	27	31	34	39	44	53
Puddings and Custards												
Chantilly Cream—1 c (120 gm)	122	183.7	295	371	48	52	56	63	69	76	86	101
Coconut custard, Puerto Rican style (flan de coco)—1 c (245 gm)	368	141.5	232	687	67	72	77	83	89	97	107	120
Custard—½ c (122 gm)	133	26.3	53	130	20	21	22	24	25	27	29	32
Custard, baked—1 c (265 gm)	278	60.8	131	305	42	45	48	51	54	58	63	70
Fresh corn custard, Puerto Rican style (Mazamorra—Mundo Nuevo)—½ c (140 gm)	0	251.8	292	532	46	51	57	65	72	81	94	112
Mousse, chocolate—½ c (85 gm)	146	71.4	126	189	30	32	34	37	40	43	48	54
Mousse, not chocolate (inc fruit mousse)—½ c (85 gm)	54	81.2	131	190	21	23	25	28	30	34	38	44
Pudding, bread (inc w/raisins)—½ c (100 gm)	82	20.2	51	168	13	14	15	16	17	18	20	22
Pudding, canned, chocolate—1 can (5 oz) (142 gm)	6	55.7	97	204	11	12	13	15	17	19	22	26

Food												
Pudding, canned, flavors other than chocolate—1 can (5 oz) (142 gm)	24	20	18	16	14	12	11	10	220	93	53.5	3
Pudding, canned, lo-cal, chocolate—1 can (5 oz) (142 gm)	2	1	1	1	1	1	1	1	71	4	2.4	3
Pudding, canned, lo-cal, flavors other than chocolate—1 can (5 oz) (142 gm)	1	1	1	1	1	1	1	1	77	3	1.5	3
Pudding, chocolate, prep fr starch base mix, made w/milk, cooked—1 c (260 gm)	22	19	17	15	14	13	11	11	322	70	38.6	31
Pudding, chocolate, starch base, prep fr home recipe—1 c (260 gm)	31	27	24	21	19	17	16	14	385	110	60.6	29
Pudding, coconut—½ c (113 gm)	13	11	10	9	8	7	6	6	136	33	24.3	12
Pudding, Indian—½ c (119 gm)	18	16	14	13	12	11	11	10	154	40	21.8	53
Pudding, lo-cal, chocolate—½ c (130 gm)	1	1	1	1	1	1	1	1	65	4	2.2	3
Pudding, lo-cal, flavors other than chocolate—½ c (130 gm)	1	1	1	1	1	1	1	1	70	2	1.4	3
Pudding pops, chocolate—1 pudding pop (1.75 fl oz) (57 gm)	9	8	7	6	6	5	4	4	99	24	21.0	1
Pudding pops, flavors other than chocolate—1 pudding pop (1.75 fl oz) (57 gm)	9	8	7	6	5	5	4	4	93	24	20.5	1
Pudding, prep fr dry mix, lo-cal, milk added, chocolate (inc D-Zerta)—½ c (125 gm)	1	1	1	1	1	1	1	1	63	3	2.1	3
Pudding, prep fr dry mix, lo-cal, milk added, flavors other than chocolate (inc D-Zerta)—½ c (125 gm)	1	1	1	1	1	1	1	1	68	2	1.4	3
Pudding, prep fr dry mix, milk added, chocolate—½ c (131 gm)	8	7	6	6	5	5	4	4	159	35	13.1	15
Pudding, prep fr dry mix, milk added, flavors other than chocolate—½ c (132 gm)	12	10	9	8	7	7	6	6	150	34	20.9	15

	1200	1500	1800	2100	2400	2800	3200	3600	TOT CAL	FAT CAL	S/FAT CAL	CHOL MG
Puddings and Custards *(cont.)*												
Pudding, prep fr starch base mix, made w/milk wo/cooking—1 c (260 gm)	19	16	14	13	12	11	10	9	325	59	32.3	29
Pudding, pumpkin—½ c (133 gm)	9	8	7	6	5	5	5	4	114	29	15.9	11
Pudding, rice—½ c (113 gm)	9	8	7	6	6	5	5	4	159	28	17.1	12
Pudding, sodium-controlled, w/milk base (inc Diet Care)—½ c (128 gm)	1	1	1	1	1	1	1	1	69	2	1.4	3
Pudding, tapioca, chocolate—½ c (83 gm)	8	7	6	5	5	4	4	4	92	24	14.6	9
Pudding, tapioca, made fr dry mix—½ c (83 gm)	8	7	6	6	5	5	4	4	99	24	14.6	11
Pudding, tapioca, made fr home recipe—½ c (103 gm)	25	23	21	20	19	17	16	15	129	42	20.6	104
Pudding, vanilla (blanc mange), starch base, prep fr home recip—1 c (255 gm)	27	24	21	19	17	16	14	13	283	89	49.1	36
Pudding, w/fruit & vanilla wafers—½ c (94 gm)	14	13	12	11	10	9	9	8	135	33	13.7	51
Rennin dessert, chocolate, made fr mix, w/milk—1 c (255 gm)	26	22	20	18	16	15	13	12	260	87	48.0	31
Rennin dessert, home prep w/tablet—1 c (255 gm)	24	21	19	17	15	14	13	12	227	80	44.1	31
Rennin dessert, prep fr mix, w/milk, other than strawberry & raspberry—1 c (250 gm)	25	21	19	17	16	14	13	12	238	81	44.5	33
Rennin dessert, prep fr mix, w/milk, strawberry & raspberry—1 c (250 gm)	25	21	19	17	16	14	13	12	238	81	44.5	33
Rice pudding, w/raisins—1 c (265 gm)	22	19	17	15	14	13	12	11	387	74	40.3	29

Food												
Tapioca desserts, apple—1 c (250 gm)	0	0	0	0	0	0	0	0	293	3	0	0
Tapioca desserts, cream pudding—1 c (165 gm)	41	37	34	31	29	27	26	24	221	76	35.2	160
Tapioca, dry—1 tbsp (8 gm)	0	0	0	0	0	0	0	0	30	0	0	0
Yogurt												
Frozen, carob-coated—1 bar (41 gm)	12	10	9	8	7	6	6	5	83	33	25.9	3
Frozen, chocolate—1 8-oz container (227 gm)	15	12	11	10	9	8	7	7	247	41	26.4	18
Frozen, chocolate-coated—1 bar (41 gm)	19	16	14	12	11	10	9	8	94	50	42.0	3
Frozen, flavors other than chocolate—1 8-oz container (227 gm)	15	12	11	10	9	8	7	7	247	41	26.4	18
Frozen, sandwich—1 sandwich (85 gm)	5	4	3	3	3	3	2	2	148	18	8.5	5
Fruit variety, low-fat milk (inc custard style)—1 8-oz container (227 gm)	8	7	6	5	5	4	4	4	231	22	14.2	10
Fruit variety, nonfat milk—1 8-oz container (227 gm)	2	2	2	1	1	1	1	1	213	4	2.5	5
Fruit variety, whole milk (inc breakfast yogurt)—1 8-oz container (227 gm)	22	19	17	15	14	12	11	10	270	66	41.7	22
Plain, low-fat milk—1 8-oz container (227 gm)	11	10	9	8	7	6	6	5	144	32	20.4	14
Plain, nonfat milk—1 8-oz container (227 gm)	2	1	1	1	1	1	1	1	127	4	2.4	4
Plain, whole milk—1 8-oz container (227 gm)	24	20	18	16	15	13	12	11	139	66	42.8	29
Vanilla, lemon, or coffee flavor, low-fat milk (inc liquid yogurt, LeShake, Tuscan)—1 8-oz container (227 gm)	9	8	7	6	6	5	5	4	194	26	16.5	11
Vanilla, lemon, or coffee flavor, whole milk—1 8-oz container (227 gm)	22	18	16	15	13	12	11	10	229	65	40.9	22

THE 2-IN-1 SYSTEM

	1200	1500	1800	2100	2400	2800	3200	3600	TOT CAL	FAT CAL	S/FAT CAL	CHOL MG
NUTS												
Acorns, dried—1 oz (28 gm)	5	4	3	3	3	2	2	2	145	80	10.4	0
Acorns, raw—1 oz (28 gm)	4	3	3	2	2	2	2	1	105	61	7.9	0
Almond meal, partially defatted, w/salt added—1 oz (28 gm)	2	2	1	1	1	1	1	1	116	47	4.4	0
Almond paste—1 oz (28 gm)	3	2	2	2	2	1	1	1	127	69	6.6	0
Almond powder, full-fat—1 oz (28 gm)	6	5	4	4	3	3	3	2	168	132	12.5	0
Almond powder, partially defatted—1 oz (28 gm)	2	1	1	1	1	1	1	1	112	41	3.9	0
Almonds, dry-roasted, unblanched, wo/salt added—1 oz (28 gm)	6	5	4	4	3	3	3	2	167	131	12.5	0
Almonds, dry-roasted, wo/salt—1 oz (approx 22 whole kernels) (28 gm)	5	5	4	4	3	3	2	2	164	130	12.3	0
Almonds, oil-roasted, blanched, wo/salt added—1 oz (24 whole kernels) (28 gm)	6	5	4	4	4	3	3	3	174	145	13.7	0
Almonds, oil-roasted, unblanched, wo/salt added—1 oz (22 whole kernels) (28 gm)	6	5	4	4	4	3	3	3	176	148	13.9	0
Almonds, toasted, unblanched—1 oz (28 gm)	5	5	4	4	3	3	2	2	167	130	12.3	0
Almonds, unroasted—1 c slivered, blanched (94 gm)	19	16	14	12	11	9	9	8	554	442	41.9	0
——1 c, sliced, unblanched (95 gm)	19	16	14	12	11	10	9	8	560	446	42.3	0
——1 c, ground (95 gm)	19	16	14	12	11	10	9	8	560	446	42.3	0
——1 c, sliced, blanched (105 gm)	21	17	15	13	12	11	10	9	618	493	46.8	0
——1 c, slivered, unblanched (115 gm)	23	19	17	15	13	12	10	9	677	540	51.2	0
——1 c whole, blanched (145 gm)	29	24	21	18	17	15	13	12	854	681	64.6	0

Food												
Beechnuts, dried—1 oz (28 gm)	6	5	5	4	4	3	3	3	164	128	14.6	0
Brazil nuts, dried, unblanched—1 oz (6–8 kernels) (28 gm)	18	15	13	12	11	9	8	8	186	169	41.3	0
Butternuts, dried—1 oz (28 gm)	1	1	1	1	1	1	1	1	174	146	3.3	0
Cashew nuts, dry-roasted, wo/salt added—1 oz (28 gm)	10	9	8	7	6	5	5	4	163	119	23.4	0
Cashew nuts, honey-roasted (28 gm)	8	7	6	5	5	4	4	3	155	106	17.7	0
Cashew nuts, oil-roasted, wo/salt added—1 oz (28 gm)	11	9	8	7	6	6	5	4	163	123	24.3	0
Chestnuts, Chinese, boiled & steamed—1 oz (28 gm)	0	0	0	0	0	0	0	0	44	2	.3	0
Chestnuts, Chinese, dried—1 oz (28 gm)	½	½	0	0	0	0	0	0	103	5	.7	0
Chestnuts, Chinese, roasted—1 oz (28 gm)	0	0	0	0	0	0	0	0	68	3	.4	0
Chestnuts, European, boiled & steamed—1 oz (28 gm)	½	0	0	0	0	0	0	0	37	4	.6	0
Chestnuts, European, dried, peeled—1 oz (28 gm)	1	1	1	1	½	½	½	½	105	10	1.9	0
Chestnuts, European, roasted—1 oz (28 gm)	½	½	½	½	½	0	0	0	70	5	1.1	0
Chestnuts, Japanese, boiled & steamed—1 oz (28 gm)	0	0	0	0	0	0	0	0	16	1	.1	0
Chestnuts, Japanese, dried—1 oz (28 gm)	0	0	0	0	0	0	0	0	102	4	.4	0
Chestnuts, Japanese, roasted—1 oz (28 gm)	0	0	0	0	0	0	0	0	57	2	.3	0
Coconut cream, canned (liquid expressed from grated meat)—1 tbsp (19 gm)	12	10	9	8	7	6	5	5	36	31	26.8	0
Coconut cream, raw (liquid expressed fr grated meat)—1 tbsp (15 gm)	18	15	13	12	11	9	8	8	49	47	41.5	0
Coconut meat, dried (desiccated), creamed—1 oz (28 gm)	69	58	51	45	40	36	32	29	194	176	156.6	0

Nuts (cont.)

	1200	1500	1800	2100	2400	2800	3200	3600	TOT CAL	FAT CAL	S/FAT CAL	CHOL MG
Coconut meat, dried (desiccated), not sweetened—1 oz (28 gm)	65	55	47	42	38	33	30	27	187	165	146.3	0
Coconut meat, dried (desiccated), sweetened, flaked, canned—4 oz (114 gm)	128	108	93	83	74	65	59	53	505	325	288.4	0
Coconut meat, dried (desiccated), sweetened, shredded—7 oz pkt (199 gm)	250	210	182	161	145	128	115	104	997	635	563.6	0
Coconut meat, dried (desiccated), toasted—1 oz (28 gm)	47	40	34	31	27	24	22	20	168	121	106.6	0
Coconut meat, raw—1 piece (2" x 2" x ½") (45 gm)	53	45	39	34	31	27	24	22	159	136	120.2	0
Coconut milk, canned (liquid expressed fr grated meat & water)—1 tbsp (15 gm)	11	10	8	7	7	6	5	5	30	29	25.6	0
Coconut milk, frozen (liquid expressed fr grated meat & water)—1 tbsp (15 gm)	11	9	8	7	6	6	5	5	30	28	24.9	0
Coconut milk, raw (liquid expressed fr grated meat and water)—1 tbsp (15 gm)	13	11	9	8	7	6	6	5	35	32	28.5	0
Coconut water (liquid fr coconuts)—1 tbsp (15 gm)	0	0	0	0	0	0	0	0	3	0	0	0
Filberts or hazelnuts, dried, blanched—1 oz (28 gm)	6	5	4	4	3	3	3	2	191	172	12.6	0
Filberts or hazelnuts, dried, unblanched—1 oz (28 gm)	5	4	4	3	3	3	2	2	179	160	11.8	0

Food												
Filberts or hazelnuts, dry-roasted, unblanched, wo/salt added—1 oz (28 gm)	6	5	4	4	3	3	3	2	188	169	12.4	0
Filberts or hazelnuts, oil-roasted, unblanched, wo/salt added—1 oz (28 gm)	5	4	3	4	3	3	3	2	187	163	12.0	0
Ginkgo nuts, canned—1 oz (14 med kernels) (28 gm)	½	½	½	½	0	0	0	0	32	5	.8	0
Ginkgo nuts, dried—1 oz (28 gm)	½	½	½	½	½	0	0	0	99	5	1.0	0
Ginkgo nuts, raw—1 oz (28 gm)	½	½	0	0	0	0	0	0	52	5	.8	0
Hickory nuts, dried—1 oz (28 gm)	8	7	6	5	5	4	4	3	187	165	18.0	0
Macadamia nuts, dried—1 oz (28 gm)	13	11	9	8	7	6	6	5	199	188	28.3	0
Macadamia nuts, oil-roasted, wo/salt added—1 oz (10–12 kernels) (28 gm)	13	11	9	8	8	7	6	5	204	195	29.3	0
Mixed nuts, wo/peanuts, oil-roasted, wo/salt added—1 oz (28 gm)	10	9	8	7	6	5	5	4	175	143	23.2	0
Mixed nuts, wo/peanuts, oil-roasted w/salt added—1 oz (28 gm)	10	9	8	7	6	5	5	4	175	143	23.2	0
Mixed nuts, w/peanuts, dry-roasted, wo/salt added—1 oz (28 gm)	8	7	6	5	5	4	4	3	169	131	17.6	0
Mixed nuts, w/peanuts, dry-roasted, w/salt added—1 oz (28 gm)	8	7	6	5	5	4	4	3	169	131	17.6	0
Mixed nuts, w/peanuts, oil-roasted, wo/salt added—1 oz (28 gm)	10	8	7	6	6	5	5	4	175	144	22.3	0
Mixed nuts, w/peanuts, oil-roasted, w/salt added—1 oz (28 gm)	10	8	7	6	6	5	5	4	175	144	22.3	0
Nut mixture w/dried fruit & seeds (inc trail mix, garp)—1 c (140 gm)	7	6	5	5	4	4	3	3	491	182	16.8	0
Nut mixture w/seeds—1 c (140 gm)	30	25	22	19	17	15	14	12	717	433	67.2	0
Peanut flour, defatted, wo/salt added—1 tbsp (4 gm)	0	0	0	0	0	0	0	0	13	0	0	0
Peanut kernels, dried—1 oz (28 gm)	8	6	6	5	4	4	3	3	159	124	17.2	0
Peanut kernels, oil-roasted, wo/salt added—1 oz (28 gm)	8	7	6	5	4	4	4	3	165	126	17.5	0

	1200	1500	1800	2100	2400	2800	3200	3600	TOT CAL	FAT CAL	S/FAT CAL	CHOL MG
Nuts (cont.)												
Peanut kernels, oil-roasted, w/salt added—1 oz (28 gm)	8	6	6	5	4	4	4	3	163	124	17.3	0
Peanuts, all types, dry-roasted, wo/salt —1 oz (28 gm)	8	6	6	5	4	4	4	3	164	125	17.4	0
Peanuts, all types, oil-roasted, w/salt —1 oz (28 gm)	8	6	6	5	4	4	4	3	163	124	17.3	0
Peanuts, boiled—1 oz (33 nuts) (28 gm)	5	4	4	3	3	2	2	2	105	79	11.0	0
Peanuts, honey-roasted—1 oz (35 nuts) (28 gm)	7	6	5	4	4	3	3	3	153	108	15.1	0
Peanuts, in shell—1 oz (32 nuts, shelled) (28 gm)	8	6	6	5	4	4	4	3	162	124	17.3	0
Peanuts, Spanish, oil-roasted, w/salt— 1 oz (28 gm)	8	7	6	5	5	4	4	4	162	123	19.0	0
Peanuts, Spanish, raw—1 oz (28 gm)	9	7	6	6	5	4	4	4	160	125	19.3	0
Peanuts, Valencia, oil-roasted, w/salt— 1 oz (28 gm)	9	7	6	6	5	5	4	4	165	130	19.9	0
Peanuts, Valencia, raw—1 oz (28 gm)	8	7	6	5	5	4	4	3	160	120	18.5	0
Peanuts, Virginia, oil-roasted, w/salt— 1 oz (28 gm)	7	6	5	5	4	4	3	3	162	122	16.0	0
Peanuts, Virginia, raw—1 oz (28 gm)	7	6	5	5	4	4	3	3	158	122	16.0	0
Pecan flour—1 oz (28 gm)	0	0	0	0	0	0	0	0	93	4	.3	0
Pecans, dried—1 oz (20 halves) (28 gm)	6	5	4	4	4	3	3	3	190	173	13.9	0
Pecans, dry-roasted, wo/salt added—1 oz (28 gm)	6	5	4	4	3	3	3	2	187	166	13.2	0
Pecans, oil-roasted, wo/salt added—1 oz (15 halves) (28 gm)	6	5	5	4	4	3	3	3	195	182	14.6	0
Pine nuts (pignoli) dried—1 oz (28 gm)	9	7	6	6	5	5	4	4	146	130	19.9	0

Food												
Pine nuts (pinyon) dried—1 oz (28 gm)	11	9	8	7	6	5	5	4	161	156	23.9	0
Pistachio nuts, dried—1 oz (47 kernels) (28 gm)	7	6	5	4	4	4	3	3	164	123	15.7	0
Pistachio nuts, dry-roasted, wo/salt added—1 oz (28 gm)	8	6	6	5	4	4	3	3	172	135	17.1	0
Pistachio nuts, dry-roasted, w/salt added—1 oz (28 gm)	8	6	6	5	4	4	3	3	172	135	17.1	0
Soybean kernels, roasted & toasted, wo/salt—1 oz (95 kernels) (28 gm)	4	3	3	2	2	2	2	1	129	61	8.1	0
Soybean kernels, roasted & toasted, w/salt added—1 oz (95 kernels) (28 gm)	4	3	3	2	2	2	2	1	129	61	8.1	0
Walnuts, black, dried—1 oz (28 gm)	4	3	3	3	2	2	2	2	172	145	9.3	0
Walnuts, English or Persian, dried—1 oz (14 halves) (28 gm)	6	5	5	4	4	3	3	3	182	158	14.3	0

Nut Butters

Food												
Almond butter, honey & cinnamon, wo/salt added—1 tbsp (16 gm)	3	3	2	2	2	2	1	1	96	76	7.1	0
Almond butter, plain, wo/salt added—1 tbsp (16 gm)	4	3	3	2	2	2	2	1	101	86	8.1	0
Cashew butter, plain, wo/salt added—1 tbsp (16 gm)	6	5	5	4	4	3	3	3	94	71	14.0	0
Peanut butter, chunk style, w/salt—2 tbsp (32 gm)	12	10	9	8	7	6	6	5	189	144	27.6	0
Peanut butter, smooth style, w/salt—2 tbsp (32 gm)	12	10	9	8	7	6	6	5	188	144	27.6	0
Peanut butter, wo/salt added—1 tbsp (16 gm)	6	5	4	4	4	3	3	3	94	72	13.8	0
Peanut spread (inc imitation peanut butter—1 tbsp (16 gm)	5	4	3	3	3	2	2	2	96	75	10.4	0

Seeds

	1200	1500	1800	2100	2400	2800	3200	3600	TOT CAL	FAT CAL	S/FAT CAL	CHOL MG
Breadfruit seeds, roasted—1 oz (28 gm)	1	1	1	1	½	½	½	½	59	7	1.9	0
Lotus seeds, dried—1 oz (42 med seeds) (28 gm)	½	½	½	0	0	0	0	0	94	5	.8	0
Pumpkin & squash seed kernels, dried—1 oz (142 kernels) (28 gm)	10	8	7	6	6	5	4	4	154	117	22.1	0
Pumpkin & squash seed kernels, roasted, wo/salt—1 oz (28 gm)	9	8	7	6	5	5	4	4	148	108	20.3	0
Pumpkin & squash seeds, whole, roasted, wo/salt—1 oz (85 seeds) (28 gm)	4	4	3	3	2	2	2	2	127	50	9.4	0
Safflower seed kernels, dried—1 oz (28 gm)	4	4	3	3	2	2	2	2	147	98	9.4	0
Safflower seed meal, partially defatted—1 oz (28 gm)	0	0	0	0	0	0	0	0	97	6	.5	0
Sesame butter, paste—1 tbsp (16 gm)	5	4	3	3	3	2	2	2	95	73	10.3	0
Sesame butter (tahini), fr raw & stone-ground kernels—1 tbsp (15 gm)	4	3	3	3	2	2	2	2	86	65	9.1	0
Sesame butter (tahini), fr roasted & toasted kernels (most common type)—1 tbsp (15 gm)	5	4	3	3	3	2	2	2	89	73	10.2	0
Sesame flour, high-fat—1 oz (28 gm)	6	5	4	4	3	3	3	2	149	95	13.3	0
Sesame flour, low-fat—1 oz (28 gm)	0	0	0	0	0	0	0	0	95	5	.5	0
Sesame flour, partially defatted—1 oz (28 gm)	2	2	1	1	1	1	1	1	109	31	4.1	0
Sesame meal, partially defatted—1 oz (28 gm)	8	6	6	5	4	4	3	3	161	122	17.2	0
Sesame seed kernels, dried—1 tbsp (8 gm)	2	2	2	2	1	1	1	1	47	40	5.5	0

Food												
Sesame seed kernels, toasted, wo/salt added—1 oz (28 gm)	8	6	5	4	4	3	3	3	161	122	17.2	0
Sesame seeds, whole, dried—1 tbsp (9 gm)	3	2	2	1	1	1	1	1	52	41	5.7	0
Sesame seeds, whole, roasted & toasted—1 oz (28 gm)	8	6	5	4	4	3	3	3	161	122	17.2	0
Sunflower seed butter, wo/salt—1 tbsp (16 gm)	3	3	2	2	2	1	1	1	93	68	7.2	0
Sunflower seed flour, partially defatted—1 c (80 gm)	½	½	½	½	0	0	0	0	261	12	1.0	0
Sunflower seed kernels, dried—1 oz (28 gm)	6	5	4	4	3	3	3	2	162	127	13.3	0
Sunflower seed kernels, dry-roasted, wo/salt—1 oz (28 gm)	6	5	4	4	3	3	3	2	165	127	13.3	0
Sunflower seed kernels, oil-roasted, wo/salt—1 oz (28 gm)	7	6	5	4	4	3	3	3	175	147	15.4	0
Sunflower seed kernels, toasted, wo/salt—1 oz (28 gm)	7	6	5	4	4	3	3	3	176	145	15.2	0
Watermelon seed kernels, dried—1 oz (28 gm)	11	9	8	7	6	6	5	5	158	121	25.0	0

PANCAKES, WAFFLES, TORTILLAS

Food												
Bread fritters, Puerto Rican style (torrejas, Galician fritters)—2 fritters w/syrup (4" x 2½" x 3¼") (110 gm)	20	18	16	15	14	13	12	12	296	88	18.5	73
Crepe, plain (inc French pancake)—1 crepe (7" dia) (28 gm)	13	12	11	10	9	9	8		66	32	10.2	57
Flour & milk patty—1 pancake (28 gm)	2	2	2	1	1	1	1	1	69	15	3.9	2
Flour & water patty (inc Chinese pancake)—1 pancake (28 gm)	0	0	0	0	0	0	0	0	58	1	.1	0
French toast, plain (inc Roman Meal)—1 sl (65 gm)	26	24	22	21	20	18	17	17	164	61	17.7	118

Pancakes, Waffles, Tortillas (cont.)

	1200	1500	1800	2100	2400	2800	3200	3600	TOT CAL	FAT CAL	S/FAT CAL	CHOL MG
Pancakes, buckwheat—2 pancakes (4" dia) (42 gm)	9	8	7	7	6	6	6	5	78	26	9.5	31
Pancakes, cornmeal—2 pancakes (4" dia) (42 gm)	6	6	5	5	4	4	4	4	85	24	6.1	22
Pancakes, made fr buckwheat mix, w/egg & milk—1 pancake (6" dia) (73 gm)	16	14	13	12	11	10	10	9	146	59	19.5	48
Pancakes, made fr home recipe—1 pancake (6" dia) (73 gm)	11	10	9	9	8	7	7	7	169	46	11.8	39
Pancakes, plain & buttermilk, made w/egg & milk—1 pancake, (6" dia) (73 gm)	16	14	13	12	11	10	10	9	164	48	16.9	54
Pancakes, plain—2 pancakes (4" dia) (42 gm)	10	9	8	8	7	7	6	6	93	29	10.5	35
Pancakes, rye—2 pancakes (4" dia) (42 gm)	8	7	6	6	5	5	5	4	122	41	10.6	21
Pancakes, sour dough—2 pancakes (4" dia) (42 gm)	2	2	1	1	1	1	1	1	95	19	4.6	0
Pancakes, whole wheat—2 pancakes (4" dia) (42 gm)	10	9	8	7	7	6	6	6	100	42	10.3	33
Pancakes, w/fruit (inc blueberry pancakes)—2 pancakes (4" dia) (42 gm)	9	8	7	7	6	6	6	5	88	27	9.5	31
Taco shell (inc tostada shell)—1 reg (5" dia) (13 gm)	2	1	1	1	1	1	1	1	59	23	3.5	0
Tortillas, corn—1 tortilla (15 gm)	0	0	0	0	0	0	0	0	32	5	.5	0
Tortillas, flour (wheat)—1 tortilla (approx 5" dia) (20 gm)	2	1	1	1	1	1	1	1	59	14	3.4	0

Food												
—1 tortilla (approx 7" dia) (40 gm)	3	3	2	2	2	2	1	1	118	27	6.8	0
Tortilla, whole wheat (inc chapati and puri)—1 tortilla (35 gm)	½	0	0	0	0	0	0	0	73	4	.6	0
Waffles, bran—2 sq waffles, inc frozen (4" sq) (74 gm)	19	17	16	15	14	13	12	11	223	82	20.2	66
Waffles, cornmeal—2 sq waffles, inc frozen (4" sq) (74 gm)	22	20	19	17	16	15	14	14	207	70	17.6	93
Waffles, fruit (inc blueberry)—2 sq waffles, include frozen (4" square) (74 gm)	13	12	11	10	9	9	8	8	184	43	16.0	41
Waffles, frozen—1 waffle (4⅝" x 3¾") (34 gm)	9	8	8	7	7	6	6	6	86	19	4.9	43
Waffles, made fr pancake & waffle mix, w/egg & milk—1 section, ¼ waffle (50 gm)	12	11	9	9	8	7	7	6	138	48	16.4	30
Waffles, plain—2 sq waffles, inc frozen (4" sq) (74 gm)	16	14	13	12	11	10	9	9	204	49	18.5	48
Waffles, whole wheat or mixed grain (inc Roman Meal)—2 sq waffles, inc frozen (4" sq) (74 gm)	27	24	22	20	19	18	17	16	203	84	27.8	94

PASTAS

Food												
Chow fun rice noodles, fat not added in cooking—1 c, ckd (152 gm)	0	0	0	0	0	0	0	0	120	1	.3	0
Macaroni, fat not added in cooking (inc lasagna noodles, orzo, ziti, rotini, shells, wagon wheels, cart wheels, manicotti, rigatoni)—1 c, ckd (140 gm)	½	½	½	0	0	0	0	0	154	5	.8	0
Noodles, chow mein, canned—5 oz can (142 gm)	37	31	27	24	22	19	18	16	694	301	76.7	17
Noodles, chow mein (inc Chinese noodles)—1 c (45 gm)	9	8	7	6	5	5	4	4	220	95	18.5	5

	1200	1500	1800	2100	2400	2800	3200	3600	TOT CAL	FAT CAL	S/FAT CAL	CHOL MG
Pastas (cont.)												
Noodles, fat not added in cooking (inc pastina, egg noodles)—1 c, ckd (160 gm)	10	9	8	8	7	7	7	6	199	21	4.7	49
Noodles, rice, fat not added in cooking—1 c, ckd (160 gm)	14	12	10	9	8	7	7	6	320	217	32.4	0
Noodles, spinach, fat not added in cooking—1 c, cooked (160 gm)	½	½	½	½	½	0	0	0	190	7	1.0	0
Noodles, whole wheat, fat not added in cooking—1 c, cooked (160 gm)	1	½	½	½	½	½	½	0	174	9	1.3	0
Spaghetti, fat not added in cooking (inc fides, linguini, vermicelli)—1 c, ckd (140 gm)	½	½	½	0	0	0	0	0	154	5	.8	0
Spaghetti, hi-prot type, fat not added—1 c, ckd (140 gm)	½	½	½	0	0	0	0	0	153	5	.8	0
Spaghetti, whole wheat, fat not added in cooking—1 cup, ckd (140 gm)	0	0	0	0	0	0	0	0	190	3	.4	0
Spaghetti, in tomato sauce w/cheese, cooked, home recipe—1 c (250 gm)	9	8	7	6	6	5	5	4	260	79	18.0	8
Spaghetti, in tomato sauce w/cheese, canned—1 c (250 gm)	1	1	1	1	1	1	1	1	190	14	0	8
POULTRY **Chicken**												
Back, w/or wo/bone, battered, fried, skin eaten (inc extra crispy fried chicken)—1 med back (yield after cooking, bone removed) (212 gm)	78	69	62	57	52	48	44	41	703	419	111.5	186

Back, w/or wo/bone, battered, fried, skin not eaten (inc extra crispy fried chicken)—1 med back (yield after cooking, bone & skin removed) (119 gm)	37	33	30	27	25	23	22	20	342	163	44.0	110
Back, w/or wo/bone, breaded, baked or fried, skin eaten (inc reg fried chicken, Shake-n-Bake)—1 med back (yield after cooking, bone removed) (212 gm)	79	69	62	57	53	48	44	42	703	413	111.6	187
Back, w/or wo/bone, breaded, baked or fried, skin not eaten (inc reg fried chicken, Shake-n-Bake)—1 med back (yield after cooking, bone & skin removed) (119 gm)	37	33	30	27	25	23	22	20	342	163	44.0	110
Back, w/or wo/bone, broiled, skin eaten—1 med back (yield after cooking, bone removed) (134 gm)	49	43	39	36	33	30	28	26	400	251	69.8	117
Back, w/or wo/bone, broiled, skin not eaten—1 med back (yield after cooking, bone and skin removed) (112 gm)	32	28	26	24	22	20	19	18	266	132	36.1	100
Back, w/or wo/bone, floured, baked or fried, skin eaten—1 med back (yield after cooking, bone removed) (128 gm)	46	41	37	33	31	28	26	25	422	237	64.3	113
Back, w/or wo/bone, floured, baked or fried, skin not eaten—1 med back (yield after cooking, bone and skin removed) (103 gm)	32	28	26	24	22	20	19	18	299	144	38.8	95
Back, w/or wo/bone, fried, no coating, skin eaten—1 med back (yield after cooking, bone removed) (128 gm)	50	44	39	36	33	31	28	26	413	256	69.3	122
Back, w/or wo/bone, fried, no coating, skin not eaten—1 med back (yield after cooking, bone & skin removed) (103 gm)	34	30	27	25	23	22	20	19	287	150	40.6	102

	1200	1500	1800	2100	2400	2800	3200	3600	TOT CAL	FAT CAL	S/FAT CAL	CHOL MG
Chicken *(cont.)*												
Back, w/or wo/bone, roasted, skin eaten—1 med back (yield after cooking, bone removed) (94 gm)	34	30	27	25	23	21	19	18	280	176	48.9	82
Back, w/or wo/bone, roasted, skin not eaten—1 med back (yield after cooking, bone & skin removed) (72 gm)	20	18	16	15	14	13	12	11	171	85	23.2	64
Back, w/or wo/bone, stewed, skin eaten—1 med back (yield after cooking, bone removed) (106 gm)	34	30	27	25	23	21	19	18	272	172	47.6	82
Back, w/or wo/bone, stewed, skin not eaten—1 med back (yield after cooking, bone & skin removed) (75 gm)	19	17	15	14	13	12	11	11	156	75	20.4	63
Breast, w/or wo/bone, battered, fried, skin eaten (inc extra crispy fried chicken)—1 med sl (approx 2" x 1½" x ¼") (14 gm)	4	3	3	3	3	2	2	2	36	17	4.4	12
—½ med breast (yield after cooking, bone removed) (140 gm)	38	34	31	29	27	25	23	22	364	166	44.3	120
Breast, w/or wo/bone, battered, fried, skin not eaten (inc extra crispy fried chicken)—1 med sl (approx 2" x 1½" x ¼") (14 gm)	3	3	2	2	2	2	2	2	26	6	1.6	13
—½ med breast (yield after cooking, bone & skin removed) (88 gm)	17	16	14	14	13	12	11	11	164	37	10.2	80
Breast, w/or wo/bone, breaded, baked or fried, skin eaten (inc reg fried chicken, Shake-n-Bake)—1 med sl (approx 2" x 1½" x ¼") (14 gm)	3	3	3	3	2	2	2	2	33	12	3.3	12

Food												
—½ med breast (yield after cooking, bone removed) (140 gm)	33	29	27	25	23	22	20	19	328	120	32.7	117
Breast, w/or wo/bone, breaded, fried or baked, skin not eaten (inc reg fried chicken, Shake-n-Bake)—1 med sl (approx 2" x 1½" x ¼") (14 gm)	3	3	2	2	2	2	2	2	26	6	1.6	13
—½ med breast (yield after cooking, bone & skin removed) (88 gm)	17	16	14	14	13	12	11	11	164	37	10.2	80
Breast, w/or wo/bone, broiled, skin eaten—1 med sl (approx 2" x 1½" x ¼") (14 gm)	3	3	3	2	2	2	2	2	27	10	2.7	12
—½ med breast (yield after cooking, bone removed) (92 gm)	20	18	17	15	14	13	13	12	180	64	18.0	77
Breast, w/or wo/bone, broiled, skin not eaten—1 med sl (approx 2" x 1½" x ¼") (14 gm)	2	2	2	2	2	2	2	2	23	4	1.3	12
—½ med breast (yield after cooking, bone & skin removed) (81 gm)	14	13	12	11	11	10	9	9	133	26	7.3	68
Breast, w/or wo/bone, floured, baked or fried, skin eaten—1 med sl (approx 2" x 1½" x ¼") (14 gm)	3	3	3	2	2	2	2	2	31	11	3.1	12
—½ med breast (yield after cooking, bone removed) (98 gm)	23	21	19	18	17	15	15	14	217	78	21.5	87
Breast, w/or wo/bone, floured, baked or fried, skin not eaten—1 med sl (approx 2" x 1½" x ¼") (14 gm)	3	3	2	2	2	2	2	2	26	6	1.6	13
—½ med breast (yield after cooking, bone & skin removed) (86 gm)	17	15	14	13	13	12	11	11	160	36	10.0	78
Breast, w/or wo/bone, fried, no coating, skin eaten (inc sautéed)—1 med sl (approx 2" x 1½" x ¼") (14 gm)	3	3	3	3	2	2	2	2	30	11	3.1	13

Chicken (cont.)

	1200	1500	1800	2100	2400	2800	3200	3600	TOT CAL	FAT CAL	S/FAT CAL	CHOL MG
——½ med breast (yield after cooking, bone removed) (98 gm)	24	21	19	18	17	16	15	14	213	79	21.9	89
Breast, w/or wo/bone, fried, no coating, skin not eaten (inc sautéed)—1 med sl (approx 2" x 1½" x ¼") (14 gm)	3	3	2	2	2	2	2	2	26	6	1.6	13
——½ med breast (yield after cooking, bone & skin removed) (86 gm)	17	15	14	13	13	12	11	11	159	37	10.0	79
Breast, w/or wo/bone, roasted, skin eaten—1 med sl (approx 2" x 1½" x ¼") (14 gm)	3	3	3	2	2	2	2	2	27	10	2.7	12
——½ med breast (yield after cooking, bone removed) (98 gm)	21	19	18	16	15	14	13	13	192	68	19.2	82
Breast, w/or wo/bone, roasted, skin not eaten—1 med sl (approx 2" x 1½" x ¼") (14 gm)	2	2	2	2	2	2	2	2	23	4	1.3	12
——½ med breast (yield after cooking, bone & skin removed) (86 gm)	15	14	13	12	11	11	10	10	141	27	7.8	73
Breast, w/or wo/bone, skinless, battered, fried, batter eaten (inc Chicken Tenders)—1 tender (piece) (12 gm)	3	2	2	2	2	2	2	2	28	10	2.7	10
Breast, w/or wo/bone, skinless, breaded, cooked, breading eaten (inc Weaver Crispy Light Fried Chicken)—½ breast (124 gm)	24	22	20	19	17	16	15	15	282	87	19.8	97

Food												
Breast, w/or wo/bone, stewed, skin eaten—1 med sl (approx 2" x 1½" x ¼") (14 gm)	3	2	2	2	2	2	2	2	26	9	2.6	10
—½ med breast (yield after cooking, bone removed) (110 gm)	22	20	18	17	16	15	14	13	201	73	20.5	82
Breast, w/or wo/bone, stewed, skin not eaten—1 med sl (approx 2" x 1½" x ¼") (14 gm)	2	2	2	2	2	2	1		21	4	1.1	11
—½ med breast (yield after cooking, bone & skin removed) (95 gm)	15	13	12	12	11	11	10	10	143	26	7.2	73
Broilers or fryers, neck, meat only, simmered—1 neck (18 gm)	4	3	3	3	3	2	2	2	32	14	3.4	14
Broilers or fryers, neck, meat & skin, simmered—1 neck (38 gm)	12	10	9	8	8	7	7	6	94	62	17.1	27
Canned, meat only, dark meat—1 can (5 oz) yields (125 gm)	23	21	19	17	16	15	14	13	206	89	24.8	78
Canned, meat only, light & dark meat—1 can (5 oz) yields (125 gm)	23	21	19	17	16	15	14	13	206	89	24.8	78
Canned, meat only, light meat—1 can (5 oz) yields (125 gm)	23	21	19	17	16	15	14	13	206	89	24.8	78
Chicken roll, roasted, dark meat—3 slices (85 gm)	14	12	11	10	9	9	8	8	135	56	15.5	43
Chicken roll, roasted, light & dark meat—3 sl (85 gm)	14	12	11	10	9	9	8	8	135	56	15.5	43
Chicken roll, roasted, light meat—3 sl (85 gm)	14	12	11	10	9	9	8	8	135	56	15.5	43
Drumstick, w/or wo/bone, battered, fried, skin eaten (inc extra crispy fried chicken)—1 med drumstick (yield after cooking, bone removed) (72 gm)	22	19	17	16	15	14	13	12	193	102	26.8	62
Drumstick, w/or wo/bone, battered, fried, skin not eaten (inc extra crispy fried chicken)—1 med drumstick (yield after cooking, bone & skin removed) (46 gm)	11	10	9	8	8	7	7	6	89	33	8.7	43

	1200	1500	1800	2100	2400	2800	3200	3600	TOT CAL	FAT CAL	S/FAT CAL	CHOL MG
Chicken *(cont.)*												
Drumstick, w/or wo/bone, breaded, baked or fried, skin eaten (inc reg fried chicken, Shake-n-Bake)—1 med drumstick (yield after cooking, bone removed) (72 gm)	20	18	16	15	14	13	12	11	184	91	24.1	61
Drumstick, w/or wo/bone, breaded, baked or fried, skin not eaten (inc reg fried chicken, Shake-n-Bake)—1 med drumstick (yield after cooking, bone & skin removed) (46 gm)	11	10	9	8	8	7	7	6	89	33	8.7	43
Drumstick, w/or wo/bone, broiled, skin eaten—1 med drumstick (yield after cooking, bone removed) (52 gm)	14	12	11	10	10	9	8	8	112	52	14.2	47
Drumstick, w/or wo/bone, broiled, skin not eaten—1 med drumstick (yield after cooking, bone & skin removed) (45 gm)	9	8	8	7	7	6	6	6	77	23	6.0	42
Drumstick, w/or wo/bone, floured, baked or fried, skin eaten—1 med drumstick (yield after cooking, bone removed) (49 gm)	14	12	11	10	10	9	8	8	119	60	16.0	44
Drumstick, w/or wo/bone, floured, baked or fried, skin not eaten—1 med drumstick (yield after cooking, bone & skin removed) (42 gm)	10	9	8	7	7	7	6	6	81	30	8.0	39
Drumstick, w/or wo/bone, fried, no coating, skin eaten—1 med drumstick (yield after cooking, bone removed) (49 gm)	14	13	12	11	10	9	9	8	118	61	16.3	45

Food												
Drumstick, w/or wo/bone, fried, no coating, skin not eaten—1 med drumstick (yield after cooking, bone & skin removed) (42 gm)	10	9	8	7	7	7	6	6	6	81	30	8.0	39
Drumstick, w/or wo/bone, roasted, skin eaten—1 med drumstick (yield after cooking, bone removed) (52 gm)	14	12	11	10	10	9	8	8	8	112	52	14.2	47
Drumstick, w/or wo/bone, roasted, skin not eaten—1 med drumstick (yield after cooking, bone & skin removed) (44 gm)	9	8	8	7	7	6	6	6	6	75	22	5.8	41
Drumstick, w/or wo/bone, skinless, breaded, cooked, breading eaten (inc Weaver Crispy Light Fried Chicken)—1 drumstick (64 gm)	17	15	14	13	12	11	10	10	10	164	75	18.0	56
Drumstick, w/or wo/bone, stewed, skin eaten—1 med drumstick (yield after cooking, bone removed) (57 gm)	14	12	11	10	10	9	8	8	8	116	54	14.8	47
Drumstick, w/or wo/bone, stewed, skin not eaten—1 medium drumstick (yield after cooking, bone & skin removed) (46 gm)	9	8	8	7	7	6	6	6	6	77	23	6.2	40
Leg (drumstick & thigh), w/or wo/bone, battered, fried, skin eaten (inc extra crispy fried chicken)—1 med leg (yield after cooking, bone removed) (158 gm)	49	43	39	36	34	31	29	27	430	229	60.7	142	
Leg (drumstick & thigh), w/or wo/bone, battered, fried, skin not eaten (inc extra crispy fried chicken)—1 med leg (yield after cooking, bone & skin removed) (102 gm)	26	23	21	20	19	17	16	15	211	85	22.8	100	

THE 2-IN-1 SYSTEM

	1200	1500	1800	2100	2400	2800	3200	3600	TOT CAL	FAT CAL	S/FAT CAL	CHOL MG
Chicken (cont.)												
Leg (drumstick & thigh), w/or wo/bone, breaded, baked or fried, skin eaten (inc reg fried chicken, Shake-n-Bake) —1 med leg (yield after cooking, bone removed) (158 gm)	47	42	38	35	33	30	28	26	416	211	56.7	141
Leg (drumstick & thigh), w/or wo/bone, breaded, baked or fried, skin not eaten (inc regular fried chicken, Shake-n-Bake) —1 med leg (yield after cooking, bone & skin removed) (102 gm)	26	23	21	20	19	17	16	15	211	85	22.8	100
Leg (drumstick & thigh), w/or wo/bone, broiled, skin eaten—1 med leg (yield after cooking, bone removed) (109 gm)	32	28	26	24	22	20	19	18	251	131	36.3	100
Leg (drumstick & thigh), w/or wo/bone, broiled, skin not eaten—1 med leg (yield after cooking, bone & skin removed) (97 gm)	23	21	19	18	17	16	15	14	184	73	19.9	91
Leg (drumstick & thigh), w/or wo/bone, floured, baked or fried, skin eaten—1 med leg (yield after cooking, bone removed) (112 gm)	34	30	27	25	23	22	20	19	283	144	39.0	105
Leg (drumstick & thigh), w/or wo/bone, floured, baked or fried, skin not eaten—1 med leg (yield after cooking, bone & skin removed) (94 gm)	24	21	20	18	17	16	15	14	194	79	21.0	92

Food											
Leg (drumstick & thigh), w/or wo/bone, fried, no coating, skin eaten—1 med leg (yield after cooking, bone removed) (112 gm)	35	31	28	26	24	22	21	20	278	148	40.1	108
Leg (drumstick & thigh), w/or wo/bone, fried, no coating, skin not eaten—1 med leg (yield after cooking, bone & skin removed) (94 gm)	24	22	20	18	17	16	15	14	193	79	21.1	93
Leg (drumstick & thigh), w/or wo/bone, roasted, skin eaten—1 med leg (yield after cooking, bone removed) (114 gm)	33	29	27	25	23	21	20	19	263	137	37.9	104
Leg (drumstick & thigh), w/or wo/bone, roasted, skin not eaten—1 med leg (yield after cooking, bone & skin removed) (95 gm)	23	20	19	17	16	15	14	14	180	72	19.5	89
Leg (drumstick & thigh), w/or wo/bone, stewed, skin eaten—1 med leg (yield after cooking, bone removed) (125 gm)	34	30	27	25	24	22	20	19	273	144	39.9	104
Leg (drumstick & thigh), w/or wo/bone, stewed, skin not eaten—1 med leg (yield after cooking, bone & skin removed) (101 gm)	23	20	19	18	16	15	14	14	186	73	19.9	89
Neck or ribs, w/or wo/bone, breaded, baked or fried, skin eaten (inc reg fried chicken, Shake-n-Bake)—1 neck (yield after cooking, bone removed) (52 gm)	21	18	16	15	14	12	12	11	174	112	29.9	47
Neck or ribs, w/or wo/bone, breaded, baked or fried, skin not eaten (inc reg fried chicken, Shake-n-Bake)—1 neck (yield after cooking, bone & skin removed) (25 gm)	7	6	6	5	5	5	4	4	57	27	6.7	26

	1200	1500	1800	2100	2400	2800	3200	3600	TOT CAL	FAT CAL	S/FAT CAL	CHOL MG
Chicken *(cont.)*												
Neck or ribs, w/or wo/bone, floured, baked or fried, skin eaten—1 neck (yield after cooking, bone removed) (36 gm)	14	12	11	10	9	9	8	7	119	76	20.4	33
Neck or ribs, w/or wo/bone, floured, baked or fried, skin not eaten—1 neck (yield after cooking, bone & skin removed) (22 gm)	6	6	5	5	4	4	4	4	50	23	5.9	23
Neck or ribs, w/or wo/bone, fried, no coating, skin eaten—1 neck (yield after cooking, bone removed) (36 gm)	15	13	12	11	10	9	8	8	117	80	21.4	35
Neck or ribs, w/or wo/bone, fried, no coating, skin not eaten—1 neck (yield after cooking, bone & skin removed) (22 gm)	6	6	5	5	4	4	4	4	49	24	6.0	23
Neck or ribs, w/or wo/bone, roasted, skin eaten—1 neck (yield after cooking, bone removed) (36 gm)	17	15	13	12	11	10	10	9	107	74	20.5	48
Neck or ribs, w/or wo/bone, roasted, skin not eaten—1 neck (yield after cooking, bone & skin removed) (24 gm)	8	7	6	6	5	5	5	4	56	32	8.4	26
Neck or ribs, w/or wo/bone, stewed, skin eaten—1 neck (yield after cooking, bone removed) (38 gm)	12	10	9	8	8	7	6	6	93	62	17.0	26
Neck or ribs, w/or wo/bone, stewed, skin not eaten—1 neck (yield after cooking, bone & skin removed) (18 gm)	4	3	3	3	3	2	2	2	32	13	3.4	14

Food												
Nuggets—6 nuggets (108 gm)	34	29	26	24	22	20	19	17	320	190	50.3	72
Patty, breaded, cooked (inc Rondelets)—1 patty (75 gm)	23	20	18	17	15	14	13	12	222	132	35.0	50
Patty w/cheese, breaded, cooked (inc Cheese Recipe Chicken Rondelets)—1 patty (3 oz) (85 gm)	22	19	17	15	14	12	11	11	236	126	37.1	33
Poultry, mechanically deboned, fr backs & necks wo/skin, raw—½ lb (227 gm)	79	70	64	59	55	50	47	44	450	316	96.0	236
Poultry, mechanically deboned, fr backs & necks w/skin, raw—½ lb (227 gm)	113	100	90	83	77	70	65	61	616	505	152.1	295
Poultry, mechanically deboned, fr mature hens, raw—½ lb (227 gm)	93	84	76	71	66	61	57	54	551	408	96.5	324
Roasting, dark meat, meat only,—(94 gm)	20	18	16	15	14	13	12	12	168	74	20.5	70
Roasting, light meat, meat only—(78 gm)	12	11	11	10	9	9	8	8	119	29	7.6	58
Roasting, meat & skin—½ chicken (480 gm)	129	114	103	95	88	81	75	71	1071	579	161.4	365
Stewing, dark meat, meat only—(73 gm)	23	20	18	17	16	14	13	13	188	101	26.7	69
Stewing, meat & skin—½ chicken (261 gm)	85	75	67	62	57	52	48	45	744	443	120.1	205
Thigh, w/or wo/bone, battered, fried, skin eaten (inc extra crispy fried chicken)—1 med thigh (yield after cooking, bone removed) (86 gm)	28	24	22	20	19	17	16	15	237	127	34.0	80
Thigh, w/or wo/bone, battered, fried, skin not eaten (inc extra crispy fried chicken)—1 med thigh (yield after cooking, bone & skin removed) (54 gm)	15	13	12	11	10	10	9	9	117	50	13.4	55
Thigh, w/or wo/bone, breaded, baked or fried, skin eaten (inc reg fried chicken, Shake-n-Bake)—1 med thigh (yield after cooking, bone removed) (86 gm)	27	24	22	20	19	17	16	15	231	120	32.4	80

	1200	1500	1800	2100	2400	2800	3200	3600	TOT CAL	FAT CAL	S/FAT CAL	CHOL MG
Chicken (cont.)												
Thigh, w/or wo/bone, breaded, baked, or fried, skin not eaten (inc reg fried chicken, Shake-n-Bake)—1 med thigh (yield after cooking, bone & skin removed) (54 gm)	15	13	12	11	10	10	9	9	117	50	13.4	55
Thigh, w/or wo/bone, broiled, skin eaten—1 med thigh (yield after cooking, bone removed) (58 gm)	18	16	15	14	13	12	11	10	142	80	22.5	54
Thigh, w/or wo/bone, broiled, skin not eaten—1 med thigh (yield after cooking, bone & skin removed) (52 gm)	14	12	11	11	10	9	9	8	108	51	14.1	49
Thigh, w/or wo/bone, floured, baked, or fried, skin eaten—1 med thigh (yield after cooking, bone removed) (62 gm)	19	17	16	14	13	12	12	11	161	83	22.7	60
Thigh, w/or wo/bone, floured, baked or fried, skin not eaten—1 med thigh (yield after cooking, bone & skin removed) (52 gm)	14	13	12	11	10	9	9	8	112	48	12.9	53
Thigh, w/or wo/bone, fried, no coating, skin eaten—1 med thigh (yield after cooking, bone removed) (62 gm)	20	18	16	15	14	13	12	11	158	86	23.5	62
Thigh, w/or w/bone, fried, no coating, skin not eaten—1 med thigh (yield after cooking, bone & skin removed) (52 gm)	14	13	12	11	10	10	9	8	111	48	13.1	54

Food												
Thigh, w/or wo/bone, roasted, skin eaten—1 med thigh (yield after cooking, bone removed) (62 gm)	20	17	16	14	13	12	12	11	152	86	24.0	57
Thigh, w/or wo/bone, roasted, skin not eaten—1 med thigh (yield after cooking, bone & skin removed) (52 gm)	14	12	11	11	10	9	9	8	108	51	14.1	49
Thigh, w/or wo/bone, skinless, breaded, cooked, breading eaten (inc Weaver Crispy Light Fried Chicken)—1 thigh (78 gm)	18	16	15	14	13	12	11	10	195	88	19.1	61
Thigh, w/or wo/bone, stewed, skin eaten—1 med thigh (yield after cooking, bone removed) (68 gm)	20	18	16	15	14	13	12	11	157	90	25.0	57
Thigh, w/or wo/bone, stewed, skin not eaten—1 med thigh (yield after cooking, bone & skin removed) (55 gm)	14	12	11	10	10	9	8	8	107	48	13.3	49
Wing, w/or wo/bone, battered, fried, skin eaten (inc extra crispy fried chicken)—1 med wing (yield after cooking, bone removed) (49 gm)	17	15	14	13	12	11	10	9	159	96	25.7	39
Wing, w/or wo/bone, battered, fried, skin not eaten (inc extra crispy fried chicken)—1 med wing (yield after cooking, bone & skin removed) (22 gm)	5	4	4	4	4	3	3	3	46	18	4.9	18
Wing, w/or wo/bone, breaded, baked or fried, skin eaten (inc reg fried chicken, Shake-n-Bake)—1 med wing (yield after cooking, bone removed) (49 gm)	18	15	14	13	12	11	10	9	159	98	26.5	37
Wing, w/or wo/bone, breaded, baked or fried, skin not eaten (inc reg fried chicken, Shake-n-Bake)—1 med wing (yield after cooking, bone & skin removed) (22 gm)	5	4	4	4	4	3	3	3	46	18	4.9	18

Chicken (cont.)

	1200	1500	1800	2100	2400	2800	3200	3600	TOT CAL	FAT CAL	S/FAT CAL	CHOL MG
Wing, w/or wo/bone, broiled, skin eaten—1 med wing (yield after cooking, bone removed) (35 gm)	12	11	10	9	8	7	7	6	101	61	17.1	29
Wing, w/or wo/bone, broiled, skin eaten—1 med wing (yield after cooking, bone & skin removed) (20 gm)	4	4	4	3	3	3	3	3	40	15	4.0	17
Wing, w/or wo/bone, floured, baked or fried, skin eaten—1 med wing (yield after cooking, bone removed) (32 gm)	12	10	9	8	8	7	7	6	102	63	17.3	26
Wing, w/or wo/bone, floured, baked or fried, skin not eaten—1 med wing (yield after cooking, bone & skin removed) (20 gm)	5	4	4	4	3	3	3	3	42	16	4.5	17
Wing, w/or wo/bone, fried, no coating, skin eaten—1 med wing (yield after cooking, bone removed) (32 gm)	12	10	9	9	8	7	7	6	101	65	17.8	26
Wing, w/or wo/bone, fried, no coating, skin not eaten—1 med wing (yield after cooking, bone & skin removed) (20 gm)	5	4	4	4	3	3	3	3	42	16	4.5	17
Wing, w/or wo/bone, roasted, skin eaten—1 med wing (yield after cooking, bone removed) (34 gm)	12	10	9	8	8	7	7	6	98	59	16.6	28
Wing, w/or wo/bone, roasted, skin not eaten—1 med wing (yield after cooking, bone & skin removed) (21 gm)	5	4	4	4	3	3	3	3	42	15	4.2	18

Food												
Wing, w/or wo/bone, stewed, skin eaten—1 med wing (yield after cooking, bone removed) (40 gm)	12	10	9	9	8	7	7	6	99	60	16.9	28
Wing, w/or wo/bone, stewed, skin not eaten—1 med wing (yield after cooking, bone and skin removed) (24 gm)	5	4	4	4	3	3	3	3	43	15	4.3	18

Duck

Food												
Battered, fried—4 oz, w/bone, ckd (yield after bone removed) (88 gm)	28	25	23	21	20	18	17	16	195	87	29.5	95
—1 drumstick (yield after cooking, bone removed) (70 gm)	22	20	18	17	16	15	14	13	155	69	23.5	76
—1 back (yield after cooking, bone removed) (96 gm)	30	27	25	23	22	20	19	18	213	95	32.2	104
Pressed, Chinese—4 oz, cooked (113 gm)	21	19	17	15	14	13	12	11	232	107	33.3	42
Roasted, skin eaten—1 wing (yield after cooking, bone removed) (30 gm)	15	13	12	11	10	9	8	8	100	76	26.0	25
—4 oz, w/bone, cooked (yield after bone removed) (72 gm)	37	32	29	26	24	21	20	18	241	183	62.3	60
—1 leg (drumstick & thigh) (yield after cooking, bone removed) (55 gm)	28	25	22	20	18	16	15	14	184	139	47.6	46
Roasted, skin not eaten—1 wing (yield after cooking, bone & skin removed) (25 gm)	8	7	6	6	5	5	4	4	50	25	9.3	22
—4 oz, w/bone, ckd (yield after bone & skin removed) (52 gm)	16	14	13	12	11	10	9	9	104	52	19.4	46
—1 leg (drumstick & thigh) (yield after cooking, bone & skin removed) (46 gm)	14	12	11	10	10	9	8	8	92	46	17.2	41

Turkey

	1200	1500	1800	2100	2400	2800	3200	3600	TOT CAL	FAT CAL	S/FAT CAL	CHOL MG
All classes, neck, meat only, simmered —1 neck (152 gm)	44	40	37	34	32	30	28	27	274	99	33.3	186
Back, cooked—4 oz, w/bone, ckd (yield after bone removed) (68 gm)	21	18	17	15	14	13	12	12	164	87	25.4	61
Canned—1 can (5 oz) yields (125 gm)	23	21	19	17	16	15	14	13	204	77	22.5	83
Canned, meat only, w/broth—½ can (2.5 oz) (71 gm)	13	12	11	10	9	9	8	8	116	44	12.8	47
Dark meat, roasted, skin eaten—1 med sl (approx 3" x 2" x ¼") (28 gm)	8	7	6	6	5	5	5	4	62	29	8.7	25
4 oz, boneless (113 gm)	31	28	25	23	22	20	19	18	249	117	35.4	100
Dark meat, roasted, skin not eaten—1 med sl (approx 3" x 2" x ¼") (28 gm)	6	6	5	5	5	4	4	4	52	18	6.1	24
4 oz, boneless (yield after skin removed) (104 gm)	24	21	20	18	17	16	15	14	193	67	22.5	88
Drumstick, cooked, skin eaten—1 med drumstick (approx 14–18 lb bird) (yield after cooking, bone removed) (216 gm)	55	49	45	41	39	36	33	31	447	190	59.1	183
Drumstick, cooked, skin not eaten—1 med drumstick (approx 14–18 lb bird) (yield after cooking, bone & skin removed) (199 gm)	45	41	37	35	33	30	28	27	370	129	43.1	168
Drumstick, roasted, skin eaten—1 med drumstick (approx 14–18 lb bird) (yield after cooking, bone removed) (204 gm)	52	46	42	39	36	34	31	30	422	179	55.9	172

Food												
Drumstick, roasted, skin not eaten—1 med drumstick (approx 14–18 lb bird) (yield after cooking, bone & skin removed) (188 gm)	43	39	35	33	31	29	27	25	349	121	40.7	159
Drumstick, smoked, cooked, skin eaten—1 med drumstick (approx 14–18 lb bird) (yield after cooking, bone removed) (204 gm)	53	47	43	40	38	35	33	31	420	176	54.9	184
Ground—1 c (110 gm)	26	24	22	20	19	17	16	15	227	96	27.9	90
Light & dark meat, roasted, skin eaten—1 med sl (approx 3" x 2" x ¼") (28 gm)	7	6	5	5	5	4	4	4	58	24	7.1	23
—4 oz, boneless (113 gm)	27	24	22	20	19	18	17	16	234	99	28.8	92
Light & dark meat, roasted, skin not eaten—1 med sl (approx 3" x 2" x ¼") (28 gm)	5	5	4	4	4	3	3	3	47	12	4.1	21
—4 oz, boneless (yield after skin removed) (104 gm)	19	17	16	15	14	13	12	12	176	46	15.3	79
Light meat, breaded, baked or fried, skin eaten—1 med sl (approx 3" x 2" x ¼") (28 gm)	6	5	5	5	4	4	4	3	63	25	6.9	19
—4 oz, boneless (113 gm)	25	22	20	18	17	16	15	14	253	103	28.0	78
Light meat, breaded, baked or fried, skin not eaten—1 med sl (approx 3" x 2" x ¼") (28 gm)	4	4	3	3	3	3	3	3	46	11	3.2	19
—4 oz, boneless (yield after skin removed) (100 gm)	16	14	13	12	12	11	10	10	165	38	11.5	67
Light meat, cooked, skin eaten (inc turkey breast from deli)—1 med slice (approx 3" x 2" x ¼") (28 gm)	6	5	5	4	4	4	4	3	55	21	5.9	21
Light meat, cooked, skin not eaten (inc turkey breast from deli, turkey cutlet)—1 med sl (approx 3" x 2" x ¼") (28 gm)	4	4	3	3	3	3	3	3	44	8	2.6	19
—4 oz, boneless (113 gm)	24	21	20	18	17	16	15	14	222	84	23.7	86

Turkey (cont.)

	1200	1500	1800	2100	2400	2800	3200	3600	TOT CAL	FAT CAL	S/FAT CAL	CHOL MG
4 oz, boneless (yield after skin removed) (100 gm)	15	14	13	12	11	10	10	9	156	29	9.2	69
Light meat, roasted, skin eaten—1 med sl (approx 3" x 2" x 1/4") (28 gm)	6	5	5	4	4	4	4	3	55	21	5.9	21
4 oz, boneless (113 gm)	24	21	20	18	17	16	15	14	222	84	23.7	86
Light meat, roasted, skin not eaten—med sl (approx 3" x 2" x 1/4") (28 gm)	4	4	3	3	3	3	3	3	44	8	2.6	19
4 oz, boneless (yield after skin removed) (100 gm)	15	14	13	12	11	10	10	9	156	29	9.2	69
Light or dark meat, battered, fried, skin eaten—1 med sl (approx 3" x 2" x 1/4") (28 gm)	8	7	6	6	5	5	4	4	79	45	11.8	17
4 oz, boneless (113 gm)	32	28	25	23	21	19	18	17	321	184	47.8	70
Light or dark meat, battered, fried, skin not eaten—1 med sl (approx 3" x 2" x 1/4") (28 gm)	5	5	4	4	4	4	3	3	51	16	4.9	20
4 oz, boneless (yield after skin removed) (104 gm)	20	18	16	15	14	13	13	12	191	58	18.0	76
Light or dark meat, smoked, cooked, skin eaten—1 med sl (approx 3" x 2" x 1/4") (28 gm)	7	6	6	5	5	4	4	4	58	24	7.2	23
4 oz, boneless (113 gm)	27	24	22	21	19	18	17	16	236	99	29.0	93
Light or dark meat, smoked, cooked, skin not eaten—1 med sl (approx 3" x 2" x 1/4") (28 gm)	5	5	4	4	4	3	3	3	48	13	4.1	21
4 oz, boneless (yield after skin removed) (104 gm)	19	17	16	15	14	13	12	12	177	47	15.3	79

Food												
Light or dark meat, stewed, skin eaten—1 med sl (approx 3" x 2" x ¼") (28 gm)	6	5	5	4	4	4	4	3	48	16	4.3	24
—4 oz, boneless (113 gm)	23	21	19	18	17	16	15	14	194	63	17.3	96
Light or dark meat, stewed, skin not eaten—1 med sl (approx 3" x 2" x ¼") (28 gm)	6	5	5	5	4	4	4	4	49	16	5.2	23
—4 oz, boneless (yield after skin removed) (104 gm)	22	20	18	17	16	15	14	13	182	61	19.3	87
Neck, cooked—1 neck (152 gm)	43	39	36	34	32	30	28	27	272	99	33.2	184
Nuggets—6 nuggets (108 gm)	23	21	19	17	16	14	13	13	279	116	31.6	60
Tail, cooked—1 tail (yield after cooking, bone removed (58 gm)	18	16	14	13	12	11	10	10	140	75	21.7	52
Thigh, cooked, skin eaten—1 med thigh (approx 14–18 lb bird) (yield after cooking, bone removed 297 gm)	75	67	61	57	53	49	46	43	614	261	81.3	251
Thigh, cooked, skin not eaten—1 med thigh (approx 14–18 lb bird) (yield after cooking, bone & skin removed 274 gm)	62	56	51	48	45	42	39	37	509	177	59.3	232
Turkey patties, breaded, battered, fried—1 patty (2.25 oz) (64 gm)	18	16	14	13	12	11	10	9	181	104	27.0	40
Turkey roll, light & dark meat—2 sl (2 oz) (57 gm)	9	8	8	7	7	6	6	5	84	36	10.4	31
Turkey roll, light meat—2 sl (2 oz) (57 gm)	8	7	7	6	6	5	5	5	83	37	10.3	24
Turkey roll, light or dark meat, cooked—3 sl (85 gm)	13	12	11	10	9	9	8	8	132	44	14.5	45
Turkey sticks, breaded, battered, fried—1 stick (2.25 oz) (64 gm)	18	15	14	13	12	11	10	9	178	97	25.3	41
Wing, cooked, skin eaten—1 med wing (yield after cooking, bone removed) (174 gm)	45	40	37	34	31	29	27	25	396	193	52.8	140

	1200	1500	1800	2100	2400	2800	3200	3600	TOT CAL	FAT CAL	S/FAT CAL	CHOL MG
Turkey *(cont.)*												
Wing, cooked, skin not eaten—1 med wing (yield after cooking, bone & skin removed) (146 gm)	22	20	18	17	16	15	14	14	228	42	13.5	100
Wing, smoked, cooked, skin eaten—1 med wing (yield after cooking, bone removed) (174 gm)	44	39	35	33	31	28	26	25	385	180	49.0	141
Other Poultry												
Cornish game hen, cooked, skin eaten—1 leg (drumstick & thigh) (yield after cooking, bone removed) (57 gm)	16	15	13	12	11	10	10	9	135	69	19.3	50
—1 breast (yield after cooking, bone removed) (110 gm)	32	28	25	23	22	20	19	18	261	134	37.3	96
—½ hen (10 oz, raw) (yield after cooking, bone removed) (153 gm)	44	39	35	33	30	28	26	25	363	186	51.9	134
Cornish game hen, cooked, skin not eaten—1 leg (drumstick & thigh) (yield after cooking, bone & skin removed) (48 gm)	10	9	9	8	8	7	7	6	91	32	8.8	42
—1 breast (yield after cooking, bone and skin removed) (90 gm)	20	18	16	15	14	13	13	12	170	60	16.4	80
—½ hen (10 oz, raw) (yield after cooking, bone and skin removed) (125 gm)	27	25	23	21	20	19	18	17	236	83	22.8	111
Cornish game hen, roasted, skin eaten—1 leg (drumstick & thigh) (yield after cooking, bone removed) (57 gm)	16	15	13	12	11	10	10	9	135	69	19.3	50

Food												
—1 breast (yield after cooking, bone removed) (110 gm)	32	28	25	23	22	20	19	18	261	134	37.3	96
—½ hen (10 oz, raw) (yield after cooking, bone removed) (153 gm)	44	39	35	33	30	28	26	25	363	186	51.9	134
Cornish game hen, roasted, skin not eaten—1 leg (drumstick & thigh) (yield after cooking, bone & skin removed) (48 gm)	10	9	9	8	8	7	7	6	91	32	8.8	42
—1 breast (yield after cooking, bone & skin removed) (90 gm)	20	18	16	15	14	13	13	12	170	60	16.4	80
—½ hen (10 oz, raw) (yield after cooking, bone and skin removed) (125 gm)	27	25	23	21	20	19	18	17	236	83	22.8	111
Dove, cooked (inc squab, pigeon)—1 dove (yield after cooking, bone removed) (111 gm)	35	32	29	27	25	23	22	21	242	129	34.8	128
Dove, fried (inc squab, pigeon)—1 dove (yield after cooking, bone removed) (111 gm)	36	32	29	27	25	24	22	21	256	139	37.1	124
Goose, domesticated, meat only, roasted—1 lb goose (143 gm)	48	42	38	35	33	30	28	26	340	163	58.7	138
Goose, domesticated, meat & skin, roasted—1 lb goose (188 gm)	78	69	62	56	52	47	44	41	574	371	116.3	172
Goose, wild, roasted—4 oz, w/bone, ckd (yield after bone removed) (76 gm)	31	28	25	23	21	19	18	16	230	149	46.7	69
Guinea hen, meat & skin, raw—½ guinea (345 gm)	64	58	53	50	47	44	41	39	545	201	55.0	255
Guinea hen, meat only, raw—½ guinea (264 gm)	33	30	28	26	25	24	22	21	292	59	15.3	166
Pheasant, cooked (inc grouse)—1 pheasant leg (drumstick & thigh) (yield after cooking, bone removed) (75 gm)	21	18	17	16	14	13	12	12	184	81	23.6	66

	1200	1500	1800	2100	2400	2800	3200	3600	TOT CAL	FAT CAL	S/FAT CAL	CHOL MG
Other Poultry (cont.)												
½ pheasant breast (yield after cooking, bone removed) (127 gm)	35	31	28	26	25	23	21	20	312	137	39.9	112
Quail, cooked (inc partridge)—1 quail (yield after cooking, bone removed) (76 gm)	22	20	18	16	15	14	13	12	177	96	26.8	65
Squab (pigeon), light meat wo/skin, raw—1 breast (101 gm)	19	17	16	15	14	14	13	12	135	41	10.7	91
Squab (pigeon), meat only, raw—1 squab (168 gm)	37	33	31	29	27	25	24	22	239	113	29.6	151
Squab (pigeon), meat & skin, raw—1 squab (199 gm)	96	84	75	68	63	57	53	49	584	427	151.0	189
Poultry Organs												
Chicken broilers or fryers, giblets, fried—1 c (145 gm)	123	113	106	100	95	90	85	82	402	176	49.5	647
Chicken broilers or fryers, giblets, simmered—1 c (145 gm)	97	91	86	81	78	74	70	68	228	62	19.4	570
Chicken, giblets, excluding liver, cooked—1 c diced (145 gm)	50	46	43	41	39	37	35	34	221	47	13.5	280
Chicken giblets, simmered—1 c (145 gm)	96	89	83	78	75	71	67	64	281	122	34.7	515
Chicken gizzard, all classes, simmered—1 c (145 gm)	50	46	44	41	39	37	36	34	222	48	13.6	281
Chicken, gizzard, cooked—1 gizzard (22 gm)	7	7	6	6	6	6	5	5	33	7	2.0	42
Chicken heart, all classes, simmered—1 c (145 gm)	68	62	58	55	52	49	47	45	268	104	29.4	350

Food												
Chicken liver, battered, fried—½ c, diced (70 gm)	64	60	56	54	51	48	46	44	150	54	16.0	368
Chicken liver, braised—½ c diced (70 gm)	74	69	65	62	59	56	54	51	109	34	11.5	439
Chicken liver, breaded, fried—½ c diced (70 gm)	70	65	61	58	55	52	50	48	159	56	16.8	400
Chicken liver, fried or sautéed, no coating—½ c diced (70 gm)	74	69	65	61	59	56	53	51	144	57	17.3	425
Duck, domesticated, liver, raw—1 liver (44 gm)	38	35	33	32	30	29	28	27	60	18	5.7	227
Goose, liver, raw—1 liver (94 gm)	81	76	72	68	65	62	59	57	125	36	13.4	484
Poultry liver, other than chicken or turkey, cooked—1 oz, raw (yield after cooking) (22 gm)	23	22	20	19	19	18	17	16	34	11	3.6	138
Turkey giblets, simmered, some giblet fat—1 c (145 gm)	103	96	91	86	83	78	75	72	243	67	20.2	606
Turkey gizzard, all classes, simmered—1 c (145 gm)	59	55	51	49	47	44	42	40	236	50	14.5	336
—1 gizzard, turkey, ckd (67 gm)	23	21	20	19	18	17	16	16	102	22	6.2	129
Turkey heart, all classes, simmered—1 c (145 gm)	61	57	53	50	48	45	43	41	257	79	22.9	327
Turkey liver, all classes, simmered—1 c (140 gm)	147	137	130	123	118	112	107	103	237	75	23.7	876
Turkey liver, cooked—½ c (70 gm)	73	68	65	61	59	56	53	51	118	37	11.8	436
SANDWICHES												
Bacon & cheese sandwich w/mayonnaise —(121 gm)	65	56	49	44	40	36	33	30	439	252	123.7	65
Bacon & egg sandwich—(177 gm)	102	92	85	80	75	71	67	63	391	196	72.5	445
Bacon, chicken, & tomato club sandwich (lettuce, tomato, mayonnaise)—(246 gm)	33	29	26	23	22	20	18	17	502	222	51.8	65

Sandwiches (cont.)

	1200	1500	1800	2100	2400	2800	3200	3600	TOT CAL	FAT CAL	S/FAT CAL	CHOL MG
Bacon, lettuce, & tomato sandwich (lettuce, tomato, mayonnaise)—(164 gm)	23	20	17	16	14	13	12	11	346	172	44.1	22
Bacon sandwich w/mayonnaise—(91 gm)	37	32	28	25	23	21	19	17	390	226	71.2	37
Beef barbecue or Sloppy Joe, on bun—(186 gm)	32	28	25	22	21	19	17	16	387	145	52.3	56
Bologna & cheese sandwich w/mayonnaise—(111 gm)	45	39	34	31	28	25	23	21	365	201	87.1	42
Bologna sandwich w/mayonnaise—(83 gm)	19	16	14	13	12	10	9	9	259	122	37.5	16
Cheeseburger, 1/4 lb meat, plain, on bun—(184 gm)	70	61	54	49	45	40	37	34	554	264	120.6	107
Cheeseburger, 1/4 lb meat, w/mayonnaise or salad dressing, on bun (inc lettuce, pickles, onions, and/or mustard)—(228 gm)	76	65	58	52	48	43	40	37	630	336	131.3	112
Cheeseburger, 1/4 lb meat, w/mayonnaise or salad dressing & tomatoes, on bun (inc lettuce, pickles, onions, and/or mustard)—(274 gm)	77	67	59	54	49	44	41	38	656	341	134.5	114
Cheeseburger, 1/4 lb meat, w/tomato and/or catsup, on bun (inc lettuce, pickles, onions, and/or mustard)—(220 gm)	70	60	54	49	44	40	37	34	575	267	120.3	106
Cheeseburger, plain, on bun—(107 gm)	33	29	26	23	21	19	18	16	323	132	58.6	48
Cheeseburger w/mayonnaise or salad dressing, tomato & bacon, on bun (inc lettuce, pickles, onions, and/or mustard)—(288 gm)	89	77	68	61	56	51	46	43	741	407	156.0	126

Food												
Cheese sandwich—1 sandwich (83 gm)	31	26	23	21	19	17	15	14	276	129	60.2	27
Cheese sandwich, grilled—1 sandwich (83 gm)	34	29	25	23	20	18	17	15	306	158	66.1	27
Cheese sandwich, hoagie—1 sandwich (156 gm)	77	66	58	52	48	43	39	36	484	240	146.6	79
Cheese spread sandwich—1 sandwich (78 gm)	19	16	14	13	12	10	9	9	216	68	37.0	16
Chicken barbecue sandwich—(119 gm)	14	13	12	11	10	9	9	8	252	54	14.1	51
Chicken fillet sandwich—(126 gm)	25	22	20	19	17	16	15	14	329	113	30.4	74
Chicken frank, plain, on bun (inc turkey frank, plain, on bun)—(85 gm)	19	17	15	14	13	12	11	10	235	99	27.4	45
Chicken salad or chicken spread sandwich—(113 gm)	15	13	12	11	10	9	9	8	279	113	20.9	37
Chicken sandwich w/mayonnaise (inc sliced, roast chicken sandwich)—(112 gm)	15	13	12	11	11	10	9	9	269	73	15.2	52
Chiliburger, on bun (inc hamburger w/chili)—(159 gm)	40	35	31	28	26	24	22	20	408	170	65.1	72
Corn dog (frankfurter w/cornbread coating)—(88 gm)	34	29	26	24	21	19	18	16	278	160	60.8	46
Corned beef sandwich—(130 gm)	21	19	17	15	14	13	12	11	250	76	30.3	51
Corny dog, w/chili, on bun—(162 gm)	37	31	28	25	23	21	19	17	432	181	66.2	46
Crab cake sandwich, on bun—(140 gm)	25	23	21	19	18	17	16	15	318	85	18.4	107
Cuban sandwich, not Puerto Rican style, w/mayonnaise—(227 gm)	53	46	40	37	33	30	28	25	672	268	93.3	74
Double cheeseburger (2 patties), plain, on bun—(158 gm)	64	55	49	44	41	37	34	31	479	236	110.3	96
Double cheeseburger (2 patties), plain, on double-decker bun—(186 gm)	65	57	50	45	41	37	34	32	562	250	113.7	96
Double cheeseburger (2 patties), w/mayonnaise or salad dressing, on bun (inc lettuce, pickles, onions, and/or mustard)—(187 gm)	67	58	51	46	42	38	35	32	518	272	115.7	99

THE 2-IN-1 SYSTEM

Sandwiches (cont.)

	1200	1500	1800	2100	2400	2800	3200	3600	TOT CAL	FAT CAL	S/FAT CAL	CHOL MG
Double cheeseburger (2 patties), w/mayonnaise or salad dressing, on double-decker bun (inc lettuce, pickles, onions, and/or mustard)—(224 gm)	71	61	54	49	45	41	37	34	637	322	124.4	102
Double cheeseburger (2 patties), w/mayonnaise or salad dressing & tomatoes, on bun (inc lettuce, pickles, onions, and/or mustard)—(225 gm)	67	58	51	46	42	38	35	33	532	274	116.1	99
Double cheeseburger (2 patties), w/tomato and/or catsup, on bun (inc lettuce, pickles, onions, and/or mustard)—(192 gm)	64	55	49	44	41	37	34	31	500	239	110.5	96
Double hamburger (2 patties), plain, on bun—(130 gm)	38	33	29	27	24	22	20	19	374	158	60.7	69
Double hamburger (2 patties), w/mayonnaise or salad dressing & tomatoes, on double-decker bun (inc lettuce, pickles, onions, and/or mustard)—(241 gm)	45	39	35	31	29	26	24	22	554	245	75.0	75
Double hamburger (2 patties), w/mayonnaise or salad dressing, on bun (inc lettuce, pickles, onions, and/or mustard)—(159 gm)	41	35	31	28	26	24	22	20	412	194	66.1	72
Double hamburger (2 patties), w/mayonnaise or salad dressing & tomatoes, on bun (inc lettuce, pickles, onions, and/or mustard)—(197 gm)	41	35	32	29	26	24	22	20	427	195	66.4	72

Food											
Double hamburger (2 patties), w/tomato and/or catsup, on bun (inc lettuce, pickles, onions, and/or mustard)—(164 gm)	38	33	29	27	24	22	20	19	395	160	60.9	69
Egg, cheese, & ham on English muffin (inc Egg McMuffin)—1 egg muffin (138 gm)	66	59	55	51	48	44	42	40	303	126	58.5	256
Egg salad sandwich—1 sandwich (124 gm)	54	49	45	43	40	38	36	34	366	188	36.0	243
Fishburger, on bun w/mayonnaise—(129 gm)	18	16	15	13	12	11	10	10	353	126	26.9	42
Fish sandwich, on bun w/mayonnaise—(140 gm)	21	18	17	15	14	13	12	11	376	137	29.7	50
Frankfurter, plain, on bun—(85 gm)	25	21	19	17	15	14	13	12	263	138	48.5	23
Frankfurter, w/catsup and/or mustard, on bun—(105 gm)	27	23	20	18	17	15	13	12	296	151	52.4	24
Frankfurter, w/cheese, plain, on bun—(118 gm)	53	46	40	36	33	29	27	25	385	227	102.5	51
Frankfurter w/chili & cheese, on bun (inc chili cheese dog)—(147 gm)	55	47	42	37	34	31	28	26	414	236	105.1	55
Frankfurter, w/chili, on bun (inc chili dog)—(152 gm)	33	28	25	23	21	18	17	15	356	177	62.9	34
—foot-long sandwich (236 gm)	52	44	39	35	32	29	26	24	552	275	97.7	53
Fried egg sandwich—1 sandwich (96 gm)	44	40	37	35	33	31	29	28	193	66	24.3	210
Gyro sandwich (pita bread, beef, lamb, condiments, tomato, mayonnaise)—(105 gm)	14	12	11	10	9	8	8	7	215	93	21.9	27
Ham & cheese sandwich, grilled w/mayonnaise—(141 gm)	46	39	35	31	29	26	24	22	417	207	82.2	59
Ham & cheese sandwich w/lettuce, mayonnaise—(155 gm)	44	38	33	30	27	25	23	21	382	184	77.6	59
Ham & cheese sandwich, on bun w/lettuce, mayonnaise (inc ham & cheese hero sandwich)—(154 gm)	44	38	34	30	28	25	23	21	367	190	79.1	59

	1200	1500	1800	2100	2400	2800	3200	3600	TOT CAL	FAT CAL	S/FAT CAL	CHOL MG
Sandwiches (cont.)												
Ham & egg sandwich—(124 gm)	56	51	47	44	42	40	37	36	269	99	33.6	262
Ham & tomato club sandwich w/lettuce, tomato, mayonnaise—(254 gm)	61	53	47	42	39	35	32	30	611	282	105.0	92
Hamburger, ¼ lb meat, plain, on bun—(156 gm)	44	38	34	31	28	26	24	22	449	185	70.9	80
Hamburger, ¼ lb meat, w/mayonnaise or salad dressing & tomatoes, on bun (inc lettuce, pickles, onions, and/or mustard)—(244 gm)	50	43	38	35	32	29	27	25	544	257	81.4	86
Hamburger, ¼ lb meat, w/mayonnaise or salad dressing, on bun (inc lettuce, pickles, onions, and/or mustard)—(200 gm)	50	43	38	35	32	29	27	25	525	257	81.6	86
Hamburger, ¼ lb meat, w/tomato and/or catsup, on bun (inc lettuce, pickles, onions, and/or mustard)—(192 gm)	44	38	34	31	28	26	24	22	470	188	70.8	80
Hamburger, 2½ oz meat, w/mayonnaise or salad dressing & tomatoes, on bun (inc lettuce, pickles, onions, and/or mustard)—(178 gm)	31	27	24	22	20	18	17	15	369	158	51.2	53
Hamburger, plain, on bun—(93 gm)	20	18	16	14	13	12	11	10	270	93	33.8	35
Ham salad sandwich—(107 gm)	16	14	13	11	10	9	9	8	258	94	29.4	22
Ham sandwich w/lettuce, mayonnaise—(127 gm)	18	15	14	12	11	10	10	9	277	105	28.0	33
Hors d'oeuvres w/mayonnaise (inc finger sandwich)—3 hors d'oeuvres (69 gm)	12	11	10	10	9	8	8	7	173	51	12.0	46
Luncheon meat sandwich w/mayonnaise—(83 gm)	19	16	14	13	12	10	9	9	259	122	37.5	16

Food											
Meatball & spaghetti sauce sandwich—(189 gm)	44	38	34	31	29	26	24	23	438	166	65.8	93
Meat spread or potted meat sandwich—(107 gm)	19	16	14	13	12	10	9	9	269	103	34.1	22
Pastrami sandwich—(134 gm)	35	31	27	25	22	20	19	17	339	168	60.1	55
Peanut butter & banana sandwich—1 sandwich (131 gm)	10	8	7	6	6	5	5	4	325	125	22.3	1
Peanut butter & jelly sandwich—1 sandwich (93 gm)	10	9	8	7	6	5	5	4	333	132	23.1	1
Peanut butter sandwich—1 sandwich (93 gm)	12	10	9	8	7	6	6	5	348	157	27.4	1
Pig in a blanket (frankfurter wrapped in dough)—(85 gm)	30	26	23	20	18	16	15	14	278	189	60.9	22
Pizzaburger (hamburger, cheese, sauce) on ½ bun—(137 gm)	52	45	40	36	33	30	28	26	355	184	87.5	86
Pizzaburger (hamburger, cheese, sauce) on whole bun—(165 gm)	54	47	41	37	34	31	28	26	439	198	90.9	86
Pochito (frankfurter & beef chili wrapped in tortilla)—(122 gm)	28	24	21	19	18	16	14	13	275	151	53.3	30
Pork barbecue or Sloppy Joe, on bun—(186 gm)	23	20	18	17	16	14	13	12	359	110	32.8	56
Pork sandwich—(136 gm)	29	26	23	22	20	18	17	16	334	115	38.0	80
Pork sandwich, w/gravy—(218 gm)	31	27	24	22	21	19	18	17	361	121	40.4	81
Reuben sandwich (corned beef sandwich w/sauerkraut & cheese) w/mayonnaise—(181 gm)	64	56	49	44	41	37	34	31	535	315	114.0	89
Roast beef sandwich—(136 gm)	32	28	25	23	21	19	18	17	341	117	46.2	73
Roast beef sandwich dipped in egg, fried in butter, w/gravy & mayonnaise,—(258 gm)	69	62	56	52	48	44	41	39	522	265	82.7	210
Roast beef sandwich w/cheese—(175 gm)	55	48	42	39	35	32	30	28	460	186	87.2	102
Roast beef sandwich, w/gravy—(222 gm)	36	32	28	26	24	22	20	19	385	135	55.1	75

Sandwiches (cont.)

	1200	1500	1800	2100	2400	2800	3200	3600	TOT CAL	FAT CAL	S/FAT CAL	CHOL MG
Salami sandwich w/mayonnaise— (82 gm)	18	15	14	12	11	10	9	8	238	106	32.8	21
Sandwich spread—1 tbsp (15.0 gm)	5	4	4	3	3	3	3	3	58	46	6.9	11
Sardine sandwich w/lettuce, mayonnaise —(214 gm)	58	53	49	45	43	40	38	36	466	242	43.9	247
Sausage sandwich—(107 gm)	34	29	26	23	21	19	18	16	345	174	58.9	48
Scrambled egg sandwich—1 sandwich (112 gm)	51	47	43	41	39	36	34	33	225	77	28.3	245
Soyburger w/cheese—1 sandwich (140 gm)	19	16	14	13	12	10	9	9	330	119	38.8	13
Steak & cheese sandwich, plain, on roll—(170 gm)	73	63	56	50	46	41	38	35	543	287	131.4	96
Steak & cheese submarine sandwich, w/tomato & lettuce, on roll—(214 gm)	71	61	54	49	45	40	37	34	543	273	127.2	95
Steak sandwich, plain, on roll—(142 gm)	47	41	36	33	30	27	25	23	438	208	81.6	69
Steak submarine sandwich, w/tomato & lettuce, on roll—(186 gm)	45	39	35	31	29	26	24	22	438	194	77.6	69
Submarine, cold cut sandwich, on bun w/lettuce (inc grinder, poorboy)—(198 gm)	74	63	55	50	45	41	37	34	560	336	139.7	74
Taco burger, on bun (inc chiliburger w/cheese)—(127 gm)	24	21	19	17	15	14	13	12	286	103	42.3	36
Tomato sandwich w/lettuce, mayonnaise —1 sandwich (134 gm)	7	6	5	5	4	4	3	3	216	84	13.7	6
Tuna salad sandwich—(172 gm)	21	19	17	15	14	13	12	11	374	159	32.4	45
Turkey salad or turkey spread sandwich —(92 gm)	9	8	7	6	6	5	5	4	219	66	16.5	13

Food												
Turkey sandwich w/mayonnaise (inc sliced, roast turkey sandwich)—(143 gm)	22	19	18	16	15	14	13	12	334	106	24.0	70
Turkey sandwich, w/gravy—(284 gm)	26	23	21	20	18	17	16	15	389	88	26.8	89
SEAFOOD (Fish and Shellfish)												
Abalone, floured or breaded, fried—4 oz, boneless, ckd (113 gm)	18	16	15	14	13	12	11	11	206	66	16.7	69
Anchovy, canned—1 oz, boneless (28 gm)	6	5	4	4	4	3	3	3	49	26	7.3	15
Anchovy, cooked—1 oz, boneless (28 gm)	6	5	4	4	4	3	3	3	49	26	7.3	15
Anchovy, pickled, w/or wo/oil, not heavy salt—5 anchovies (20 gm)	5	4	4	4	3	3	3	3	35	19	7.2	11
Barracuda, baked or broiled—4 oz, boneless, (113 gm)	18	16	15	14	13	12	12	11	180	61	15.4	73
Barracuda, floured or breaded, fried—4 oz, boneless, (113 gm)	21	19	17	16	15	14	13	12	225	90	24.2	69
Bass, black sea, baked, stuffed—1 piece w/⅓ c stuffing (205 gm)	56	48	43	39	36	32	30	28	531	292	92.3	94
Bass, striped, oven-fried—1 fillet (200 gm)	44	39	35	32	30	28	26	24	392	153	54.0	126
Bluefish, baked or broiled—1 fillet (155 gm)	23	21	20	18	17	16	16	15	246	73	13.9	109
Bluefish, fried—1 fillet (195 gm)	34	31	28	26	24	22	21	20	400	172	35.1	119
Bonito, (inc Atlantic, Pacific and striped), raw—4 oz (113 gm)	23	20	18	17	16	14	13	13	190	74	30.5	62
Burbot, fried—1 oz (28 gm)	6	6	5	5	5	5	4	4	46	16	2.5	33
Butterfish, raw, fr Gulf waters—4 oz (113 gm)	14	13	12	11	11	10	9	9	107	29	10.2	62
Butterfish, raw, fr northern waters—4 oz (113 gm)	19	17	15	14	13	12	11	11	191	104	20.3	62
Carp, baked or broiled (inc bream, buffalofish, chub, sucker)—4 oz, boneless (113 gm)	22	19	18	17	15	14	13	13	214	102	21.0	79

	1200	1500	1800	2100	2400	2800	3200	3600	TOT CAL	FAT CAL	S/FAT CAL	CHOL MG
Seafood (Fish & Shellfish) *(cont.)*												
Carp, floured or breaded, fried (inc buffalofish, chub, sucker)—4 oz, boneless, cooked (113 gm)	33	30	27	25	24	22	20	19	301	147	35.8	112
Carp, smoked (inc bream, buffalofish, sucker)—1 oz (28 gm)	5	5	5	4	4	4	4	3	51	17	3.6	24
Catfish, baked, broiled (inc bullhead)—4 oz, boneless (113 gm)	21	19	17	16	15	14	13	13	199	89	20.0	79
Catfish, battered, fried (inc bullhead)—4 oz, boneless (113 gm)	32	28	25	23	22	20	18	17	301	182	44.8	79
Catfish, floured or breaded, fried (inc bullhead)—4 oz, boneless (113 gm)	33	30	27	25	23	22	20	19	291	138	35.3	112
Catfish, steamed or poached—4 oz, boneless (113 gm)	17	15	14	13	13	12	11	11	146	40	10.2	78
Chub, raw, fish (carp family)—4 oz (113 gm)	19	17	15	14	13	12	11	11	164	89	20.3	62
Clam fritters—1 fritter (2" dia) (40 gm)	13	12	11	10	9	9	8	8	124	54	10.8	52
Clams, baked or broiled—1 c (8 lg clams, 12 med clams, 15 sm clams) (150 gm)	19	17	15	14	13	12	11	11	170	84	17.5	70
Clams, battered, fried—1 c (102 gm)	30	27	25	23	22	20	19	18	225	112	28.6	114
Clams, canned—1 c, solids only (160 gm)	19	17	16	15	15	14	13	13	152	29	7.2	100
Clams, floured or breaded, fried (inc baked)—1 c (8 lg clams, 12 med clams, 15 sm clams) (150 gm)	33	29	27	25	24	22	21	19	278	118	30.2	124
Clams, raw—1 clam (16 gm)	2	1	1	1	1	1	1	1	12	2	.6	8

Food												
Clams, smoked, in oil—1 oz (28 gm)	6	5	5	4	4	4	3	3	50	31	7.7	16
Clams, steamed or boiled—1 c (150 gm)	18	16	15	14	13	12	12	11	163	77	16.2	71
Cod, baked, broiled—4 oz, boneless (113 gm)	14	12	12	11	10	10	9	9	134	39	7.4	66
Cod, battered, fried—4 oz, boneless (113 gm)	20	18	16	15	14	13	12	12	192	82	19.5	71
Cod, breaded or battered, baked—4 oz, boneless (113 gm)	22	20	19	17	16	15	14	14	219	94	19.4	89
Cod, dried, salted—1 oz, boneless (28 gm)	4	3	3	3	3	3	3	3	36	2	.3	23
Cod, dried, salted, salt removed in water—1 oz, dried, soaked in water (34 gm)	4	3	3	3	3	3	3	3	36	2	.4	22
Cod, floured or breaded, fried—4 oz, boneless (113 gm)	26	23	21	19	18	17	16	15	233	111	27.7	87
Cod, smoked—4 oz boneless (113 gm)	14	13	12	12	11	11	10	10	117	4	.7	86
Cod, steamed or poached—4 oz, boneless (113 gm)	12	11	10	10	10	9	9	8	110	9	1.7	71
Crab, baked or broiled—1 blue Atlantic, about 6" (shell removed) (48 gm)	9	9	8	7	7	7	6	6	62	28	5.2	45
—1 snow crab, legs, body section w/3 legs (cooked, shell removed) (54 gm)	11	10	9	8	8	8	7	7	70	32	5.9	51
—1 Pacific (1 lb live weight) (yield after cooking, shell removed 113 gm)	22	20	19	18	17	16	15	14	146	66	12.3	106
Crab, canned (inc white or king meat)—4 oz (113 gm)	23	21	19	18	17	16	15	15	152	69	12.8	110
—1 c, flaked & pieces (118 gm)	20	18	17	16	16	15	14	14	114	25	4.1	114
Crab, deviled—1 c (240 gm)	64	58	53	49	46	43	40	38	451	203	58.3	245
Crab, hard shell, steamed (inc boiled)—1 blue Atlantic, about 6" (shell removed) (48 gm)	8	8	7	7	6	6	6	6	44	8	1.3	48

Seafood (Fish & Shellfish) (cont.)

	1200	1500	1800	2100	2400	2800	3200	3600	TOT CAL	FAT CAL	S/FAT CAL	CHOL MG
—1 snow crab, legs, body section w/3 legs (cooked, shell removed) (54 gm)	9	8	8	8	7	7	7	6	50	9	1.4	54
—1 Pacific (1 lb live weight) (yield after cooking, shell removed) (113 gm)	19	18	17	16	15	14	14	13	104	19	3.0	112
—1 c, flaked & pieces (118 gm)	20	18	17	16	16	15	14	14	109	20	3.2	117
Crab, imperial—1 c (220 gm)	83	75	68	64	60	55	52	49	323	150	79.2	308
Crab, soft shell, floured or breaded, fried—1 crab (65 gm)	27	24	22	20	19	17	16	15	213	118	29.2	87
Crayfish, boiled or steamed—4 oz, wo/shell (113 gm)	21	20	19	18	17	16	15	15	118	17	2.6	127
Crayfish, floured or breaded, fried—4 oz, wo/shell (113 gm)	48	44	41	38	36	34	32	31	265	118	29.4	224
Croaker, baked, broiled (inc angelfish, butterfly fish, drumfish, goatfish, kingfish)—4 oz, boneless (113 gm)	21	19	17	16	15	14	13	13	205	89	18.3	83
Croaker, floured or breaded, fried (inc angelfish, butterfly fish, drumfish, goatfish, kingfish)—4 oz, boneless (113 gm)	35	31	29	26	25	23	21	20	310	146	36.7	120
Dogfish, spiny, grayfish, raw—4 oz (113 gm)	19	17	15	14	13	12	11	11	176	92	20.3	62
Eel, smoked—4 oz, boneless (113 gm)	41	35	32	29	26	24	22	20	374	283	66.3	73
Eel, steamed or poached—4 oz, boneless (113 gm)	36	32	29	26	24	22	20	19	330	233	54.8	78
Fish cakes, fried—1 reg cake (5-bite size) (60 gm)	11	10	9	8	7	7	6	6	103	43	16.2	25

Food												
Fish cakes, fried, frozen, reheated—1 reg cake (5-bite size) (60 gm)	19	16	14	13	12	11	10	9	162	96	37.8	16
Fish sticks, baked or broiled—2 sticks (48 gm)	10	9	8	8	7	7	6	6	101	41	8.1	40
Fish sticks, battered, fried—2 sticks (48 gm)	10	9	8	7	7	6	6	6	100	49	11.6	31
Fish sticks, breaded or battered, baked —2 sticks (48 gm)	10	9	8	8	7	7	6	6	101	41	8.1	40
Fish sticks, floured or breaded, fried—2 sticks (48 gm)	11	10	9	9	8	7	7	7	106	48	11.6	39
Flounder, baked or broiled (inc dab, fluke, halibut, sole, turbot)—4 oz, boneless (113 gm)	14	13	12	11	11	10	9	9	135	40	8.3	66
Flounder, battered, fried (inc dab, fluke, halibut, sole, turbot)—4 oz, boneless (113 gm)	18	16	15	14	13	12	11	11	204	74	14.4	72
Flounder, breaded or battered, baked—4 oz, boneless (113 gm)	20	18	17	16	15	14	13	13	190	59	13.1	92
Flounder, floured or breaded, fried (inc dab, fluke, halibut, sole, turbot)—4 oz, boneless (113 gm)	26	23	21	20	18	17	16	15	234	112	28.4	87
Flounder, smoked (inc dab, fluke, halibut, sole, turbot)—4 oz, boneless (113 gm)	21	18	17	15	14	13	12	11	254	153	26.5	57
Flounder, steamed or poached (inc dab, fluke, halibut, sole, turbot)—4 oz, boneless (113 gm)	12	11	11	10	10	9	9	8	112	10	2.5	71
Frog legs, cooked—2 legs fried (yield after bone removed) (48 gm)	18	16	15	14	13	13	12	11	121	52	13.4	78
Haddock, baked or broiled (inc burbot, cusk, hake, ling, monkfish, pollack, scrod)—4 oz, boneless (113 gm)	16	14	13	13	12	11	11	10	136	38	7.3	79
Haddock, battered, fried—4 oz, boneless (113 gm)	21	19	18	16	15	14	13	13	193	82	19.4	81

Seafood (Fish & Shellfish) (cont.)

	1200	1500	1800	2100	2400	2800	3200	3600	TOT CAL	FAT CAL	S/FAT CAL	CHOL MG
Haddock, breaded, baked or broiled (inc burbot, cusk, hake, ling, monkfish, pollack, scrod)—4 oz, boneless (113 gm)	30	27	25	23	22	20	19	18	258	117	29.4	110
Haddock, floured or breaded, fried (inc burbot, cusk, hake, ling, monkfish, pollack, scrod)—4 oz, boneless (113 gm)	30	27	25	23	22	20	19	18	258	117	29.4	110
Haddock, smoked (inc burbot, cusk, hake, ling, monkfish, pollack, scrod, white fish)—4 oz, boneless (113 gm)	14	13	12	12	11	11	10	10	117	4	.7	86
Halibut (Atlantic & Pacific), broiled—1 fillet, (6½" x 2¾") (125 gm)	22	19	18	16	15	14	13	13	214	79	22.5	75
Herring, baked or broiled (inc alewife, milkfish, shad)—4 oz, boneless (113 gm)	44	39	35	33	30	28	26	24	297	185	54.8	127
Herring, dried, salted (inc alewife, milkfish, shad)—1 oz, boneless (28 gm)	11	10	9	8	8	7	7	6	72	45	14.5	31
Herring, floured or breaded, fried (inc alewife, milkfish shad)—4 oz, boneless (113 gm)	51	45	41	38	35	32	30	28	372	220	63.4	146
Herring, pickled (inc in wine sauce)—1 oz, boneless (28 gm)	10	9	8	7	7	6	6	5	62	38	12.3	28
Herring, pickled, in cream sauce—1 oz boneless (28 gm)	11	10	9	8	7	7	6	6	62	41	16.5	25
Herring, plain, canned, solid & liquid—15 oz can (425 gm)	108	97	89	83	78	73	68	64	884	520	98.7	412

Food												
Herring, smoked, kippered—4 oz, boneless (113 gm)	38	34	31	29	27	25	23	22	239	132	42.3	124
Herring, w/tomato sauce, canned, solid & liquid—1 herring w/tbsp sauce (55 gm)	11	10	9	9	8	7	7	7	97	52	9.9	43
Kingfish, (southern Gulf & northern whiting), raw—4 oz (113 gm)	14	13	12	11	11	10	9	9	119	31	10.2	62
Lobster, baked or broiled—1 sm lobster (1 lb live weight) (yield after cooking, shell removed, 118 gm)	32	28	26	24	23	21	19	18	154	65	33.4	109
—1 med lobster (2½ lb live weight) (yield after cooking, shell removed, 295 gm)	79	71	65	60	56	52	49	46	386	163	83.4	272
Lobster, canned—4 oz (113 gm)	16	15	14	13	13	12	12	11	108	15	2.0	96
Lobster floured or breaded, fried—4 oz, wo/shell, ckd (113 gm)	31	28	25	24	22	21	19	18	247	123	28.4	117
Lobster Newburg—1 c (250 gm)	131	117	107	99	93	86	80	76	485	239	135.0	455
Lobster salad—approx ½ c (4 oz) (260 gm)	29	26	24	23	21	20	19	18	286	149	23.4	120
Lobster, steamed or boiled—1 sm lobster (1 lb live weight) (yield after cooking, shell removed, 118 gm)	17	15	15	14	13	13	12	12	112	16	2.1	100
—1 med lobster (2½ lb live weight) (yield after cooking, shell removed, 295 gm)	41	39	37	35	33	32	30	29	280	40	5.3	251
Mackerel, Atlantic, broiled, w/butter or margarine—1 fillet (8½" x 2½") (105 gm)	38	33	30	28	26	24	22	21	248	149	47.3	106
Mackerel, baked or broiled (inc enenui, garfish, ono, needlefish, wahoo)—4 oz, boneless (113 gm)	39	35	32	29	27	25	24	22	283	175	44.0	126
Mackerel, breaded or floured, fried (inc enenui, garfish, ono, needlefish, wahoo)—4 oz, boneless (113 gm)	47	41	38	35	32	30	28	26	355	223	57.7	136

Seafood (Fish & Shellfish) (cont.)	1200	1500	1800	2100	2400	2800	3200	3600	TOT CAL	FAT CAL	S/FAT CAL	CHOL MG
Mackerel, canned, drained—4 oz, boneless (113 gm)	34	30	28	26	24	23	21	20	243	121	31.6	127
Mackerel, dried (inc enenui, garfish, ono, needlefish, wahoo)—1 oz, boneless (28 gm)	11	10	9	8	8	7	6	6	85	63	16.4	27
Mackerel, pickled (inc enenui, garfish, ono, needlefish, wahoo)—1 oz, boneless (28 gm)	9	8	8	7	7	6	6	5	65	37	9.8	32
Mackerel, salted—1 oz, boneless, dried (28 gm)	11	10	9	8	8	7	6	6	85	63	16.4	27
Mackerel, smoked (inc enenui, garfish, ono, needlefish, wahoo)—4 oz, boneless (113 gm)	32	29	26	24	23	21	20	19	248	133	34.7	108
Mullet, baked or broiled—4 oz, boneless (113 gm)	29	26	23	21	20	18	17	16	261	138	38.1	79
Mullet, breaded or floured, fried—4 oz, boneless (113 gm)	39	35	32	29	27	25	23	22	339	176	49.4	112
Mussels, cooked—4 oz (113 gm)	18	16	15	14	13	12	11	11	183	85	17.6	66
Ocean perch, baked or broiled (inc bocaccio, menpachi, redfish, rockfish)—4 oz, boneless (113 gm)	15	14	13	12	11	11	10	10	147	44	8.4	73
Ocean perch, battered, fried (inc bocaccio, menpachi, redfish, rockfish)—4 oz, boneless (113 gm)	24	21	19	18	17	15	14	14	230	119	28.3	74
Ocean perch, breaded or battered, baked (inc bocaccio, menpachi, redfish, rockfish)—4 oz, boneless (113 gm)	21	19	18	17	16	15	14	13	200	63	13.2	98

Food												
Ocean perch, floured or breaded, fried (inc bocaccio, menpachi, redfish, rockfish)—4 oz, boneless (113 gm)	27	24	22	21	19	18	17	16	244	116	28.5	93
Octopus, cooked—4 oz boneless (113 gm)	32	29	27	25	24	22	21	20	200	80	20.6	146
Octopus, smoked—4 oz boneless, cooked (113 gm)	32	30	28	27	26	24	23	22	141	14	3.8	193
Oysters, baked or broiled—1 eastern oyster (12 gm)	2	1	1	1	1	1	1	1	12	7	1.6	6
—1 Pacific oyster (38 gm)	5	5	4	4	4	3	3	3	39	22	5.1	18
—4 oz, wo/shell (113 gm)	15	13	12	11	11	10	9	9	118	65	15.2	53
Oysters, battered, fried—1 eastern oyster (8 gm)	2	2	2	2	2	2	2	1	17	9	2.4	9
—1 Pacific oyster (26 gm)	8	7	7	6	6	5	5	5	55	29	7.7	29
—4 oz, wo/shell (113 gm)	35	31	29	26	25	23	22	20	241	127	33.5	127
Oysters, canned—4 oz, (113 gm)	14	13	12	11	10	10	9	9	108	28	8.9	64
Oysters, floured or breaded, fried—1 eastern oyster (8 gm)	2	2	2	1	1	1	1	1	16	9	2.3	6
—1 Pacific oyster (26 gm)	6	6	5	5	4	4	4	4	50	28	7.4	20
—4 oz, wo/shell (113 gm)	28	25	23	21	19	18	17	16	220	122	32.3	87
Oysters, raw—1 eastern oyster (15 gm)	2	1	1	1	1	1	1	1	10	2	.8	8
—1 Pacific oyster (48 gm)	5	4	4	4	4	4	3	3	32	8	2.6	24
—4 oz, wo/shell (113 gm)	12	11	10	9	9	8	8	8	75	18	6.1	57
Oysters, smoked—4 oz (113 gm)	19	17	16	15	14	13	13	12	121	30	9.9	92
Oysters, steamed—1 eastern oyster (12 gm)	1	1	1	1	1	1	1	1	8	2	.6	6
—1 Pacific oyster (38 gm)	4	4	3	3	3	3	3	3	25	6	2.1	19
—4 oz, wo/shell (113 gm)	12	11	10	9	9	8	8	8	75	18	6.1	57
Perch, baked or broiled (inc freshwater bass, bluegill, crappie, sunfish, walleye)—4 oz, boneless (113 gm)	16	14	13	12	12	11	10	10	151	41	9.5	73

	1200	1500	1800	2100	2400	2800	3200	3600	TOT CAL	FAT CAL	S/FAT CAL	CHOL MG
Seafood (Fish & Shellfish) (cont.)												
Perch, battered, fried (inc freshwater bass, bluegill, crappie, sunfish, walleye)—4 oz, boneless (113 gm)	21	19	17	16	15	14	13	12	205	84	21.1	76
Perch, breaded or battered, baked (inc freshwater bass, bluegill, crappie, sunfish, walleye)—4 oz, boneless (113 gm)	22	20	18	17	16	15	14	14	204	60	14.2	98
Perch, floured or breaded, fried (inc freshwater bass, bluegill, crappie, sunfish, walleye)—4 oz, boneless (113 gm)	33	30	27	25	23	22	20	19	295	127	33.1	117
Perch, steamed or poached—4 oz, boneless (113 gm)	14	13	12	11	11	10	10	9	129	11	3.8	78
Pike, baked or broiled (inc muskellunge, pickerel)—4 oz, boneless (113 gm)	15	14	13	12	11	11	10	10	147	43	8.3	73
Pike, floured or breaded, fried (inc muskellunge, pickerel)—4 oz, boneless (113 gm)	27	24	22	20	19	18	17	16	244	115	28.4	93
Pollack, creamed—1 c (250 gm)	44	39	35	32	29	27	25	23	320	133	67.5	93
Pompano, baked or broiled (inc akule, blackfish, bluefish, butterfish, dolphinfish, jack, mahimahi, paplo, sablefish, scad, tilefish, ulva, yellowtail)—4 oz, boneless (113 gm)	38	33	30	27	25	23	21	20	291	173	57.9	80
Pompano, battered, fried (inc akule, blackfish, bluefish, butterfish, dolphinfish, jack, mahimahi, paplo, sablefish, scad, tilefish, ulva, yellowtail)—4 oz, boneless (113 gm)	37	33	29	27	24	22	20	19	305	191	58.3	74

Food												
Pompano, floured or breaded, fried (inc akule, blackfish, bluefish, butterfish, dolphinfish, jack, mahimahi, paplo, sablefish, scad, tilefish, ulva, yellowtail)—4 oz, boneless (113 gm)	46	40	36	33	31	28	26	24	361	202	64.4	112
Pompano, smoked (inc akule, blackfish, bluefish, butterfish, dolphinfish, jack, mahimahi, paplo, sablefish, scad, tilefish, ulva, yellowtail)—4 oz, boneless (113 gm)	31	27	24	22	21	19	17	16	220	113	44.1	73
Porgy, baked or broiled (inc scup, sea bream, marine sheepshead, snapper)—4 oz, boneless (113 gm)	23	21	19	17	16	15	14	13	229	105	22.4	83
Porgy, battered, fried (inc scup, sea bream, marine sheepshead, snapper)—4 oz, boneless (113 gm)	18	16	15	14	13	12	12	11	184	51	11.0	81
Porgy, breaded or battered, baked (inc scup, sea bream, marine sheepshead, snapper)—4 oz, boneless (113 gm)	31	28	25	23	22	20	19	18	305	158	33.8	103
Porgy, floured or breaded, fried (inc scup, sea bream, marine sheepshead, snapper)—4 oz, boneless (113 gm)	36	32	30	27	26	24	22	21	328	158	39.7	120
Porgy, steamed or poached (inc scup, sea bream, marine sheepshead, snapper)—4 oz, boneless (113 gm)	17	15	14	13	13	12	11	11	159	43	10.2	78
Ray, baked or broiled (inc skate)—4 oz, boneless (113 gm)	16	15	14	13	12	12	11	11	181	43	7.7	82
Ray, floured or breaded, fried (inc skate)—4 oz, boneless (113 gm)	30	27	25	23	21	20	19	18	285	114	28.0	112
Roe, caviar, sturgeon, granular—1 oz (28 gm)	16	15	14	13	13	12	11	11	73	38	7.6	84

	1200	1500	1800	2100	2400	2800	3200	3600	TOT CAL	FAT CAL	S/FAT CAL	CHOL MG
Seafood (Fish & Shellfish) (cont.)												
Roe, caviar, sturgeon, pressed—1 oz (28 gm)	20	19	17	17	16	15	14	14	88	42	7.6	108
Roe, cod & shad, cooked—1 oz (28 gm)	19	18	17	16	15	15	14	13	52	16	3.0	115
Roe, herring—1 tbsp (14 gm)	8	8	7	7	7	6	6	6	18	3	.5	50
Sablefish, raw—4 oz (113 gm)	32	28	25	23	21	19	18	16	215	152	50.9	62
Salmon, Atlantic, canned, solid & liquid —7.75 oz can (220 gm)	39	34	30	28	26	23	22	20	447	241	59.4	81
Salmon, baked or broiled (inc saltwater trout)—4 oz, boneless (113 gm)	14	13	12	11	10	9	9	8	199	81	15.6	48
Salmon, canned—4 oz (113 gm)	11	10	9	9	8	8	7	7	173	73	11.7	40
Salmon, chinook (king), canned, solid & & liquid, wo/salt—7.75 oz can (220 gm)	49	43	38	34	31	28	26	24	462	277	85.1	75
Salmon chinook, (king), canned, solid & liquid, w/salt—7.75 oz can (220 gm)	49	43	38	34	31	28	26	24	462	277	85.1	75
Salmon, chub, canned, solid & liquid, w/salt—7.75 oz can (220 gm)	21	19	18	16	15	14	13	13	306	103	19.8	81
Salmon, coho (silver), canned, solid & liquid, w/salt—7.75 oz can (220 gm)	30	26	24	22	20	19	17	16	337	140	39.6	79
Salmon, dried—1 oz, boneless (28 gm)	3	3	2	2	2	2	2	2	42	12	2.2	12
Salmon, floured or breaded, fried (inc saltwater trout)—4 oz, boneless (113 gm)	25	22	20	19	17	16	15	14	271	121	28.5	79
Salmon, pink (humpback), canned, solid & liquid, w/salt—7.75 oz can (220 gm)	26	23	21	19	18	16	15	14	310	117	30.7	77
Salmon rice loaf—1 piece, ⅛ of loaf (174 gm)	13	11	10	9	9	8	8	7	212	70	15.7	37
Salmon, smoked (inc lox)—4 oz, boneless (113 gm)	17	15	14	12	12	11	10	9	199	95	23.5	43

Food												
Salmon, sockeye (red), canned, solid & liquid, w/salt—7.75 oz can (220 gm)	38	33	30	27	25	23	21	20	376	185	59.4	77
Salmon, steamed or poached (inc saltwater trout)—4 oz, boneless (113 gm)	12	11	10	9	9	8	8	7	169	47	8.9	50
Sardines, Atlantic, canned, in oil, drained solids—3.25 oz can, drained (92 gm)	27	25	23	22	21	19	18	18	187	92	16.6	129
Sardines, Atlantic, canned, in oil, solid & liquid—3.75 can (106 gm)	41	36	33	31	28	26	25	23	330	233	47.7	127
Sardines, cooked—4 oz (113 gm)	34	31	29	27	26	24	23	22	230	113	21.4	159
Sardines, Pacific, canned, in brine or mustard, solid & liquid—4 oz (113 gm)	28	26	24	22	21	20	19	18	221	122	20.3	124
Sardines, Pacific, canned, in tomato sauce, solid & liquid—4 oz (113 gm)	28	26	24	22	21	20	19	18	223	124	20.3	124
Scallops, baked or broiled—4 oz (113 gm)	11	10	9	9	8	8	7	7	141	41	7.6	48
Scallops, floured or breaded, fried—4 oz (113 gm)	23	21	19	17	16	15	14	13	236	112	27.6	71
Scallops, steamed or boiled—4 oz (113 gm)	9	8	7	7	6	6	6	5	113	33	6.0	38
Sea bass, baked or broiled (inc grouper, striped bass, weakfish)—4 oz, boneless (113 gm)	15	14	13	12	12	11	10	10	155	42	8.3	75
Sea bass, flonred or breaded, fried (inc grouper, striped bass, weakfish)—4 oz, boneless (113 gm)	26	23	21	20	19	17	16	15	235	89	22.4	101
Sea bass, pickled (mero en escabeche)—4 oz, boneless (113 gm)	22	19	17	16	14	13	12	11	322	259	35.3	42
Sea urchin (roe)—1 oz (28 gm)	15	14	13	13	12	11	11	10	41	21	3.8	87
Seaweed, dulse, raw—4 oz (113 gm)	0	0	0	0	0	0	0	0	0	0	0	0
Shad, American, baked—1 oz (28 gm)	7	7	6	5	5	5	4	4	56	29	10.1	19
Shark, baked or broiled (inc dogfish, grayfish)—4 oz, boneless (113 gm)	25	22	20	18	17	16	15	14	237	137	29.7	73

	1200	1500	1800	2100	2400	2800	3200	3600	TOT CAL	FAT CAL	S/FAT CAL	CHOL MG
Seafood (Fish & Shellfish) *(cont.)*												
Shrimp, battered, fried (inc shrimp scampi, Arthur Treacher's shrimp, shrimp tempura, prawn)—1 sm shrimp (6 gm)	3	2	2	2	2	2	2	2	14	6	1.6	12
—1 med shrimp (11 gm)	5	4	4	4	4	3	3	3	26	11	2.9	22
—1 lg shrimp (17 gm)	7	7	6	6	5	5	5	5	40	18	4.4	34
—1 extra-lg shrimp (25 gm)	11	10	9	8	8	7	7	7	58	26	6.5	49
—4 oz, wo/shell (113 gm)	48	44	41	38	36	34	32	30	265	118	29.4	223
Shrimp, broiled (inc prawn)—4 oz (113 gm)	35	33	31	29	28	27	25	24	155	41	7.7	205
Shrimp, canned—4 oz, cooked (113 gm)	27	26	24	23	22	21	20	19	131	11	1.7	170
Shrimp, dried—1 oz (28 gm)	17	16	15	14	14	13	13	12	82	7	1.1	106
Shrimp, floured or breaded, fried (inc prawn)—1 sm shrimp (6 gm)	3	2	2	2	2	2	2	2	14	6	1.6	12
—1 med shrimp (11 gm)	5	4	4	4	4	3	3	3	26	11	2.9	22
—1 lg shrimp (17 gm)	7	7	6	6	5	5	5	5	40	18	4.4	34
—1 extra-lg shrimp (25 gm)	11	10	9	8	8	7	7	7	58	26	6.5	49
—1 jumbo shrimp (30 gm)	13	12	11	10	10	9	8	8	70	31	7.8	59
—4 oz, wo/shell (113 gm)	48	44	41	38	36	34	32	30	265	118	29.4	223
Shrimp or lobster paste, canned—1 tsp (7 gm)	3	3	3	3	2	2	2	2	13	6	3.2	12
Shrimp, steamed or boiled (inc prawn)—4 oz, wo/shell (113 gm)	36	33	32	30	29	28	26	25	136	11	1.7	224
Smelt, baked or broiled (inc capelin)—4 oz, boneless (113 gm)	19	18	16	15	14	13	12	12	193	76	15.5	80
Smelt, battered, fried—4 oz, boneless (113 gm)	31	27	25	22	21	19	18	17	297	173	41.9	79

Food											
Smelt, floured or breaded, fried (inc capelin)—4 oz, boneless (113 gm)	32	28	26	24	22	21	20	18	285	128	31.8	112
Snails, cooked—1 oz, wo/shell (28 gm)	5	5	4	4	4	3	3	3	37	11	5.4	19
Squid, baked, broiled (inc cuttlefish, calamari)—1 c (140 gm)	52	49	46	43	42	39	38	36	180	52	10.5	304
Squid, breaded, fried—4 oz (113 gm)	46	42	39	37	35	33	32	30	213	81	20.4	236
Squid, canned (inc calameres en su tinta, squid in its own ink)—1 can (4 oz) (115 gm)	38	36	34	32	31	30	28	27	111	11	2.4	238
Squid, dried (inc cuttlefish, calamari)—1 oz, boneless (28 gm)	30	28	26	25	24	23	22	21	87	8	1.9	185
Squid, pickled—1 oz boneless (28 gm)	8	7	7	7	6	6	5	5	23	2	.5	48
Squid, raw (inc cuttlefish)—4 oz boneless (113 gm)	26	24	23	22	21	20	19	18	89	9	2.1	159
Squid, steamed or boiled—4 oz, boneless (113 gm)	38	35	33	32	30	29	28	27	109	11	2.3	234
Sturgeon, smoked—4 oz, boneless (113 gm)	20	18	17	16	16	15	14	13	180	33	7.8	106
Sturgeon, steamed—4 oz, boneless (113 gm)	15	13	13	12	11	11	10	10	132	24	5.7	77
Swordfish, baked or broiled (inc marlin)—4 oz, boneless (113 gm)	20	18	17	15	15	14	13	12	198	85	19.3	75
Swordfish, floured or breaded, fried (inc marlin)—4 oz, boneless (113 gm)	31	27	25	23	21	20	18	17	276	143	36.2	93
Terrapin, cooked (inc turtle)—4 oz, boneless (113 gm)	14	13	12	11	11	10	9	9	152	37	7.7	68
Trout, baked or broiled (inc cisco, lake herring, steelhead, whitefish)—4 oz, boneless (113 gm)	28	24	22	20	19	17	16	15	289	166	36.8	73
Trout, battered, fried (inc cisco, cisco, lake herring, steelhead, whitefish)—4 oz, boneless (113 gm)	21	19	17	16	15	14	13	12	215	95	22.5	70

Seafood (Fish & Shellfish) (cont.)

	1200	1500	1800	2100	2400	2800	3200	3600	TOT CAL	FAT CAL	S/FAT CAL	CHOL MG
Trout, breaded, baked (inc cisco, lake herring, steelhead, whitefish)—4 oz, boneless (113 gm)	19	17	16	15	14	13	13	12	184	35	8.5	95
Trout, floured or breaded, fried (inc cisco, lake herring, steelhead, whitefish)—4 oz, boneless (113 gm)	26	24	22	20	19	17	16	15	251	105	26.1	95
Trout, smoked (inc chub, cisco, lake herring steelhead, whitefish)—4 oz, boneless (113 gm)	19	18	16	16	15	14	13	13	204	38	8.4	99
Tuna, canned, oil pack—4 oz (113 gm)	24	21	19	18	17	15	14	13	223	84	27.5	74
Tuna, canned, water pack—4 oz (113 gm)	15	14	13	13	12	11	11	11	180	10	2.8	89
Tuna, fresh, baked or broiled (inc ahi, aku, bonito)—4 oz, boneless (56 gm)	9	8	8	7	7	6	6	6	102	32	7.6	37
Tuna, fresh, dried (inc ahi, aku, bonito)—1 c (42 gm)	9	8	7	7	6	6	5	5	88	34	9.4	29
Tuna, fresh, floured or breaded, fried (inc ahi, aku, bonito)—4 oz, boneless (113 gm)	30	27	24	22	21	19	18	17	293	133	34.2	94
Tuna, fresh, raw (inc ahi, aku, bonito)—4 oz, boneless (113 gm)	13	12	11	11	10	10	9	9	151	31	8.2	63
Tuna, fresh, smoked (inc ahi, aku, bonito)—4 oz, boneless (113 gm)	32	29	26	24	23	21	20	19	248	133	34.7	108
Tuna salad—1 c (205 gm)	41	36	33	31	29	26	25	23	349	194	45.4	133
Weakfish, broiled—4 oz (112 gm)	32	29	26	24	22	20	19	18	232	115	40.3	92
Whale meat, raw—4 oz (113 gm)	13	12	11	11	10	9	9	8	176	76	10.2	57

Food												
Whitefish, baked, stuffed—4 oz (113 gm)	26	23	20	18	17	15	14	13	243	142	40.7	50
Whitefish, smoked—4 oz (113 gm)	20	18	16	15	14	13	12	11	175	74	20.3	69
Whiting, baked or broiled—4 oz, boneless (113 gm)	16	14	13	12	12	11	10	10	129	45	9.5	73
Whiting, battered, fried—4 oz, boneless (113 gm)	21	19	17	16	15	14	13	12	188	88	21.1	76
Whiting, floured or breaded, fried—4 oz, boneless (113 gm)	28	25	22	21	19	18	17	16	229	117	29.4	93

SOUPS

Food												
Barley (inc beef barley, chicken barley, mushroom barley soup)—1 c (244 gm)	1	1	1	1	1	1	1	½	96	6	2.5	0
Bean w/bacon, dehydrated, prep w/water—1 c (8 fl oz) (265 gm)	4	4	3	3	3	2	2	2	105	19	8.6	3
Bean w/frankfurters, canned, prep w/equal volume water—1 c (8 fl oz) (250 gm)	10	9	8	7	7	6	5	5	187	63	19.1	13
Bean w/ham, canned, chunky, ready-to-serve—1 c (8 fl oz) (243 gm)	17	14	13	12	11	9	9	8	231	77	30.0	22
Bean, w/macaroni & meat (inc pasta fazool)—1 c (253 gm)	18	15	13	12	11	10	9	8	186	101	36.0	12
Bean w/pork, canned, prep w/equal volume water—1 c (8 fl oz) (253 gm)	7	6	5	4	4	3	3	3	173	53	13.8	3
Beef & rice, Puerto Rican style—1 c (250 gm)	14	13	12	11	10	9	8	8	143	43	17.5	42
Beef broth, cube, prep w/water—1 c (8 fl oz) (241 gm)	½	½	½	½	0	0	0	0	8	2	.9	0
Beef broth or bouillon, powder, prep w/water—1 c (8 fl oz) (244 gm)	2	1	1	1	1	1	1	1	19	6	3.1	1
Beef broth or boullion, canned, ready-to-serve—1 c (8 fl oz) (240 gm)	1	1	1	1	1	1	½	½	16	5	2.3	0

417

	1200	1500	1800	2100	2400	2800	3200	3600	TOT CAL	FAT CAL	S/FAT CAL	CHOL MG
Soups (cont.)												
Beef, canned, chunky, ready-to-serve—1 c (8 fl oz) (240 gm)	12	11	9	8	8	7	6	6	170	46	22.9	14
Beef dumpling—1 c (241 gm)	44	39	36	33	31	28	26	25	330	139	51.6	137
Beef mushroom, canned, prep w/equal volume water—1 c (8 fl oz) (244 gm)	7	6	5	5	4	4	4	3	73	27	13.4	7
Beef noodle, canned, prep w/equal volume water—1 c (8 fl oz) (244 gm)	5	5	4	4	3	3	3	2	84	28	10.4	5
Beef noodle, chunky style—1 c (240 gm)	13	12	11	10	9	9	8	8	145	38	14.6	43
Beef noodle, dehydrated, prep w/water—1 c (8 fl oz) (251 gm)	1	1	1	1	1	1	1	1	41	7	2.3	3
Beef rice—1 c (241 gm)	8	7	6	6	5	5	4	4	106	30	12.9	16
Beef vegetable w/potato, stew type (inc chunky style)—1 c (240 gm)	7	6	5	5	4	4	4	3	146	40	13.9	7
Beer soup, made w/milk—1 c (245 gm)	28	26	24	22	21	20	19	18	134	39	17.1	131
Beet (borscht)—1 c (245 gm)	11	10	8	7	7	6	5	5	71	37	22.4	8
Bird's nest (chicken, ham, & noodles)—1 c (244 gm)	7	7	6	6	5	5	4	4	111	25	7.2	26
Black bean, canned, prep w/equal volume water—1 c (8 fl oz) (247 gm)	2	1	1	1	1	1	1	1	117	14	3.6	0
Broccoli (inc cream of broccoli soup)—1 c (237 gm)	27	23	21	18	17	15	14	13	235	143	53.9	23
Cabbage—1 c (245 gm)	2	2	2	1	1	1	1	1	47	22	5.1	0
Carrot, w/rice & milk—1 c (245 gm)	4	3	3	2	2	2	2	2	90	16	7.1	3
Cauliflower, cream of, prep w/milk—1 c (248 gm)	30	26	23	20	19	17	15	14	252	156	59.3	26
Cauliflower, dehydrated, prep w/water—1 c (8 fl oz) (256 gm)	1	1	1	1	1	1	½	½	68	15	2.3	0

Food												
Cheddar cheese soup—1 c (251 gm)	44	38	33	30	27	24	22	20	231	131	82.0	48
Cheese, canned, prep w/equal volume milk—1 c (8 fl oz) (251 gm)	44	38	33	30	27	24	22	21	230	131	82.1	48
Cheese, canned, prep w/equal volume water—1 c (8 fl oz) (247 gm)	31	27	24	21	19	17	16	14	155	95	59.9	30
Chicken broth, canned, prep w/equal volume water—1 c (8 fl oz) (244 gm)	2	1	1	1	1	1	1	1	39	13	3.5	1
Chicken broth, cube, dehydrated, prep w/water—1 c (8 fl oz) (243 gm)	½	½	½	½	½	0	0	0	13	3	.5	1
Chicken broth or bouillon, dehydrated, prep w/water—1 c (8 fl oz) (244 gm)	1	1	1	1	1	1	1	1	21	10	2.4	1
Chicken, canned, chunky, ready-to-serve—1 c (8 fl oz) (251 gm)	13	11	10	9	8	8	7	7	178	59	17.8	30
Chicken, cream of, prepared w/water (inc instant)—1 c (244 gm)	10	8	7	7	6	5	5	5	117	66	18.7	10
Chicken gumbo, canned, prep w/equal volume water—1 c (8 fl oz) (244 gm)	2	2	2	2	1	1	1	1	56	13	3.0	5
Chicken mushroom, canned, prep w/equal volume water—1 c (8 fl oz) (244 gm)	11	10	8	8	7	6	6	5	132	83	21.6	10
Chicken noodle, canned, chunky, ready-to-serve—1 c (8 fl oz) (240 gm)	8	7	7	6	6	5	5	4	175	54	12.5	18
Chicken noodle, canned, prep w/equal volume water—1 c (8 fl oz) (241 gm)	4	3	3	3	2	2	2	2	75	23	5.8	7
Chicken noodle, dehydrated, prep w/water—1 c (8 fl oz) (252 gm)	1	1	1	1	1	1	1	1	53	11	2.3	3
Chicken noodle, w/carrots—1 c (241 gm)	12	11	10	10	9	9	8	8	129	25	6.2	60
Chicken noodle, w/meatballs, canned, chunky, ready-to-serve—1 c (8 fl oz) (248 gm)	6	5	4	4	4	3	3	3	99	32	9.6	10
Chicken rice—1 cup (241 gm)	3	3	2	2	2	2	2	2	60	17	4.1	7
Chicken rice, canned, chunky, ready-to-serve—1 c (8 fl oz) (240 gm)	6	5	4	4	4	3	3	3	127	29	8.6	12

Soups (cont.)

	1200	1500	1800	2100	2400	2800	3200	3600	TOT CAL	FAT CAL	S/FAT CAL	CHOL MG
Chicken rice, dehydrated, prep w/water—1 c (8 fl oz) (253 gm)	2	2	1	1	1	1	1	1	60	13	3.0	3
Chicken rice, Puerto Rican style (sopa de pollo con arroz)—1 c (220 gm)	16	14	13	12	11	10	9	9	170	66	19.0	46
Chicken vegetable, canned, chunky, ready-to-serve—1 c (8 fl oz) (240 gm)	8	7	7	6	6	5	5	4	167	43	13.0	17
Chicken vegetable, canned, prep w/ equal volume water—1 c (8 fl oz) (241 gm)	5	4	4	4	3	3	3	3	74	25	7.7	10
Chicken vegetable, dehydrated, prep w/water—1 c (8 fl oz) (251 gm)	1	1	1	1	1	1	1	1	49	7	1.6	3
Chicken w/dumplings, canned, prep w/ equal volume water—1 c (8 fl oz) (241 gm)	11	9	9	8	7	7	6	6	97	50	11.8	34
Chicken w/dumplings—1 c (241 gm)	10	9	9	8	7	7	6	6	96	50	11.7	34
Chicken w/dumplings & potatoes—1 c (251 gm)	10	9	8	7	7	6	6	6	113	46	11.0	31
Chicken w/rice, canned, condensed—1 c (8 fl oz) (246 gm)	6	5	4	4	4	3	3	3	120	34	8.2	12
Chili beef, canned, prep w/equal volume water—1 c (8 fl oz) (250 gm)	15	13	12	10	9	8	8	7	169	59	30.1	13
Chili beef, chunky style—1 c (240 gm)	15	13	12	11	10	9	8	8	177	55	21.9	33
Clam chowder, Manhattan, canned, w/tomato, prep w/equal volume water —1 c (8 fl oz) (244 gm)	2	2	2	1	1	1	1	1	77	21	4.0	2
Clam chowder, Manhattan (inc chunky style)—1 c (240 gm)	6	5	5	4	4	4	3	3	106	25	11.4	8

Food												
Clam chowder, Manhattan, canned, chunky, ready-to-serve—1 c (8 fl oz) (240 gm)	11	9	8	7	7	6	5	5	133	31	18.9	14
Clam chowder, New England, canned, prep w/equal volume milk—1 c (8 fl oz) (248 gm)	15	13	12	11	10	9	8	7	163	59	26.5	22
Clam chowder, New England, canned, prep w/equal volume water—1 c (8 fl oz) (244 gm)	2	2	2	2	2	1	1	1	95	26	3.7	5
Consomme, w/gelatin, dehydrated, prep w/water—1 c (8 fl oz) (249 gm)	—	—	—	—	—	—	—	—	17	0	—	0
Corn, cream of, prep w/milk—1 c (248 gm)	29	25	22	19	18	16	14	13	276	152	56.5	24
Crab, canned, ready-to-serve—1 c (8 fl oz) (244 gm)	3	3	2	2	2	2	2	2	76	14	3.4	10
Crab, prep w/milk (inc crab bisque, seafood bisque)—1 c (248 gm)	33	29	26	24	23	21	19	18	247	123	41.8	93
Crab, tomato-base—1 c (244 gm)	15	13	12	12	11	10	10	9	164	66	12.5	60
Cream of asparagus, canned, prep w/equal volume milk—1 c (8 fl oz) (248 gm)	17	14	13	12	11	9	9	8	161	74	30.0	22
Cream of asparagus, canned, prep w/equal volume water—1 c (8 fl oz) (244 gm)	5	4	4	3	3	3	2	2	87	37	9.4	5
Cream of asparagus, dehydrated, prep w/water—1 c (8 fl oz) (251 gm)	0	0	0	0	0	0	0	0	58	15	.4	0
Cream of celery, canned, prep w/equal volume milk—1 c (8 fl oz) (248 gm)	21	18	16	14	13	12	11	10	165	87	35.5	32
Cream of celery, canned, prep w/equal volume water—1 c (8 fl oz) (244 gm)	8	7	6	6	5	5	4	4	90	50	12.6	15
Cream of celery, dehydrated, prep w/water—1 c (8 fl oz) (254 gm)	1	1	1	1	1	1	1	1	63	14	2.2	1
Cream of chicken, canned, prep w/equal volume water—1 c (8 fl oz) (244 gm)	10	8	7	7	6	5	5	5	116	67	18.7	10

Soups *(cont.)*

	1200	1500	1800	2100	2400	2800	3200	3600	TOT CAL	FAT CAL	S/FAT CAL	CHOL MG
Cream of chicken, dehydrated, prep w/water—1 c (8 fl oz) (261 gm)	14	12	10	9	8	7	7	6	107	48	30.5	3
Cream of chicken, canned, prep w/equal volume milk—1 c (8 fl oz) (248 gm)	23	20	17	16	14	13	12	11	191	103	41.7	27
Cream of mushroom, canned, prep w/equal volume milk—1 c (8 fl oz) (248 gm)	24	20	18	16	14	13	12	11	203	122	46.1	20
Cream of mushroom, canned, prep w/equal volume water—1 c (8 fl oz) (244 gm)	10	9	7	7	6	5	5	4	129	81	22.0	2
Cream of onion, canned, prep w/equal volume milk—1 c (8 fl oz) (248 gm)	21	18	16	15	13	12	11	10	186	85	36.3	32
Cream of onion, canned, prep w/equal volume water—1 c (8 fl oz) (244 gm)	8	7	6	6	5	5	4	4	107	48	13.2	15
Cream of potato, canned, prep w/equal volume milk—1 c (8 fl oz) (248 gm)	18	16	14	13	11	10	9	9	148	58	33.8	22
Cream of potato, canned, prep w/equal volume water—1 c (8 fl oz) (244 gm)	6	5	4	4	3	3	3	3	73	22	11.0	5
Cream of shrimp, canned, prep w/equal volume milk—1 c (8 fl oz) (248 gm)	29	25	22	20	18	16	15	14	165	84	52.1	35
Cream of shrimp, canned, prep w/equal volume water—1 c (8 fl oz) (244 gm)	16	13	12	11	10	9	8	7	90	47	29.1	17
Cream of shrimp, prep w/milk (inc shrimp bisque)—1 c (248 gm)	43	38	34	32	29	27	25	24	285	143	52.1	125
Cream of vegetable, dehydrated, prep w/water—1 c (8 fl oz) (260 gm)	6	5	4	4	3	3	3	2	105	51	12.9	0
Cucumber, cream of, prep w/milk—1 c (248 gm)	30	25	22	20	18	16	15	13	241	152	57.8	25

Food												
Egg drop soup—¾ c (183 gm)	19	18	16	16	15	14	13	13	55	27	7.9	100
Escarole, canned, ready-to-serve—1 c (8 fl oz) (248 gm)	2	2	2	2	2	1	1	1	27	16	4.9	2
Escarole soup—1 c (245 gm)	34	29	26	23	21	19	17	16	279	178	67.4	29
Fish broth—1 c (244 gm)	2	1	1	1	1	1	1	1	39	13	3.7	0
Fish chowder (inc fisherman's, seafood chowder)—1 c (244 gm)	22	20	18	17	15	14	13	12	192	54	26.6	67
Garbanzo or chickpea—1 c (253 gm)	1	1	1	1	1	1	½	½	189	22	2.3	0
Gazpacho—1 c (244 gm)	1	1	1	1	1	1	1	½	56	20	2.6	0
Gazpacho, canned, ready-to-serve—1 c (8 fl oz) (244 gm)	1	1	1	1	1	1	1	½	57	20	2.6	0
Ham, noodle & vegetable, Puerto Rican-style—1 c (250 gm)	11	10	9	8	8	7	7	6	141	36	11.8	36
Ham, pasta & vegetable—1 c (244 gm)	7	6	6	5	5	4	4	4	182	26	7.7	22
Hot & sour (inc hot and spicy Chinese soup)—1 c (244 gm)	12	11	10	9	8	7	7	6	135	60	19.9	22
Leek, cream of, prep w/milk—1 c (248 gm)	21	18	16	15	13	12	11	10	186	84	36.4	32
Leek, dehydrated, prep w/water—1 c (8 fl oz) (254 gm)	5	4	3	3	3	2	2	2	71	19	9.2	3
Lentil—1 c (248 gm)	11	10	8	8	7	6	5	5	217	74	23.9	5
Lentil w/ham, canned, ready-to-serve—1 c (8 fl oz) (248 gm)	6	5	4	4	3	3	3	3	139	25	10.1	7
Lima bean—1 c (253 gm)	5	4	3	3	3	3	2	2	117	29	10.1	3
Lobster bisque—1 c (248 gm)	36	32	28	26	24	22	20	19	272	146	52.3	82
Lobster gumbo—1 c (244 gm)	10	9	8	7	7	6	6	5	174	68	13.0	27
Macaroni & potato—1 c (244 gm)	10	9	8	7	6	6	5	5	216	39	18.6	11
Matzo ball—1 c (241 gm)	18	16	15	14	13	13	12	11	120	51	12.1	81
Minestrone, canned, prep w/equal volume water—1 c (8 fl oz) (241 gm)	2	2	2	2	2	1	1	1	83	23	4.9	2
Minestrone, canned, chunky, ready-to-serve—1 c (8 fl oz) (240 gm)	7	6	5	5	4	4	3	3	127	25	13.4	5
Minestrone, dehydrated, prep w/water—1 c (8 fl oz) (254 gm)	4	3	3	3	2	2	2	2	79	15	7.4	3

	1200	1500	1800	2100	2400	2800	3200	3600	TOT CAL	FAT CAL	S/FAT CAL	CHOL MG
Soups (cont.)												
Mushroom barley, canned, prep w/equal volume water—1 c (8 fl oz) (244 gm)	2	1	1	1	1	1	1	1	73	21	4.0	0
Mushroom, dehydrated, dry—1 reg pkt (74 gm)	11	10	8	7	7	6	5	5	328	149	25.3	1
Mushroom w/beef stock, canned, prep w/equal volume water—1 c (8 fl oz) (244 gm)	7	6	5	5	4	4	4	3	85	36	13.9	7
Onion, canned, prep w/equal volume water—1 c (8 fl oz) (241 gm)	1	1	1	1	1	1	½	½	57	15	2.3	0
Onion, dehydrated, prep w/water—1 c (8 fl oz) (246 gm)	1	½	½	½	½	½	0	0	28	5	1.2	0
Onion, French—1 c (241 gm)	1	1	1	1	1	1	½	½	58	16	2.4	0
Oxtail—1 c (244 gm)	5	4	4	3	3	3	2	2	68	22	11.0	2
Oxtail, dehydrated, prep w/water—1 c (8 fl oz) (253 gm)	6	5	4	4	3	3	3	2	71	23	11.4	3
Oyster stew, canned, prepared w/equal volume milk—1 c (8 fl oz) (245 gm)	25	22	19	17	16	14	13	12	134	71	45.5	32
Oyster stew, canned, prepared w/equal volume water—1 c (8 fl oz) (241 gm)	12	10	9	8	8	7	6	6	59	34	22.5	14
Oyster stew, home prep, 1 part oyster to 2 parts milk—1 c (240 gm)	42	37	33	30	28	25	23	22	233	139	64.8	86
Oyster stew, home prep, 1 part oyster to 3 parts milk—1 c (240 gm)	40	34	31	28	26	23	21	20	206	114	64.8	70
Pea, green, canned, prep w/equal volume milk—1 c (8 fl oz) (254 gm)	19	16	14	13	12	10	9	9	239	63	36.0	18
Pea, green, canned, prep w/equal volume water—1c (8 fl oz) (250 gm)	6	5	4	4	3	3	3	2	164	26	12.7	0

Food												
Pea, green, mix, dehydrated, prep w/water—1 c (8 fl oz) (271 gm)	2	2	2	2	1	1	1	1	133	14	3.9	3
Pea, split w/ham, canned, chunky, ready-to-serve—1 c (8 fl oz) (240 gm)	7	6	5	5	5	4	4	3	184	36	14.3	7
Pea, split w/ham, canned, prep w/equal volume water—1 c (8 fl oz) (253 gm)	8	7	6	6	5	5	4	4	189	40	15.8	8
Pepperpot, canned, prep w/equal volume water—1 c (8 fl oz) (241 gm)	10	8	7	7	6	5	5	5	103	41	18.5	10
Pinto bean—1 c (253 gm)	½	½	½	0	0	0	0	0	184	6	.8	0
Pork, rice & vegetable—1 c (244 gm)	12	10	9	9	8	8	7	7	121	37	12.7	38
Portuguese bean—1 c (253 gm)	8	7	6	6	5	5	4	4	181	38	12.9	16
Potato chowder (inc corn chowder)—1 c (248 gm)	29	25	22	20	18	16	15	14	234	136	57.0	27
Potato, prep w/milk—1 c (248 gm)	18	16	14	13	12	10	9	9	149	58	33.9	22
Potato, prep w/water—1 c (244 gm)	6	5	4	4	3	3	3	3	73	21	11.0	5
Salmon, cream style—1 c (248 gm)	20	17	16	14	13	12	11	10	298	149	26.4	50
Scotch broth, canned, prep w/equal volume water—1 c (8 fl oz) (241 gm)	5	4	4	4	3	3	3	2	80	23	10.0	5
Scotch broth (lamb, vegetables & barley)—1 c (241 gm)	5	4	4	4	3	3	3	2	80	24	10.0	5
Shav—1 c (240 gm)	35	32	30	29	27	26	25	24	58	36	10.8	192
Shrimp gumbo—1 c (244 gm)	10	9	9	8	8	7	7	6	149	42	8.1	43
Soup, mostly noodles (inc spaghetti, Top Ramen, Oriental Noodle Soup, meat flavors, saimin soup)—1 c (244 gm)	8	7	6	6	5	5	5	5	158	17	3.7	38
Soybean, made w/milk—1 c (253 gm)	15	13	11	10	9	8	7	7	198	106	29.1	13
Soybean, miso broth—1 c (240 gm)	2	2	2	1	1	1	1	1	78	29	5.1	0
Spanish vegetable, Puerto Rican style (caldo gallego)—1 c (250 gm)	39	34	31	28	25	23	21	20	302	176	64.7	69
Spinach soup—1 c (245 gm)	27	23	21	18	17	15	14	12	233	142	53.8	23
Split pea—1 c (250 gm)	6	5	4	4	3	3	3	2	165	26	12.6	0
Stockpot, canned, prep w/equal volume water—1 c (8 fl oz) (247 gm)	4	4	3	3	2	2	2	2	100	35	7.7	5
Sweet and sour—1 c (244 gm)	2	2	1	1	1	1	1	1	73	7	2.3	5

	1200	1500	1800	2100	2400	2800	3200	3600	TOT CAL	FAT CAL	S/FAT CAL	CHOL MG
Soups (cont.)												
Tomato beef, prep w/water—1 c (244 gm)	7	6	5	5	4	4	3	3	139	39	14.3	5
Tomato beef rice, prep w/water—1 c (244 gm)	3	2	2	2	2	1	1	1	126	26	5.7	1
Tomato beef w/noodle, canned, prep w/equal volume water—1 c (8 fl oz) (244 gm)	7	6	5	5	4	4	3	3	140	39	14.3	5
Tomato bisque, canned, prep w/equal volume milk—1 c (8 fl oz) (251 gm)	16	14	12	11	10	9	8	8	198	59	28.3	23
Tomato bisque, canned, prep w/equal volume water—1 c (8 fl oz) (247 gm)	3	3	2	2	2	2	2	1	123	23	4.9	5
Tomato, canned, prep w/equal volume milk—1 c (8 fl oz) (248 gm)	14	12	11	10	9	8	7	7	160	54	26.1	17
Tomato, canned, prep w/equal volume water—1 c (8 fl oz) (244 gm)	1	1	1	1	1	1	1	1	86	17	3.3	0
Tomato, cream of (inc tomato bisque, canned tomato soup prep w/milk)—1 c (248 gm)	14	12	11	10	9	8	7	7	161	54	26.1	17
Tomato, dehydrated, prep w/water—1 c (8 fl oz) (265 gm)	4	4	3	3	3	2	2	2	102	22	9.7	1
Tomato noodle, prep w/water (inc tomato macaroni soup)—1 c (244 gm)	4	4	3	3	3	3	3	2	130	21	4.1	15
Tomato rice, canned, prep w/equal volume water—1 c (8 fl oz) (247 gm)	2	2	2	2	1	1	1	1	120	24	4.7	2
Tomato vegetable, dehydrated, prep w/water—1 c (8 fl oz) (253 gm)	1	1	1	1	1	1	1	1	55	8	3.3	0
Tomato vegetable w/noodles, prep w/water—1 c (241 gm)	1	1	1	1	1	1	1	½	72	17	2.6	0

Food												
Turkey, chunky, ready-to-serve—1 c (8 fl oz) (236 gm)	6	5	5	4	4	4	3	3	136	40	11.0	9
Turkey noodle, chunky style—1 c (236 gm)	15	13	12	11	11	10	9	9	173	49	12.8	58
Turkey noodle, canned, prep w/equal volume water—1 c (8 fl oz) (244 gm)	3	3	2	2	2	2	2	1	69	18	5.0	5
Turkey vegetable, canned, prep w/equal volume water—1 c (8 fl oz) (241 gm)	4	3	3	3	2	2	2	2	74	27	8.1	2
Turtle & vegetable (inc snapper)—1 c (244 gm)	15	14	13	12	11	11	10	9	112	38	8.3	71
Vegetable bean, prep w/water (inc minestrone)—1 c (241 gm)	3	2	2	2	2	1	1	1	82	23	5.0	2
Vegetable beef, dehydrated, prep w/water—1 c (8 fl oz) (253 gm)	2	2	2	2	1	1	1	1	53	10	4.9	1
Vegetable beef, canned, prep w/equal volume water—1 c (8 fl oz) (244 gm)	4	4	3	3	3	2	2	2	79	17	7.7	5
Vegetable broth, bouillon (inc onion broth, pot liquor)—1 c (240 gm)	0	0	0	0	0	0	0	0	17	0	0	0
Vegetable, canned, chunky, ready-to-serve—1 c (8 fl oz) (240 gm)	2	2	2	1	1	1	1	1	122	33	4.9	0
Vegetable, cream of, prep w/milk—1 c (248 gm)	29	25	22	20	18	16	15	14	242	118	54.7	31
Vegetable, made fr dry mix—1 c (253 gm)	2	2	2	1	1	1	1	1	53	10	5.0	0
Vegetable noodle, prep w/water (inc vegetable w/dumplings, alphabet vegetable soup)—1 c (241 gm)	1	1	1	1	1	1	1	½	72	17	2.6	0
Vegetable rice, prep w/water—1 c (241 gm)	1	1	1	1	1	1	½	½	98	15	2.3	0
Vegetable, Spanish style, stew type—1 c (227 gm)	14	12	11	10	9	9	8	7	191	60	21.2	31
Vegetable w/beef broth, canned, prep w/equal volume water—1 c (8 fl oz) (241 gm)	2	2	2	1	1	1	1	1	81	17	4.0	2

THE 2-IN-1 SYSTEM

	1200	1500	1800	2100	2400	2800	3200	3600	TOT CAL	FAT CAL	S/FAT CAL	CHOL MG
Soups *(cont.)*												
Vegetarian vegetable, canned, prep w/equal volume water—1 c (8 fl oz) (241 gm)	1	1	1	1	1	1	1	½	72	17	2.6	0
Vichyssoise—1 c (248 gm)	18	16	14	13	12	10	9	9	149	58	33.9	22
Watercress soup—1 c (245 gm)	5	5	4	4	4	4	4	4	25	2	.3	31
White bean, Puerto Rican style (sopon de habichuelas blancas)—1 c (275 gm)	10	9	8	7	6	6	5	5	258	64	20.4	9
Won ton—1 c (241 gm)	23	21	19	17	16	15	14	13	210	89	29.4	65
Zucchini, cream of—1 c (248 gm)	27	23	20	18	16	15	13	12	223	137	52.0	23
SPICES and HERBS												
Allspice, ground—1 tsp (2 gm)	0	0	0	—	0	0	0	0	5	2	.4	0
Anise seed—1 tsp (2 gm)	—	—	—	—	—	—	—	—	7	3	—	0
Basil, ground—1 tsp (1 gm)	0	0	0	0	0	0	0	0	4	1	—	0
Bay leaf, crumbled—1 tsp (1 gm)	0	0	0	0	0	0	0	0	2	1	.1	0
Caraway seed—1 tsp (2 gm)	0	0	0	0	0	0	0	0	7	3	.1	0
Cardamom, ground—1 tsp (2 gm)	0	0	0	0	0	0	0	0	6	1	.1	0
Celery seed—1 tsp (2 gm)	0	0	0	0	0	0	0	0	8	5	.4	0
Chervil, dried—1 tsp (1 gm)	—	—	—	—	—	—	—	—	1	0	—	0
Chili powder—1 tsp (3 gm)	0	0	0	0	0	0	0	0	8	4	.2	0
Cinnamon, ground—1 tsp (2 gm)	—	—	—	—	—	—	—	—	6	1	.2	0
Cloves, ground—1 tsp (2 gm)	½	½	½	0	0	0	0	0	7	4	.8	0
Coriander leaf, dried—1 tsp (1 gm)	0	0	0	0	0	0	0	0	2	0	0	0
Coriander, fresh, raw—¼ cup (4 gm)	0	0	0	0	0	0	0	0	1	0	0	0
Coriander seed—1 tsp (2 gm)	0	0	0	0	0	0	0	0	5	3	.2	0
Cumin seed—1 tsp (2 gm)	—	—	—	—	—	—	—	—	8	5	—	0

Food												
Curry powder—1 tsp (2 gm)									7	3	—	0
Dill seed—1 tsp (2 gm)	0	0							6	3	.2	0
Dill weed, dried—1 tsp (1 gm)									3	0	—	0
Fennel seed—1 tsp (2 gm)	0	0							7	3	.1	0
Fenugreek seed—1 tsp (4 gm)									12	2	—	0
Garlic powder—1 tsp (3 gm)									9	0	.4	0
Ginger, ground—1 tsp (2 gm)	0	0							6	1	1.4	0
Mace, ground—1 tsp (2 gm)	1	1	½	½	½	½			8	5	—	0
Marjoram, dried—1 tsp (1 gm)									2	8	.4	0
Mustard seed, yellow—1 tsp (3 gm)	0	0	½						15	8	5.1	0
Nutmeg, ground—1 tsp (2 gm)	2	2	1	1					12	7	—	0
Onion powder—1 tsp (2 gm)									7	0	.4	0
Oregano, ground—1 tsp (2 gm)	0	0	0						5	2	.4	0
Paprika—1 tsp (2 gm)	0	0	0						6	3	.4	0
Parsley, dried—1 tsp (0 gm)									1	0	.2	0
Pepper, black—1 tsp (2 gm)	0	0	½						5	1	.5	0
Pepper, red or cayenne—1 tsp (2 gm)	0	0							6	3	—	0
Pepper, white—1 tsp (2 gm)									7	1	1.3	0
Poppy seed—1 tsp (3 gm)	1	1	½	½	½	½			15	12	—	0
Poultry seasoning—1 tsp (2 gm)									5	1	—	0
Pumpkin pie spice—1 tsp (2 gm)									6	2	—	0
Rosemary, dried—1 tsp (1 gm)	0								4	2	.4	0
Saffron—1 tsp (1 gm)									2	0	—	0
Sage, ground—1 tsp (1 gm)	0								4	1	.4	0
Savory, ground—1 tsp (1 gm)									5	1	—	0
Tarragon, ground—1 tsp (2 gm)	0								4	1	.4	0
Thyme, ground—1 tsp (1 gm)									5	1	—	0
Turmeric, ground—1 tsp (2 gm)									8	2	—	0

SUGARS and SWEETS
Candies and Chewing Gum

Food												
$100,000 bar—1 bar (1.5 oz) (43 gm)	20	17	15	13	12	10	9	9	206	75	44.5	3
3 Musketeer bar—1 bar (8 oz) (23 gm)	6	5	5	4	4	3	3	3	101	26	13.7	2
Almond Joy— 1 bar (1.6 oz) (45 gm)	20	17	14	13	11	10	9	8	184	71	44.0	1
Almond Roca— 1 piece (11 gm)	5	4	4	4	3	3	2	2	50	20	10.4	1

Candies and Chewing Gum (cont.)

	1200	1500	1800	2100	2400	2800	3200	3600	TOT CAL	FAT CAL	S/FAT CAL	CHOL MG
Almonds, chocolate-covered—5 almonds (15 gm)	8	6	6	5	4	4	4	3	85	59	16.3	2
Almonds, sugar-coated—4 pieces (14 gm)	1	1	1	1	½	½	½	½	64	23	1.9	0
Andes Mint Wafers—1 piece (5 gm)	4	3	3	2	2	2	2	1	27	14	8.1	0
A-OK Space Energy Stix—1 stick (10 gm)	1	1	1	1	½	½	½	½	41	8	1.8	0
Applause—1 oz (28 gm)	14	12	10	9	8	7	6	6	133	51	30.5	1
Ayds—1 ind wrapped piece (7 gm)	1	½	½	½	½	½	0	0	29	6	1.2	0
Baby Ruth—1 bar (1.2 oz) (34 gm)	13	11	10	9	8	7	6	6	162	67	25.1	11
Bit-O-Honey—1 bar, 6 candy chews (1.7 oz) (48 gm)	1	1	1	1	1	1	1	½	171	13	2.6	0
Blow-Pop—1 lollipop (21 gm)	0	0	0	0	0	0	0	0	80	0	0	0
Bonbon (inc Russell Stover)—1 bonbon (14 gm)	7	6	5	5	4	4	3	3	51	18	15.9	0
Bonkers!—1 pkt, 8 pieces (1.45 oz) (41 gm)	1	1	1	1	1	1	½	½	146	11	2.2	0
Boston Baked Beans—1 box (1.75 oz) (50 gm)	11	9	8	7	6	6	5	5	258	145	25.2	0
Bottle Caps—1 pkt (69 oz) (20 gm)	0	0	0	0	0	0	0	0	76	0	0	0
Bounty—1 bar (1.75 oz) (50 gm)	22	18	16	14	13	11	10	9	205	79	48.9	1
Brach's Royal—1 oz (28 gm)	8	7	6	5	5	4	4	3	112	26	18.0	1
Breath Mints—1 pack (18 gm)	0	0	0	0	0	0	0	0	69	0	0	0
Bridge Mix (inc Brach's)—9 pieces (27 gm)	19	16	14	13	11	10	9	8	146	91	42.2	5
Bubble gum—1 piece (4 gm)	0	0	0	0	0	0	0	0	13	0	0	0
Butter creams—1 butter cream (20 gm)	3	3	3	2	2	2	2	1	79	19	7.8	0

Item												
Butterfinger—1 bar (1.6 oz) (45 gm)	14	12	10	9	8	7	6	6	211	74	31.2	0
Butternut bar—1 bar (1.6 oz) (45 gm)	12	10	9	8	7	6	6	5	195	73	27.6	1
Butterscotch disc—1 piece (6 gm)	0	0	0	0	0	0	0	0	23	0	0	0
Butterscotch morsels—1 oz (28 gm)	12	10	9	8	7	6	5	5	132	45	26.9	0
Candy cane—1 cane (15 gm)	0	0	0	0	0	0	0	0	57	0	0	0
Candy corn—1 oz (28 gm)	0	0	0	0	0	0	0	0	105	0	0	0
Candy hearts—5 pieces (30 gm)	0	0	0	0	0	0	0	0	115	0	0	0
Candy necklace—1 necklace (.75 oz) (21 gm)	0	0	0	0	0	0	0	0	80	0	0	0
Caramel, chocolate-covered—2 pieces (18 gm)	7	6	5	5	4	4	3	3	77	25	16.1	1
Caramel, chocolate-flavored roll—2 pieces (14 gm)	3	3	2	2	2	2	2	1	55	10	7.6	0
Caramel creams—6 pieces (30 gm)	9	7	6	6	5	4	4	4	120	28	19.3	1
Caramel, flavor other than chocolate—2 pieces (18 gm)	5	4	4	3	3	3	2	2	72	17	11.6	0
Caramello—1 bar (1.2 oz) (34 gm)	14	12	10	9	8	7	6	6	145	47	30.3	2
Caramel, w/nuts—2 pieces (12 gm)	4	3	3	2	2	2	2	2	51	18	8.2	0
Caramel w/nuts & cereal, chocolate-covered—1 oz (28 gm)	8	7	6	5	5	4	4	3	121	46	17.2	1
Caramel w/nuts, chocolate-covered—1 oz (28 gm)	9	7	6	6	5	5	4	4	139	53	19.9	0
Certs—1 cert (7 oz roll) (2 gm)	0	0	0	0	0	0	0	0	8	0	0	0
Charleston Chew—1 bar (1.875 oz) (53 gm)	19	16	14	12	11	10	9	8	216	63	41.4	3
Charm pop—1 pop (21 gm)	0	0	0	0	0	0	0	0	80	0	0	0
Cherries, chocolate-covered—1 cherry (14 gm)	2	2	2	2	1	1	1	1	55	13	5.5	0
Chewing gum, sugared—1 piece (4 gm)	0	0	0	0	0	0	0	0	13	0	0	0
Chewing gum, uncoated, sugarless—1 piece (2 gm)	0	0	0	0	0	0	0	0	5	0	0	0
Chew-Its—1 bar (45 gm)	18	15	13	12	11	9	9	8	192	62	40.2	3
Chick-o-Stick—1 stick (1.55 oz) (44 gm)	9	8	7	6	6	5	4	4	204	81	20.5	2

Candies and Chewing Gum (cont.)

	1200	1500	1800	2100	2400	2800	3200	3600	TOT CAL	FAT CAL	S/FAT CAL	CHOL MG
Chocolate, coated, coconut center—1 oz (28 gm)	11	10	8	7	7	6	5	5	123	44	25.6	0
Chocolate-coated fondant—1 mint (2½" dia) (35 gm)	5	4	3	3	3	2	2	2	144	33	10.3	0
Chocolate-coated vanilla creams—1 oz (28 gm)	6	5	4	4	3	3	3	2	122	43	12.4	1
Chocolate coins—5 pieces (25 gm)	20	17	15	13	12	11	9	9	130	73	43.2	6
Chocolate jelly—1 piece (14 gm)	3	3	2	2	2	2	2	1	55	12	6.9	1
Chocolate Jots—1 oz (28 gm)	5	4	4	3	3	2	2	2	111	26	11.0	0
Chocolate, milk, plain—4 miniature bars (28 gm)	22	19	16	15	13	12	11	10	146	81	48.4	6
Chocolate, milk, w/almonds—1 piece (41 gm)	27	23	20	18	16	14	13	12	218	131	58.3	7
Chocolate, milk, w/cereal—1 piece (34 gm)	24	20	18	16	14	13	11	10	171	86	51.4	7
Chocolate, milk, w/fruit & nuts—1 piece (33 gm)	17	15	13	11	10	9	8	7	160	82	37.3	4
Chocolate, milk, w/nuts, not almond or peanuts—1 piece (33 gm)	24	20	18	16	14	12	11	10	178	111	51.6	6
Chocolate, milk, w/peanuts—1 piece (43 gm)	27	23	20	18	16	14	13	12	233	147	58.6	7
Chocolate, semisweet morsel—1 oz (28 gm)	23	20	17	15	14	12	11	10	142	90	52.7	0
Chocolate, sweet or dark—1 bar (1.45 oz) (41 gm)	35	29	25	22	20	18	16	14	216	130	78.6	0
Chocolate, white (inc summer coating)—5 pieces (25 gm)	18	15	13	12	11	9	8	8	133	68	40.7	1

Food												
Chocolate, white, w/almonds (inc Nestle Alpine White with Almonds)—1 bar (1.3 oz) (37 gm)	24	20	18	16	14	12	11	10	200	112	53.7	1
Chocolate Payday—1 bar (1.8 oz) (51 gm)	15	12	11	9	8	7	7	6	238	117	32.5	1
Choco-Lite—1 bar (1.125 oz) (32 gm)	24	20	17	15	14	12	11	10	172	86	51.2	6
Chuckles—1 pkt (2 oz) (57 gm)	0	0	0	0	0	0	0	0	198	4	.5	0
Chunky, w/fruit and nuts (inc Chunky Original)—1 ind wrapped piece (1.15 oz) (33 gm)	17	15	13	11	10	9	8	7	160	82	37.3	4
Chunky, w/peanuts—1 ind wrapped piece (1.15 oz) (33 gm)	21	18	15	14	12	11	10	9	179	113	44.9	6
Chunky, w/pecans—1 ind wrapped piece (1.15 oz) (33 gm)	24	20	18	16	14	12	11	10	178	111	51.6	6
Cinnamon hearts—½ c (108 gm)	0	0	0	0	0	0	0	0	413	0	0	0
Circus peanuts (Brach's)—1 oz (28 gm)	0	0	0	0	0	0	0	0	89	0	0	0
Clark bar—1 bar (1.2 oz) (34 gm)	12	10	8	7	7	6	5	5	157	60	26.2	0
Coconut candy, chocolate-covered—1 oz (28 gm)	12	10	9	8	7	6	6	5	115	44	27.4	1
Coconut candy, no chocolate covering—1 oz (28 gm)	14	12	10	9	8	7	6	6	103	36	31.7	0
Coconut candy, Puerto Rican style—1 oz (28 gm)	14	11	10	9	8	7	6	6	114	34	30.5	0
Conversation hearts—1 oz (28 gm)	0	0	0	0	0	0	0	0	107	0	0	0
Cotton candy—1 cone (21 gm)	0	0	0	0	0	0	0	0	80	0	0	0
Cough drops—1 piece (1 gm)	0	0	0	0	0	0	0	0	4	0	0	0
Date candy—1 oz (28 gm)	2	1	1	1	1	1	1	1	104	43	3.7	0
Dietetic or lo-cal candy, chocolate-covered—1 oz (28 gm)	36	30	26	23	21	18	17	15	154	99	80.9	1
Dietetic or lo-cal candy—5 pieces (15 gm)	0	0	0	0	0	0	0	0	56	0	0	0
Dietetic or lo-cal gumdrops—3 pieces (15 gm)	0	0	0	0	0	0	0	0	24	0	0	0
Dietetic or lo-cal hard candies—5 pieces (15 gm)	0	0	0	0	0	0	0	0	56	0	0	0

THE 2-IN-1 SYSTEM

	1200	1500	1800	2100	2400	2800	3200	3600	TOT CAL	FAT CAL	S/FAT CAL	CHOL MG
Candies and Chewing Gum (cont.)												
Dietetic or lo-cal mints—1 mint (2 gm)	0	0	0	0	0	0	0	0	6	0	0	0
Dots—1 box (1.25 oz) (35 gm)	0	0	0	0	0	0	0	0	121	2	.3	0
Easter egg—1 oz (28 gm)	22	19	16	15	13	12	11	10	146	81	48.4	6
Easter eggs—2 medium eggs (22 gm)	6	5	4	4	3	3	3	2	84	21	12.6	0
Easter eggs, candy-coated marshmallow—1 oz (28 gm)	0	0	0	0	0	0	0	0	96	0	0	0
Easter eggs, chocolate-covered coconut—1 oz (28 gm)	12	10	9	8	7	6	6	5	115	44	27.4	1
Easter eggs, chocolate-covered creme—1 oz (28 gm)	5	4	4	3	3	2	2	2	111	26	11.0	0
Easter eggs, peanut butter—1 egg (1.2 oz) (34 gm)	14	12	10	9	8	7	6	6	171	89	29.7	3
Fifth Avenue—1 bar (1.125 oz) (32 gm)	11	9	8	7	6	6	5	5	148	56	24.6	0
Fondant—1 c candy corn (200 gm)	4	3	3	3	2	2	2	2	728	36	9.0	0
Fondant—1 c mints, uncoated (110 gm)	2	2	2	1	1	1	1	1	400	20	4.9	0
Fondant, chocolate-covered (inc chocolate-covered boxed candy)—2 pieces (22 gm)	4	3	3	2	2	2	2	2	87	21	8.6	0
Fondant, mints (inc homemade mints)—2 pieces (22 gm)	0	0	0	0	0	0	0	0	82	0	0	0
Food Stix, fortified (inc caramel, chewy, chocolate, peanut butter, space food)—1 stick (10 gm)	1	1	1	1	½	½	½	½	41	8	1.8	0
Fruit & Nut bar (Cadbury)—1 oz (28 gm)	14	12	11	9	9	8	7	6	136	70	31.6	3
Fruit, chocolate-covered—1 oz (28 gm)	6	5	5	4	4	3	3	3	111	24	13.7	2

Food												
Fruit leather—1 oz (28 gm)	½	½	0	0	0	0	0	0	82	5	.7	0
Fruit nut bar—1 oz (28 gm)	2	1	1	1	1	1	1	1	104	43	3.7	0
Fruit peel, candied—3 pieces (9 gm)	0	0	0	0	0	0	0	0	28	0	0	0
Fruit roll-up—1 roll-up (14 gm)	0	0	0	0	0	0	0	0	41	3	.3	0
Fruit slices, jellied—1 oz (28 gm)	0	2	0	0	0	0	0	0	97	2	.2	0
Fudge, brown sugar (penuche)—1 cu in (30 gm)	2	2	1	1	1	1	1	1	121	32	4.6	1
Fudge, caramel & nut, chocolate-coated—1 cu in (34 gm)	9	8	7	6	6	5	5	4	147	55	20.9	1
Fudge, chocolate, chocolate-coated—1 cu in (30 gm)	13	11	9	8	7	7	6	5	129	43	27.7	2
Fudge, chocolate, chocolate-coated, w/nuts—1 cu in (30 gm)	12	10	9	8	7	6	6	5	136	56	26.9	1
Fudge, chocolate, plain—1 cu in (30 gm)	9	8	7	6	5	5	4	4	120	33	20.9	0
Fudge, chocolate, w/nuts—1 cu in (30 gm)	10	8	7	6	6	5	5	4	128	47	22.4	0
Fudge, divinity—2 pieces (24 gm)	½	½	½	½	½	0	0	0	95	16	1.0	0
Fudge, peanut butter (inc chocolate fudge w/peanut butter)—1 cu in (30 gm)	9	8	7	6	5	5	4	4	134	57	20.4	0
Fudge, peanut butter, w/nuts—1 cu in (30 gm)	9	8	7	6	6	5	4	4	140	68	20.4	0
Fudge, vanilla (inc fruit-flavored)—1 cu in (30 gm)	9	7	6	6	5	5	4	4	119	30	19.6	1
Fudge, vanilla, w/nuts—1 cu in (30 gm)	9	7	6	6	5	5	4	4	127	44	19.5	1
Fun Fruits—1 pouch (9 oz) (26 gm)	½	0	0	0	0	0	0	0	75	5	.6	0
Funs and Dips—1 oz (28 gm)	0	0	0	0	0	0	0	0	107	0	0	0
Goobers—1 oz (28 gm)	19	16	14	12	11	10	9	8	157	104	41.3	3
Good and Fruity—1 box (1.5 oz) (43 gm)	0	0	0	0	0	0	0	0	149	3	.3	0
Good and Plenty—1 box (1.8 oz) (51 gm)	1	1	½	½	½	½	½	½	187	2	1.5	0
Goo Goo Cluster—1 bar (1.75 oz) (50 gm)	16	13	11	10	9	9	8	7	248	95	35.5	0

Candies and Chewing Gum (cont.)

	1200	1500	1800	2100	2400	2800	3200	3600	TOT CAL	FAT CAL	S/FAT CAL	CHOL MG
Gumdrops—10 gumdrops (36 gm)	0	0	0	0	0	0	0	0	125	2	.3	0
Halvah, chocolate-covered—1 oz (28 gm)	10	9	8	7	6	5	5	4	142	76	22.8	2
Halvah, plain—1 oz (28 gm)	6	5	4	4	3	3	3	2	141	74	12.9	0
Hard candy—3 pieces (18 gm)	0	0	0	0	0	0	0	0	69	0	0	0
Hazelnut, w/chocolate (Cadbury)—1 bar (2 oz) (57 gm)	41	35	30	27	24	21	19	18	308	192	89.1	10
Heath bar—1 bar (1.125 oz) (32 gm)	14	12	10	9	8	7	6	6	140	47	30.2	2
Hershey bar—1 bar (1.45 oz) (41 gm)	33	28	24	21	19	17	15	14	213	119	70.8	9
Hershey kiss or star—6 pieces (30 gm)	24	20	18	16	14	13	11	10	156	87	51.8	7
Hershey bar w/almonds—1 bar (1.45 oz) (41 gm)	27	23	20	18	16	14	13	12	218	131	58.3	7
Honey-combed hard candy, w/peanut butter—5 pieces (30 gm)	6	5	5	4	4	3	3	3	139	55	14.0	1
Honey-combed hard candy, w/peanut butter, chocolate-covered—3 pieces (33 gm)	11	9	8	7	7	6	5	5	153	58	25.4	0
Hot Tamales—1 box (1.625 oz) (46 gm)	0	0	0	0	0	0	0	0	160	3	.4	0
Irish Cream mints—1 piece (58 oz)	3	3	2	2	2	2	1	1	67	16	6.7	0
Italian nougat candy—1 oz (28 gm)	5	4	4	3	3	3	2	2	102	15	11.6	0
Jawbreakers—1 oz (28 gm)	0	0	0	0	0	0	0	0	107	0	0	0
Jelly beans—1 oz (28 gm)	0	0	0	0	0	0	0	0	97	2	.2	0
Jimmies (inc chocolate-flavored sprinkles)—1 c (176 gm)	180	151	131	116	104	92	82	75	896	475	405.0	0
Jordon almonds—1 oz (28 gm)	2	1	1	1	1	1	1	1	128	47	3.8	0
Jujubes—1 box (1.5 oz) (43 gm)	0	0	0	0	0	0	0	0	149	3	.3	0
Juiy Fruits—1 oz (28 gm)	0	0	0	0	0	0	0	0	97	2	.2	0

Junior mints—1 box (1.6 oz) (45 gm)	0	17.7	43	178	3	4	5	5	5	6	7	8
Just Juice—1 piece (2 gm)	0	0	0	8	0	0	0	0	0	0	0	0
Kit Kat—1 bar (1.5 oz) (43 gm)	3	67.0	103	217	13	14	16	18	20	22	25	30
Kits—1 pkt (4 pieces) (8 gm)	0	.4	2	29	0	0	0	0	0	0	0	
Krackel bar—1 bar (1.2 oz) (34 gm)	7	51.4	86	171	10	11	13	14	16	18	20	24
Licorice—2 sticks (22 gm)	0	.7	1	81	0	0	0	0	0	0	½	½
Licorice, dietetic—2 pieces (20 gm)	0	0	0	75	0	0	0	0	0	0	0	0
Life Savers—1 piece (2 gm)	0	0	0	8	0	0	0	0	0	0	0	0
Lollipops—1 lollipop (5 oz) (14 gm)	0	0	0	53	0	0	0	0	0	0	0	0
M & M's—1 box (1.48 oz) (42 gm)	5	48.4	81	207	9	10	12	13	15	16	19	22
M & M's peanut chocolate candies—1 oz (28 gm)	4	27.2	69	143	5	6	7	8	8	9	11	13
Mallo Cup—1 piece (1.2 oz) (34 gm)	0	19.5	32	130	4	4	4	5	6	6	7	9
Malted milk balls—10 balls (30 gm)	6	45.4	76	151	9	10	11	12	14	16	18	21
Marathon bar—1 bar (1.38 oz) (39 gm)	3	34.8	54	166	7	7	8	9	10	12	13	16
Mars bar—1 bar (1.76 oz) (50 gm)	3	61.2	103	242	12	13	14	16	18	20	23	28
Marshmallow—2 lg reg (14 gm)	0	0	0	45	0	0	0	0	0	0	0	0
Marshmallow, candy-coated—1 oz (28 gm)	0	0	0	96	0	0	0	0	0	0	0	0
Marshmallow chicken—1 chicken (8 gm)	0	0	0	26	0	0	0	0	0	0	0	0
Marshmallow, chocolate-covered—2 pieces (22 gm)	0	12.6	21	84	2	3	3	3	4	4	5	6
Marshmallow, marshmallow creme—2 pieces (16 gm)	0	0	0	51	0	0	0	0	0	0	0	0
Marshmallow rabbit, chocolate-covered—1 rabbit (14 gm)	0	8.0	13	53	1	2	2	2	2	3	3	4
Mary Jane—1 oz (28 gm)	0	1.5	8	100	½	½	½	½	½	½	1	1
Merri-Mints (Delson)—3 mints (27 gm)	0	0	0	101	0	0	0	0	0	0	0	0
Mike and Ike—1 box (1.25 oz) (35 gm)	0	.3	2	121	0	0	0	0	0	0	0	0
Milk Duds—1 box (1.25 oz) (35 gm)	2	31.2	48	149	6	7	7	8	9	10	12	14
Milky Way bar—1 bar (1.9 oz) (54 gm)	3	38.9	72	247	8	8	9	10	12	13	15	18
Mint Leaves—1 oz (28 gm)	0	.2	2	97	0	0	0	0	0	0	0	0

THE 2-IN-1 SYSTEM

	1200	1500	1800	2100	2400	2800	3200	3600	TOT CAL	FAT CAL	S/FAT CAL	CHOL MG
Candies and Chewing Gum *(cont.)*												
Mints (Brach's)—3 pieces (33 gm)	6	5	4	4	3	3	3	2	131	31	12.9	0
Mints—1 oz (28 gm)	0	0	0	0	0	0	0	0	105	0	0	0
Mint toffee—1 oz (28 gm)	8	7	6	5	5	4	4	3	113	26	18.3	1
Mounds—1 bar (1.65 oz) (47 gm)	21	17	15	13	12	11	9	9	192	74	45.9	1
Mr. Goodbar—1 bar (1.5 oz) (43 gm)	27	23	20	18	16	14	13	12	233	147	58.6	7
Necco wafer—1 oz (28 gm)	0	0	0	0	0	0	0	0	107	0	0	0
Nerds—1 box (1.8 oz) (51 gm)	0	0	0	0	0	0	0	0	195	0	0	0
Nestlé bar—1 bar (1.45 oz) (41 gm)	33	28	24	21	19	17	15	14	213	119	70.8	9
Nestlé chocolate bar, w/almonds—1 bar (1.45 oz) (41 gm)	27	23	20	18	16	14	13	12	218	131	58.3	7
Nestlé Crunch—1 bar (1.313 oz) (37 gm)	26	22	19	17	15	14	12	11	186	94	55.9	7
Nonpareils—1 oz (28 gm)	14	11	10	9	8	7	6	6	130	50	29.6	3
Nougat, chocolate-covered—1 bar (1.875 oz) (53 gm)	19	16	14	12	11	10	9	8	216	63	41.4	3
Nougat, plain—1 oz (28 gm)	5	4	4	3	3	3	2	2	102	15	11.6	0
Nougat, w/caramel, chocolate-covered—1 bar (2 oz) (57 gm)	16	14	12	11	10	9	8	7	237	71	36.1	3
Now and Later—5 pieces (30 gm)	1	1	1	½	½	½	½	½	107	8	1.6	0
Nut roll, fudge or nougat, caramel & nuts—1 roll (25 gm)	7	6	5	5	4	4	3	3	117	57	15.9	0
Nuts, carob-coated (inc peanuts, cashews, walnuts, almonds)—1 oz (28 gm)	4	3	3	3	2	2	2	2	127	57	8.8	0
Nuts, chocolate-covered, not almonds or peanuts—1 oz (28 gm)	18	15	13	12	10	9	8	8	161	117	38.9	3
Oh Henry—1 bar (1.25 oz) (35 gm)	10	8	7	6	6	5	4	4	152	57	21.5	1

438

Food												
Oompas—1 pkt (1.375 oz) (39 gm)	19	16	14	12	11	10	9	8	182	69	41.2	5
Payday—1 bar (1.9 oz) (53 gm)	15	13	11	10	9	8	7	6	248	121	33.8	1
Peanut bar (inc peanut candy)—1 oz (28 gm)	6	5	5	4	4	3	3	3	144	81	14.1	0
Peanut brittle—1 oz (28 gm)	2	2	1	1	1	1	1	1	118	26	4.6	0
Peanut Butter Boppers—1 bar (30 gm)	12	10	9	8	7	6	6	5	159	91	28.0	0
Peanut butter, chocolate-covered (inc peanut butter treats, peanut butter candy)—1 piece (34 gm)	14	12	10	9	8	7	6	6	171	89	29.7	3
Peanut butter egg—1 egg (1.2 oz) (34 gm)	14	12	10	9	8	7	6	6	171	89	29.7	3
Peanut butter log (Clark)—4 logs (24 gm)	5	4	4	3	3	3	2	2	111	44	11.2	1
Peanut Butter Meltaway bar (Brach's)—1 bar (1.5 oz) (43 gm)	17	14	13	11	10	9	8	7	217	113	37.6	3
Peanut butter morsels—1 oz (28 gm)	7	6	5	4	4	3	3	3	148	75	15.3	0
Peanut Chews—1 bar (2 oz) (56 gm)	18	15	13	11	10	9	8	7	277	106	39.8	0
Peanut Kisses—5 pieces (30 gm)	1	1	1	½	½	½	½	½	107	8	1.6	0
Peanut Munch—1 bar (1.4 oz) (40 gm)	9	8	7	6	5	5	4	4	206	116	20.2	0
Peanut Pillows—1 oz (28 gm)	6	5	4	4	3	3	3	3	130	52	13.0	1
Peanuts, chocolate-covered—1 oz (28 gm)	19	16	14	12	11	10	9	8	157	104	41.3	3
Peanuts, sugar-coated—1 oz (28 gm)	6	5	5	4	4	3	3	3	144	81	14.1	0
Peanuts, yogurt-covered—1 oz (28 gm)	7	6	5	4	4	3	3	3	127	77	14.9	1
Pecan roll—1 roll (25 gm)	7	6	5	5	4	4	3	3	117	57	15.9	0
Peppermint Pattie (York)—1 Peter Paul bar (1.5 oz) (43 gm)	7	6	5	5	4	4	3	3	170	41	16.9	0
Pixy Stix—1 giant stick (1 oz) (28 gm)	0	0	0	0	0	0	0	0	107	0	0	0
Planters peanut bar—1 piece (16 gm)	4	3	3	2	2	2	2	2	84	48	8.4	0
Pom Poms—1 box (1.2 oz) (34 gm)	14	12	10	9	8	7	6	6	145	47	30.3	2
Pop Rocks—1 pkt (5 gm)	0	0	0	0	0	0	0	0	19	0	0	0
Powerhouse—1 bar (1.35 oz) (38 gm)	10	9	8	7	6	5	5	4	165	62	23.3	1
Pralines—1 cu in (30 gm)	2	2	2	1	1	1	1	1	121	32	4.6	1

Candies and Chewing Gum (cont.)

	1200	1500	1800	2100	2400	2800	3200	3600	TOT CAL	FAT CAL	S/FAT CAL	CHOL MG
Rainbow coconut—1 bar (1.4 oz) (40 gm)	20	17	15	13	12	10	9	8	147	51	45.3	0
Raisin clusters—1 cluster (22 gm)	8	7	6	5	5	4	4	4	94	34	17.8	2
Raisinets—1 box (1.25 oz) (35 gm)	13	11	10	9	8	7	6	6	149	54	28.4	4
Raisins, carob-covered—1 oz (28 gm)	13	11	9	8	7	6	6	5	111	36	28.6	0
Raisins, chocolate-covered—1 oz (28 gm)	11	9	8	7	6	6	5	5	119	43	22.7	3
Raisins, yogurt-covered—1 oz (28 gm)	9	7	6	6	5	4	4	4	111	33	19.4	0
Raspberry-filled drops—3 pieces (27 gm)	0	0	0	0	0	0	0	0	103	0	0	0
Red Hot Dollars—1 box (1 oz) (28 gm)	0	0	0	0	0	0	0	0	107	0	0	0
Red Hots—4 balls (32 gm)	0	0	0	0	0	0	0	0	122	0	0	0
Reese's Peanut Butter Cup—1 pkt (1.6 oz) (45 gm)	20	17	15	13	12	10	9	8	243	125	43.3	4
Reese's Pieces—1 oz (28 gm)	8	7	6	5	5	4	4	4	134	52	17.9	2
Rolo—9 pieces (27 gm)	15	13	11	10	9	8	7	6	134	55	33.0	2
Root beer barrel—3 barrels (27 gm)	0	0	0	0	0	0	0	0	103	0	0	0
Royals mint chocolate—1 bar (1.5 oz) (43 gm)	23	19	17	15	13	12	11	10	211	82	48.8	6
Saltwater taffy—1 pkt (1 oz) (28 gm)	1	1	½	½	½	½	½	½	100	8	1.5	0
Sesame Crunch (Sahadi)—5 pieces (10 gm)	2	2	1	1	1	1	1	1	54	31	4.3	0
Skittles—1 bar (1.25 oz) (35 gm)	1	1	½	½	½	½	½	½	142	9	1.4	0
Sky bar—1 bar (1.06 oz) (30 gm)	8	7	6	5	5	4	4	4	130	49	18.4	1
Smarty—1 oz (28 gm)	0	0	0	0	0	0	0	0	107	0	0	0
Snickers bar—1 bar (2 oz) (57 gm)	22	18	16	14	13	11	10	9	276	114	45.4	9
Space Dust—1 pkt (5 gm)	0	0	0	0	0	0	0	0	19	0	0	0

Item												
Special Dark—1 bar (1.45 oz) (41 gm)	29	24	21	19	17	15	13	12	221	110	65.3	0
Spun Sugar—1 cone (21 gm)	0	0	0	0	0	0	0	0	80	0	0	0
Squirrel Nuts—5 pieces (30 gm)	7	6	5	4	4	3	3	3	155	87	15.1	0
Starburst—6 pieces (30 gm)	1	1	1	½	½	½	½	½	107	8	1.6	0
Sugar Babies—1 oz (28 gm)	8	7	6	5	5	4	4	3	112	26	18.0	1
Sugar-coated chocolate discs—1 oz (28 gm)	14	11	10	9	8	7	6	6	130	50	29.6	3
Sugar Daddy—1 sucker (2 oz) (59 gm)	17	14	12	11	10	9	8	7	235	54	37.9	1
Sugar Momma—1 sucker (8 oz) (23 gm)	9	8	7	6	6	5	4	4	98	32	20.5	2
Summit cookie bars—1 bar (1.37 oz) (39 gm)	29	24	21	19	17	15	13	12	211	107	63.9	2
Taffy, plain (inc taffy, unspecified, fruit-flavored)—6 pieces (30 gm)	1	1	1	½	½	½	½	½	107	8	1.6	0
Take Five—1 bar (1.4 oz) (40 gm)	28	24	20	18	16	14	13	12	202	95	62.3	2
Tangy taffy—1 pkt (1.92 oz) (54 gm)	1	1	1	1	1	1	1	1	193	15	2.9	0
Thin Mints—1 oz (28 gm)	5	4	4	3	3	2	2	2	111	26	11.0	0
Toffee, chocolate-coated, w/nuts—3 pieces (33 gm)	14	12	10	9	8	7	7	6	149	60	31.1	2
Toffee, chocolate-covered—1 oz (28 gm)	12	10	9	8	7	6	6	5	122	41	26.4	2
Toffee, plain—1 oz (28 gm)	8	7	6	5	5	4	4	3	113	26	18.3	1
Toffifay—1 bar (1.16 oz) (33 gm)	10	9	8	7	6	5	5	4	163	62	23.5	0
Toll House morsels—1 oz (28 gm)	23	20	17	15	14	12	11	10	142	90	52.7	0
Tootsie Roll—1 roll (1.25 oz) (35 gm)	8	7	6	5	5	4	4	4	139	26	19.1	0
Truffles—1 piece (13 gm)	12	10	9	8	7	7	6	5	63	41	25.3	7
Turtles—2 turtles (38 gm)	12	10	9	8	7	6	5	5	188	72	27.0	0
Twix cookie bars—1 pkt, 2 bars (1.89 oz) (54 gm)	31	26	22	20	18	16	14	13	268	114	67.8	3
Twix peanut butter cookie bars—1 pkt, 2 bars (1.89 oz) (54 gm)	39	33	28	25	22	20	18	16	288	144	86.0	3
Twizzlers—2 sticks (22 gm)	½	½	0	0	0	0	0	0	81	1	.7	0

	1200	1500	1800	2100	2400	2800	3200	3600	TOT CAL	FAT CAL	S/FAT CAL	CHOL MG
Candies and Chewing Gum (cont.)												
Whatchamacallit—1 bar (1.4 oz) (40 gm)	20	17	15	13	12	11	9	9	216	107	44.3	4
Whoppers—1 pack (2 oz) (57 gm)	40	34	29	26	24	21	19	17	287	145	86.2	11
Zagnut—1 bar (1.75 oz) (50 gm)	11	9	8	7	6	6	5	5	232	92	23.3	2
Zero—1 bar (2 oz) (57 gm)	16	14	12	11	10	9	8	7	237	71	36.1	3
Gelatin Desserts												
Cheese, cottage cheese, w/gelatin dessert—1 c (240 gm)	18	16	14	12	11	10	9	9	201	54	34.3	20
Cheese, cottage cheese, w/gelatin dessert & fruit—1 c (240 gm)	14	12	11	9	9	8	7	7	215	42	26.2	15
Cheese, cottage cheese, w/gelatin dessert & vegetables—1 c (240 gm)	22	19	17	15	14	12	11	10	208	67	41.9	24
Danish dessert pudding—½ c (137 gm)	0	0	0	0	0	0	0	0	214	1	.2	0
Gelatin dessert—½ c (120 gm)	0	0	0	0	0	0	0	0	71	0	0	0
Gelatin dessert, dietetic, sweetened w/lo-cal sweetener—½ c (120 gm)	0	0	0	0	0	0	0	0	7	0	0	0
Gelatin dessert, dietetic, w/fruit, sweetened w/lo-cal sweetener—½ c (120 gm)	½	½	0	0	0	0	0	0	39	2	.7	0
Gelatin dessert, w/cream cheese—½ c (120 gm)	9	8	7	6	6	5	5	4	96	27	17.3	10
Gelatin dessert, w/fruit—½ c (120 gm)	½	½	0	0	0	0	0	0	78	2	.7	0
Gelatin dessert, w/fruit & cream cheese—½ c (120 gm)	6	5	5	4	4	3	3	3	94	20	12.0	6

Food												
Gelatin dessert, w/fruit & sour cream—½ c (124 gm)	9	8	7	6	5	5	4	4	134	29	17.7	7
Gelatin dessert, w/fruit & whipped cream—½ c (114 gm)	6	5	5	4	4	3	3	3	92	18	14.6	0
Gelatin dessert, w/fruit, vegetable & nuts—½ c (120 gm)	1	1	1	1	1	1	½	½	92	24	2.4	0
Gelatin dessert, w/sour cream—½ c (114 gm)	15	13	11	10	9	8	7	7	106	48	29.6	11
Gelatin dessert, w/whipped cream—½ c (114 gm)	6	5	5	4	4	3	3	3	86	16	14.0	0
Gelatin, frozen, whipped, on a stick (inc Jello Gelatin Pops)—1 bar (pop) (53 gm)	0	0	0	0	0	0	0	0	31	0	0	0
Gelatin salad, w/vegetables—½ c (122 gm)	0	0	0	0	0	0	0	0	62	1	.1	0
Haupia (coconut pudding)—½ c (107 gm)	77	65	56	50	45	40	35	32	289	196	174.3	0
Yookan—1 oz (28 gm)	0	0	0	0	0	0	0	0	49	0	0	0
Ices and Popsicles												
Ices, fruit—½ c (97 gm)	0	0	0	0	0	0	0	0	124	0	0	0
Popsicle—1 single stick (88 gm)	0	0	0	0	0	0	0	0	65	0	0	0
Snow cone, slurps—½ c (97 gm)	0	0	0	0	0	0	0	0	124	0	0	0
Jellies, Jams, and Preserves												
Chinese preserved sweet vegetable—1 sl (12 gm)	0	0	0	0	0	0	0	0	30	0	.1	0
Fruit butter, all flavors (inc apple butter)—½ tbsp (9 gm)	0	0	0	0	0	0	0	0	17	1	.1	0
Green papaya preserve, Puerto Rican style (dulce de lechoza)—5 sl w/syrup (6" x ¼" x ¾") (125 gm)	0	0	0	0	0	0	0	0	383	1	.4	0
Guava paste—1 med piece (32 gm)	0	0	0	0	0	0	0	0	90	0	0	0

	1200	1500	1800	2100	2400	2800	3200	3600	TOT CAL	FAT CAL	S/FAT CAL	CHOL MG
Jellies, Jams, and Preserves (cont.)												
Jam, preserves, all flavors—½ tbsp (10 gm)	0	0	0	0	0	0	0	0	27	0	0	0
Jams, preserves, marmalades, dietetic, all flavors, sweetened w/artificial sweetener—½ tbsp (10 gm)	0	0	0	0	0	0	0	0	1	0	0	0
Jams, preserves, marmalades, reduced sugar, all flavors—½ tbsp (10 gm)	0	0	0	0	0	0	0	0	15	0	0	0
Jellies, all flavors—½ tbsp (9 gm)	0	0	0	0	0	0	0	0	25	0	0	0
Jellies, dietetic, all flavors, sweetened w/artificial sweetener—½ tbsp (9 gm)	0	0	0	0	0	0	0	0	3	0	0	0
Marmalade, all flavors—½ tbsp (10 gm)	0	0	0	0	0	0	0	0	26	0	0	0
Sugar and Sweeteners												
Aspartame sweetener (inc Equal)—1 ind pkt (1 gm)	0	0	0	0	0	0	0	0	4	0	0	0
Brown—3 tsp unpacked (9 gm)	0	0	0	0	0	0	0	0	34	0	0	0
Cinnamon—2 tsp (8 gm)	0	0	0	0	0	0	0	0	30	0	0	0
Fructose sweetener—1 ind pkt (3 gm)	0	0	0	0	0	0	0	0	11	0	0	0
Maple—1 tsp (3 gm)	0	0	0	0	0	0	0	0	10	0	0	0
Saccharin (inc Necta Sweet)—1 tablet (½ grain) (0 gm)	0	0	0	0	0	0	0	0	0	0	0	0
Sprinkle Sweet—1 tsp (1 gm)	0	0	0	0	0	0	0	0	2	0	0	0
Sugar replacement or substitute, saccharin-based, liquid (inc Fasweet, Sweet 'n Low, Sweet 10)—1 tsp (5 gm)	0	0	0	0	0	0	0	0	0	0	0	0

Food												
Sugar Twin—1 ind pkt (1 gm)	0	0	0	0	0	0	0	0	3	0	0	
Sugar Twin, brown—1 tsp (0 gm)	0	0	0	0	0	0	0	0	1	0	0	
Sugar, white, granulated or lump (inc rock sugar, rock candy)—2 tsp (8 gm)	0	0	0	0	0	0	0	0	31	0	0	
Sweet'ner, Sweet 'n Low—1 ind pkt (1 gm)	0	0	0	0	0	0	0	0	4	0	0	
Wee Cal—1 ind pkt (1 gm)	0	0	0	0	0	0	0	0	2	0	0	
White, confectioners', powdered—1 c (120 gm)	0	0	0	0	0	0	0	0	462	0	0	
Syrups and Toppings												
Buttered blends syrup (inc Mrs. Butterworth's, Log Cabin w/butter)—2 tbsp (39 gm)	2	2	1	1	1	1	1	1	116	6	3.6	2
Cane & corn syrup—2 tbsp (41 gm)	0	0	0	0	0	0	0	0	113	0	0	0
Cane & maple syrup—2 tbsp (39 gm)	0	0	0	0	0	0	0	0	98	0	0	0
Cane, corn & maple syrup—2 tbsp (39 gm)	0	0	0	0	0	0	0	0	103	0	0	0
Cane syrup—2 tbsp (41 gm)	0	0	0	0	0	0	0	0	108	0	0	0
Carob chips—1 c (173 gm)	186	156	135	120	107	95	85	77	919	514	418.2	0
Carob powder or flour—1 c (140 gm)	1	1	1	1	1	1	½	½	252	18	2.5	0
Carob syrup—1 c (300 gm)	½	½	½	½	½	0	0	0	657	8	1.1	0
Chocolate gravy—2 tbsp (34 gm)	5	4	4	3	3	3	2	2	87	15	9.2	5
Chocolate syrup, fudge type—1 fl oz (2 tbsp) (38 gm)	11	10	8	7	7	6	5	5	124	46	25.7	0
Chocolate syrup, thin type—2 tbsp (38 gm)	1	1	1	1	½	½	½	½	83	3	1.8	0
Chocolate syrup, w/added nutrients, prep w/milk—1 c milk & 1 tbsp syrup (263 gm)	26	22	20	18	16	15	13	12	196	76	47.0	33
Chocolate syrup, wo/added nutrients, prep w/milk—1 c milk & 2 tbsp syrup (282 gm)	26	23	20	18	16	15	13	12	232	77	47.5	33

Syrups and Toppings (cont.)

	1200	1500	1800	2100	2400	2800	3200	3600	TOT CAL	FAT CAL	S/FAT CAL	CHOL MG
Corn & maple syrup (2% maple) (inc pancake syrup, Log Cabin brand)—2 tbsp (39 gm)	0	0	0	0	0	0	0	0	113	0	0	0
Corn syrup (inc Karo brand, light or dark)—2 tbsp (41 gm)	0	0	0	0	0	0	0	0	119	0	0	0
Dessert topping, nondairy, powdered—1.5 oz (43 gm)	62	52	45	40	36	32	28	26	245	153	140.2	0
Dessert topping, nondairy, powdered, prep w/milk—1.5 oz w/½ c milk (1 c, 80 gm)	35	30	26	23	21	18	17	15	151	89	76.9	8
Dessert topping, nondairy, pressurized—1 c (70 gm)	53	44	39	34	31	27	24	22	184	140	119.2	0
Dessert topping, nondairy, semisolid, frozen—1 c (75 gm)	65	55	48	42	38	33	30	27	239	171	147.1	0
Fruit syrup—2 tbsp (39 gm)	0	0	0	0	0	0	0	0	103	0	0	0
Hard sauce—2 tbsp (21 gm)	15	13	11	10	9	8	7	7	95	45	28.3	14
Honey (inc pear honey, raw honey)—2 tbsp (42 gm)	0	0	0	0	0	0	0	0	128	0	0	0
Honey, strained or extracted—1 tbsp (21 gm)	0	0	0	0	0	0	0	0	64	0	0	0
Icing, chocolate—2 tbsp (34 gm)	3	3	2	2	2	2	2	1	133	33	7.5	0
Icing, white (inc creme filling; icing w/added flavors, e.g., lemon icing, etc.)—2 tbsp (40 gm)	4	3	3	2	2	2	2	1	161	39	8.0	0
Maple syrup (100% maple) (inc Maple Cream)—2 tbsp (39 gm)	0	0	0	0	0	0	0	0	98	0	0	
Molasses—2 tbsp (41 gm)	0	0	0	0	0	0	0	0	103	0	0	0
Molasses, cane, third extraction or blackstrap—1 c (328 gm)	0	0	0	0	0	0	0	0	699	0	0	0

Food												
Sorghum syrup—2 tbsp (41 gm)	0	0	0	0	0	0	0	0	105	0	0	0
Sugar (brown) & water syrup—2 tbsp (30 gm)	0	0	0	0	0	0	0	0	36	0	0	0
Sugar, brown, liquid—2 tbsp (42 gm)	0	0	0	0	0	0	0	0	109	0	0	0
Sugar, carmelized—2 tbsp (30 gm)	0	0	0	0	0	0	0	0	112	0	0	0
Sugar (white) & water syrup (inc simple syrup)—2 tbsp (30 gm)	0	0	0	0	0	0	0	0	38	0	0	0
Syrup, dietetic—2 tbsp (30 gm)	0	0	0	0	0	0	0	0	12	0	0	0
Syrup, grenadine—2 tbsp (40 gm)	0	0	0	0	0	0	0	0	105	0	0	0
Syrup, reduced-calorie—2 tbsp (30 gm)	0	0	0	0	0	0	0	0	51	0	0	0
Topping, butterscotch or caramel—2 tbsp (42 gm)	3	2	2	2	2	2	1	1	124	8	7.0	0
Topping, chocolate, thick, fudge type—2 tbsp (42 gm)	15	13	11	10	9	8	7	6	139	52	31.7	5
Topping, dietetic—2 tbsp (28 gm)	1	1	1	1	½	½	½	½	16	4	2.1	0
Topping, fruit—2 tbsp (42 gm)	0	0	0	0	0	0	0	0	114	0	.1	0
Topping, fruit, unsweetened (inc Sorrell Ridge brand)—2 tbsp (40 gm)	0	0	0	0	0	0	0	0	24	0	0	0
Topping, marshmallow—2 tbsp (38 gm)	0	3	2	2	2	2	2	2	108	0	0	0
Topping, nut (wet)—2 tbsp (42 gm)	4	3	2	2	2	2	2	2	173	91	8.2	0
Topping, peanut butter, thick, fudge type—2 tbsp (42 gm)	3	2	2	2	2	2	1	1	125	32	7.7	0

VEGETABLES

Food												
Alfalfa sprouts, raw—½ c (17 gm)	0	0	0	0	0	0	0	0	5	1	.1	0
Algae, dried (inc spirulina)—1 c (128 gm)	10	9	7	6	5	5	4	4	376	68	23.1	0
Aloe vera juice—½ c (126 gm)	0	0	0	0	0	0	0	0	50	2	.3	0
Amaranth, boiled, drained, wo/salt—½ c (66 gm)	0	0	0	0	0	0	0	0	14	1	.3	0
Amaranth, raw—1 c (28 gm)	0	0	0	0	0	0	0	0	7	1	.2	0
Artichoke, globe (French), cooked, fat not added in cooking—1 med globe (120 gm)	0	0	0	0	0	0	0	0	52	2	.4	0

	1200	1500	1800	2100	2400	2800	3200	3600	TOT CAL	FAT CAL	S/FAT CAL	CHOL MG
Vegetables (cont.)												
Artichoke, Jerusalem, raw (inc sunchoke)—½ c (75 gm)	0	0	0	0	0	0	0	0	57	0	0	0
Artichoke salad in oil—½ c (65 gm)	3	3	2	2	2	2	2	1	78	54	7.4	0
Artichokes, (globe or French), frozen, boiled, drained, wo/salt—⅓ of 9 oz pkg (80 gm)	½	½	½	0	0	0	0	0	36	4	.8	0
Artichokes, stuffed—10 stuffed leaves (42 gm)	3	2	2	2	1	1	1	1	66	22	5.3	1
Asparagus, canned, drained solids—½ c spears (121 gm)	1	1	1	½	½	½	½	½	24	7	1.6	0
Asparagus, canned, special dietary pack, solid & liquid—½ c (122 gm)	0	0	0	0	0	0	0	0	17	2	.4	0
Asparagus, cooked, fat not added in cooking—1 med spear (15 gm)	0	0	0	0	0	0	0	0	4	0	.1	0
—1 c, fresh or frozen (180 gm)	1	½	½	½	½	½	0	0	45	5	1.2	0
—1 c, canned (242 gm)	1	1	½	½	½	½	½	½	60	7	1.5	0
Asparagus, creamed or w/cheese sauce—½ c (118 gm)	15	13	11	10	9	8	7	7	123	78	29.4	12
Asparagus, frozen, boiled, drained, wo/salt—4 spears (60 gm)	0	0	0	0	0	0	0	0	17	3	.5	0
Asparagus, raw—1 med spear (16 gm)	0	0	0	0	0	0	0	0	4	0	.1	0
—1 c (134 gm)	½	0	0	0	0	0	0	0	29	3	.6	0
Bamboo shoots, canned, drained solids—1 c ⅛" sl (131 gm)	½	½	½	½	½	0	0	0	25	5	1.1	0
Bamboo shoots, canned, cooked, fat not added in cooking—¼ c (30 gm)	0	0	0	0	0	0	0	0	8	1	.2	0
Bamboo shoots, fried—¼ c (39 gm)	3	3	3	2	2	2	2	1	42	31	7.8	0

Food												
Bamboo shoots, raw—½ c ½" sl (76 gm)	0	0	0	0	0	0	0	0	21	2	.4	0
Bean salad, yellow and/or green string beans (inc three-bean salad)—½ c (75 gm)	2	2	2	2	2	1	1	1	71	38	5.5	0
Beans, green & potatoes, cooked, fat added in cooking—½ c (72 gm)	2	2	2	2	1	1	1	1	56	14	5.2	1
Beans, green, & potatoes, cooked, fat not added in cooking—½ c (69 gm)	0	0	0	0	0	0	0	0	41	1	.3	0
Beans, green string, w/almonds, cooked, fat not added in cooking—½ c (61 gm)	3	2	2	2	2	2	1	1	96	66	6.4	0
Beans, green string, w/chickpeas, cooked, fat not added in cooking—½ c (67 gm)	0	0	0	0	0	0	0	0	42	4	.5	0
Beans, green string, w/onions, cooked, fat not added in cooking—½ c (73 gm)	0	0	0	0	0	0	0	0	19	1	.2	0
Beans, green string, w/onions, fat added in cooking—½ c (76 gm)	2	1	1	1	1	1	1	1	39	17	3.4	0
Beans, green string, w/tomato, cooked, fat not added in cooking (inc beans, green string, w/tomato-base sauce)—½ c (74 gm)	0	0	0	0	0	0	0	0	25	2	.3	0
Beans, green, w/pinto beans, cooked, fat not added in cooking (inc w/shellie beans)—½ c (68 gm)	0	0	0	0	0	0	0	0	20	1	.1	0
Beans, green, w/spaetzel, cooked, fat not added in cooking—½ c (74 gm)	1	1	1	1	1	1	1	1	39	3	.8	5
Beans, lima & corn (succotash), cooked, fat not added in cooking—½ c (96 gm)	1	½	½	½	½	½	½	½	89	8	1.4	0
Beans, lima, immature, creamed or w/cheese sauce—½ c (114 gm)	8	6	5	5	4	4	3	3	139	40	14.8	6

	1200	1500	1800	2100	2400	2800	3200	3600	TOT CAL	FAT CAL	S/FAT CAL	CHOL MG
Vegetables (cont.)												
Beans, lima, immature, canned, fat not added in cooking—½ c (87 gm)	½	0	0	0	0	0	0	0	106	2	.6	0
Beans, lima, immature, w/mushroom sauce—½ c (114 gm)	4	4	3	3	3	2	2	2	111	31	9.3	2
Beans, lima, raw—½ c (78 gm)	1	1	½	½	½	½	½	½	88	6	1.4	0
Bean sprouts, cooked, fat not added in cooking—½ c (63 gm)	1	1	½	½	½	½	½	½	32	13	1.4	0
Bean sprouts, raw (soybean or mung)—½ c (52 gm)	0	0	0	0	0	0	0	0	16	1	.2	0
Beans, string, green, canned, low-sodium, fat not added in cooking (inc pole beans, Italian beans, snap beans, French cut beans) (68 gm)	0	0	0	0	0	0	0	0	14	1	.1	0
Beans, string, green, frozen or canned, cooked, fat not added in cooking (inc pole beans, Italian beans, snap beans, French cut beans)—½ c (68 gm)	0	0	0	0	0	0	0	0	24	2	.4	0
Beans, string, green, creamed or w/cheese sauce (inc string beans in sauce)—½ c (114 gm)	16	13	12	10	9	8	8	7	131	80	30.5	13
Beans, string, green, raw—½ c (55 gm)	0	0	0	0	0	0	0	0	17	1	.1	0
Beans, string, green, w/mushroom sauce (inc green bean casserole)—½ c (114 gm)	5	4	4	3	3	3	2	2	69	35	10.6	2
Beans, string, yellow, cooked, fat not added in cooking (inc wax beans)—½ c (68 gm)	0	0	0	0	0	0	0	0	24	2	.4	0

Food											
Beans, string, yellow, creamed or w/cheese sauce—½ c (114 gm)	16	13	12	10	9	8	8	7	131	80	30.5	13
Beet greens, cooked, fat not added in cooking—½ c (72 gm)	0	0	0	0	0	0	0	0	19	1	.2	0
Beet greens, raw—½ c (19 gm)	0	0	0	0	0	0	0	0	4	0	0	0
Beets, canned, low-sodium, fat not added in cooking—½ c, diced (85 gm)	0	0	0	0	0	0	0	0	26	1	.2	0
Beets, boiled, drained, wo/salt—½ c sl (85 gm)	0	0	0	0	0	0	0	0	26	1	.1	0
Beets, canned, special dietary pack, solid & liquid—½ c (123 gm)	0	0	0	0	0	0	0	0	36	1	.1	0
Beets, cooked, fat not added in cooking—½ c (85 gm)	0	0	0	0	0	0	0	0	26	0	.1	0
Beets, Harvard, canned, solid & liquid—½ c sl (123 gm)	0	0	0	0	0	0	0	0	89	1	.1	0
Beets, pickled, canned, solid & liquid—½ c sl (114 gm)	0	0	0	0	0	0	0	0	75	1	.2	0
Beets, raw—½ c (68 gm)	3	0	0	0	0	0	0	0	30	1	.1	0
Beets w/Harvard sauce—½ c (123 gm)	3	3	2	2	2	1	1	1	135	35	6.9	0
Bitter melon, cooked, fat not added in cooking (inc Balsam pear)—½ c (62 gm)	0	0	0	0	0	0	0	0	12	1	.2	0
Bitter melon leaves, horseradish leaves, jute leaves, or radish leaves, cooked, fat not added in cooking—½ c (62 gm)	0	0	0	0	0	0	0	0	27	2	.2	0
Breadfruit, cooked, fat not added in cooking (inc pana)—½ c (126 gm)	½	½	0	0	0	0	0	0	144	3	.7	0
Breadfruit, fried (tostones)—½ c (85 gm)	5	5	4	3	3	2	2	2	189	94	12.2	0
Broccoli, batter-dipped & fried—½ c (43 gm)	4	4	3	3	3	2	2	2	62	40	6.1	10
Broccoli casserole (broccoli, noodles, & cream sauce)—½ c (114 gm)	16	13	12	11	10	9	8	8	147	70	26.6	24

Vegetables (cont.)

	1200	1500	1800	2100	2400	2800	3200	3600	TOT CAL	FAT CAL	S/FAT CAL	CHOL MG
Broccoli casserole (broccoli, rice, cheese, & mushroom sauce)—½ c (114 gm)	18	16	14	12	11	10	9	8	151	70	34.9	18
Broccoli, cooked, fat not added in cooking—1 stalk (5" long) (37 gm)	0	0	0	0	0	0	0	0	11	1	.1	0
—1 c, fresh, cut stalks (156 gm)	½	0	0	0	0	0	0	0	45	4	.6	0
—1 c, frozen, chopped (184 gm)	½	½	0	0	0	0	0	0	53	5	.7	0
Broccoli, cooked, w/cheese sauce—½ c (114 gm)	16	14	12	11	10	9	8	7	111	66	31.1	15
Broccoli, cooked, w/cream sauce—½ c (114 gm)	8	7	6	5	5	4	4	3	93	52	15.9	5
Broccoli, cooked, w/mushroom sauce—½ c (114 gm)	4	3	3	2	2	2	2	2	66	32	8.6	0
Broccoli, raw—½ c, chopped (44 gm)	0	0	0	0	0	0	0	0	12	1	.2	0
—1 c, chopped (88 gm)	0	0	0	0	0	0	0	0	25	3	.4	0
Brussels sprouts, cooked, fat not added in cooking—½ c (78 gm)	½	½	0	0	0	0	0	0	30	4	.7	0
Brussels sprouts, creamed—1 sprout (23 gm)	2	2	1	1	1	1	1	1	22	12	3.7	1
—½ c (114 gm)	9	8	7	6	5	5	4	4	107	60	18.2	6
Brussels sprouts, frozen, boiled, drained, wo/salt—½ c (78 gm)	0	0	0	0	0	0	0	0	33	3	.5	0
Brussels sprouts, raw—½ c (44 gm)	0	0	0	0	0	0	0	0	19	1	.2	0
Burdock, cooked, fat not added in cooking (inc gobo)—½ c (63 gm)	0	0	0	0	0	0	0	0	55	1	.1	0
Butterbur, canned—3 stalks (45 gm)	0	0	0	0	0	0	0	0	1	1	0	0
Cabbage, Chinese, cooked, fat not added in cooking (inc bok choy)—½ c (85 gm)	0	0	0	0	0	0	0	0	12	1	.2	0

Food												
Cabbage, Chinese, raw—½ c (38 gm)	0	0	0	0	0	0	0	0	5	1	.1	0
Cabbage, Chinese, salad, w/dressing—½ c (38 gm)	3	3	2	2	2	2	2	1	33	25	5.7	4
Cabbage, common (Danish, domestic & pointed types) raw—½ c shredded (35 gm)	0	0	0	0	0	0	0	0	8	1	.1	0
Cabbage, creamed—½ c (100 gm)	7	6	6	5	4	4	4	3	81	49	15.0	5
Cabbage, green, cooked, fat not added in cooking—½ c (75 gm)	0	0	0	0	0	0	0	0	16	2	.2	0
Cabbage, raw—½ c (45 gm)	0	0	0	0	0	0	0	0	11	1	.1	0
Cabbage, red, cooked, fat not added in cooking—½ c (75 gm)	0	0	0	0	0	0	0	0	16	1	.2	0
Cabbage, red, raw—½ c (45 gm)	0	0	0	0	0	0	0	0	12	1	.1	0
Cabbage salad or coleslaw, w/apples and/or raisins, w/dressing—½ c (68 gm)	7	6	5	5	4	4	4	3	127	90	13.5	7
Cabbage salad or coleslaw, w/dressing—½ c (92 gm)	6	5	4	4	3	3	3	3	71	19	10.6	6
Cabbage salad or coleslaw, w/pineapple, w/dressing—½ c (68 gm)	2	2	2	2	2	1	1	1	65	31	4.5	3
Cabbage, savoy, cooked, fat not added in cooking—½ c (73 gm)	0	0	0	0	0	0	0	0	20	1	.1	0
Cactus, cooked, fat not added in cooking—½ c (75 gm)	½	½	0	0	0	0	0	0	31	3	.7	0
Cactus, raw—½ c (59 gm)	0	0	0	0	0	0	0	0	24	3	.5	0
Caesar salad (w/romaine)—1 c (108 gm)	44	39	35	32	30	28	26	24	390	309	57.5	119
Carrot juice—½ c (118 gm)	0	0	0	0	0	0	0	0	47	2	.3	0
Carrot juice, canned—½ c (123 gm)	0	0	0	0	0	0	0	0	49	2	.3	0
Carrots, canned, low-sodium, fat not added in cooking—½ c sl (73 gm)	0	0	0	0	0	0	0	0	17	1	.2	0
Carrots, cooked, creamed—½ c (114 gm)	9	8	7	6	5	5	4	4	113	60	18.4	6
Carrots, cooked, fat not added in cooking—1 med (56 gm)	0	0	0	0	0	0	0	0	25	1	.2	0

	1200	1500	1800	2100	2400	2800	3200	3600	TOT CAL	FAT CAL	S/FAT CAL	CHOL MG
Vegetables (cont.)												
—1 c, frozen, sl (146 gm)	0	0	0	0	0	0	0	0	65	2	.4	0
—1 c, canned, sl (146 gm)	0	0	0	0	0	0	0	0	65	2	.4	0
—1 c, fresh, sl (156 gm)	0	0	0	0	0	0	0	0	70	3	.5	0
Carrots, cooked, glazed—½ c (80 gm)	4	4	3	3	3	2	2	2	116	50	9.9	0
Carrots, cooked, w/cheese sauce—½ c (114 gm)	9	8	7	6	6	5	5	4	87	32	16.9	11
Carrots in tomato sauce—½ c (88 gm)	4	4	3	3	2	2	2	2	134	63	9.5	0
Carrots, raw—1 sm (5½" long) (50 gm)	0	0	0	0	0	0	0	0	22	1	.1	0
—½ c, grated (55 gm)	0	0	0	0	0	0	0	0	24	1	.1	0
—1 med (61 gm)	0	0	0	0	0	0	0	0	26	1	.2	0
—1 lg (7½" long) (72 gm)	0	0	0	0	0	0	0	0	31	1	.2	0
Carrots, raw, salad (inc carrot-raisin salad)—½ c (88 gm)	11	9	8	7	7	6	5	5	210	133	20.0	11
Carrots, raw, salad w/apples—½ c (86 gm)	6	5	5	4	4	3	3	3	116	79	11.8	6
Cassava, cooked, fat not added in cooking—½ c (103 gm)	½	½	½	½	½	0	0	0	124	4	1.0	0
Cauliflower, batter-dipped, fried (inc breaded, fried)—½ c (43 gm)	8	7	6	6	5	5	4	4	83	60	14.2	11
Cauliflower, cooked, fat not added in cooking—½ c, frozen (90 gm)	0	0	0	0	0	0	0	0	20	1	.3	0
—1 c, fresh (124 gm)	0	0	0	0	0	0	0	0	28	2	.4	0
Cauliflower, creamed (inc w/cheese sauce)—1 floweret (32 gm)	4	3	3	3	3	2	2	2	29	17	7.8	4
—½ c (114 gm)	14	12	11	10	9	8	7	7	103	61	27.8	13
Cauliflower, frozen, boiled, drained, wo/salt—½ c 1" pieces (90 gm)	0	0	0	0	0	0	0	0	17	2	.3	0

Food												
Cauliflower, raw—½ c (50 gm)	0	0	0	0	0	0	0	0	12	1	.1	0
Celery, cooked, fat not added in cooking—½ c, diced (75 gm)	0	0	0	0	0	0	0	0	11	1	.2	0
Celery, creamed—½ c (114 gm)	9	8	7	6	5	5	4	4	93	59	18.1	6
Celery juice—½ c (118 gm)	0	0	0	0	0	0	0	0	18	1	.3	0
Celery, raw—1 sm stalk (5" long) (17 gm)	0	0	0	0	0	0	0	0	3	0	0	0
—1 med stalk (7½"–8" long) (40 gm)	0	0	0	0	0	0	0	0	6	0	.1	0
—1 lg stalk (11"–12" long) (64 gm)	0	0	0	0	0	0	0	0	10	1	.2	0
Chard, cooked, fat not added in cooking—½ c (73 gm)	0	0	0	0	0	0	0	0	15	1	.1	0
Chard, raw—½ c (18 gm)	0	0	0	0	0	0	0	0	3	0	0	0
Chicory, greens, raw—½ c chopped (90 gm)	½	0	0	0	0	0	0	0	21	3	.6	0
Chiles rellenos—1 chili (143 gm)	98	85	76	69	64	58	54	50	427	314	153.1	193
Chives, dried or dehydrated—1 c (15 gm)	1	1	½	½	½	½	½	½	40	9	1.4	0
Chives, freeze-dried—1 tbsp (0 gm)	0	0	0	0	0	0	0	0	1	0	0	0
Chives, raw—1 tbsp (3 gm)	0	0	0	0	0	0	0	0	1	0	0	0
Christophine, cooked, fat not added in cooking (inc chayote)—½ c (80 gm)	½	0	0	0	0	0	0	0	21	2	.7	0
Collards, cooked, fat not added in cooking—1 c, canned (162 gm)	0	0	0	0	0	0	0	0	23	2	.4	0
—1 c, frozen (170 gm)	0	0	0	0	0	0	0	0	24	2	.5	0
—1 c, fresh (190 gm)	0	0	0	0	0	0	0	0	26	3	.5	0
Collards, frozen, chopped, boiled, drained, wo/salt—½ c chopped (85 gm)	0	0	0	0	0	0	0	0	31	4	0	0
Collards, raw—½ c (93 gm)	0	0	0	0	0	0	0	0	18	2	.3	0
Corn, dried, cooked—1 oz (28 gm)	2	1	1	1	1	1	1	1	37	17	3.5	0
Corn fritter—1 fritter (35 gm)	13	11	10	9	9	8	7	7	132	68	17.7	31
Corn salad, raw—½ c (28 gm)	—	—	—	—	—	—	—	—	6	1	—	0

	1200	1500	1800	2100	2400	2800	3200	3600	TOT CAL	FAT CAL	S/FAT CAL	CHOL MG
Vegetables (cont.)												
Corn, scalloped or pudding (inc corn soufflé)—½ c (107 gm)	23	21	19	18	17	16	15	14	134	55	16.0	100
Corn, sweet, white, canned, cream style, special dietary pack—½ c (128 gm)	½	½	0	0	0	0	0	0	93	5	.7	0
Corn, sweet, white, canned—½ c (105 gm)	½	½	0	0	0	0	0	0	83	5	.7	0
Corn, sweet, white, canned, special dietary pack—½ c (105 gm)	½	½	0	0	0	0	0	0	83	5	.7	0
Corn, sweet, white, frozen, kernels cut off cob, boiled, drained, wo/salt—½ c (82 gm)	0	0	0	0	0	0	0	0	67	1	.1	0
Corn, sweet, white, raw—½ c cut off cob (77 gm)	1	½	½	½	½	½	½	0	66	8	1.3	0
Corn, sweet, yellow, canned, brine pack type, special dietary pack, solid & liquid—½ c (128 gm)	½	½	½	0	0	0	0	0	79	5	.8	0
Corn, sweet, yellow, canned, cream style, special dietary pack—½ c (128 gm)	½	½	0	0	0	0	0	0	93	5	.7	0
Corn, sweet, yellow, canned, vacuum pack—½ c (105 gm)	½	½	0	0	0	0	0	0	83	5	.7	0
Corn, sweet, yellow, canned, special dietary pack—½ c (105 gm)	½	½	0	0	0	0	0	0	83	5	.7	0
Corn, sweet, yellow, frozen, kernels cut off cob, boiled, drained, wo/salt—½ c (82 gm)	0	0	0	0	0	0	0	0	67	1	.1	0
Corn, sweet, yellow, raw—½ c cut off cob (77 gm)	1	½	½	½	½	½	½	0	66	8	1.3	0

Corn, white, cooked, fat not added in cooking—½ c (82 gm)	1	1	½	½	½	½	½	½	88	9	1.4	0
Corn, white, cream style—½ c (128 gm)	½	½	0	0	0	0	0	0	92	5	.7	0
Corn, w/cream sauce—½ c (114 gm)	9	8	7	6	5	5	4	4	149	63	18.1	6
Corn, w/peppers, red or green, cooked, fat not added in cooking (inc Mexican style corn)—½ c (89 gm)	1	1	½	½	½	½	½	½	88	9	1.5	0
Corn, w/red & green peppers, canned, solid & liquid—½ c (114 gm)	½	½	½	½	0	0	0	0	86	5	.9	0
Corn, yellow, canned, low-sodium, fat not added in cooking—½ c (82 gm)	1	1	½	½	½	½	½	½	89	9	1.5	0
Corn, yellow, cooked, fat not added in cooking—1 med ear (5″ long) (77 gm)	1	1	½	½	½	½	½	½	83	9	1.4	0
———½ c (82 gm)	1	1	½	½	½	½	½	½	88	9	1.4	0
Corn, yellow, cream style—½ c (128 gm)	½	½	0	0	0	0	0	0	92	5	.7	0
Cowpeas, common (black-eyed, crowder, southern), mature seeds, canned w/pork—½ c (120 gm)	4	4	3	3	3	2	2	2	99	17	6.6	8
Cowpeas, common (black-eyed, crowder, southern), mature seeds, canned, plain—½ c (120 gm)	1	1	½	½	½	½	½	½	92	6	1.5	0
Cowpeas, immature seeds, frozen, boiled, drained, wo/salt—½ c (85 gm)	1	1	½	½	½	½	½	½	112	5	1.4	0
Cowpeas, leafy tips, boiled, drained, wo/salt—½ c chopped (26 gm)	0	0	0	0	0	0	0	0	6	1	.1	0
Cowpeas w/snap beans, cooked, fat not added in cooking—½ c (69 gm)	½	0	0	0	0	0	0	0	39	2	.6	0
Cowpeas, young pods w/seeds, boiled, drained, wo/salt—½ c (47 gm)	0	0	0	0	0	0	0	0	16	1	.4	0
Cress, cooked, fat not added in cooking—½ c (68 gm)	0	0	0	0	0	0	0	0	16	4	.1	0

Vegetables (cont.)

	1200	1500	1800	2100	2400	2800	3200	3600	TOT CAL	FAT CAL	S/FAT CAL	CHOL MG
Cress, raw—6 sprigs (6 gm)	0	0	0	0	0	0	0	0	2	0	0	0
Cucumber & sour cream salad—½ c (67 gm)	7	6	5	5	4	4	3	3	33	23	13.6	5
Cucumber & vegetable namasu—½ c (78 gm)	0	0	0	0	0	0	0	0	26	1	.2	0
Cucumber, cooked, fat not added in cooking—½ c (90 gm)	0	0	0	0	0	0	0	0	14	1	.3	0
Cucumber, raw—½ c, sl (60 gm)	0	0	0	0	0	0	0	0	10	1	.2	0
—1 med (201 gm)	0	½	0	0	0	0	0	0	32	4	.7	0
Dandelion greens, cooked, fat not added in cooking—½ c, chopped (53 gm)	0		0	0	0	0	0	0	17	3	.4	0
Dandelion greens, raw—½ c (28 gm)	0	0	0	0	0	0	0	0	13	2	.2	0
Eggplant, batter-dipped, fried—½ c (110 gm)	15	13	11	10	9	8	8	7	165	106	25.8	20
Eggplant, cooked, fat not added in cooking—½ c (48 gm)	0	0	0	0	0	0	0	0	13	1	.2	0
Eggplant in tomato sauce, cooked, fat not added in cooking—½ c (116 gm)	0	0	0	0	0	0	0	0	33	2	.4	0
Eggplant Parmesan casserole, lo-cal—½ c (99 gm)	13	11	10	9	8	7	6	6	90	39	24.3	12
Eggplant Parmesan casserole, reg—½ c (99 gm)	24	20	18	16	15	13	12	11	160	99	43.5	29
Eggplant, raw—½ c, cubes (41 gm)	0	0	0	0	0	0	0	0	11	0	.1	0
Endive, chicory, escarole, or romaine, raw—1 c, escarole (28 gm)	0	0	0	0	0	0	0	0	5	1	.1	0
—1 c, endive (29 gm)	0	0	0	0	0	0	0	0	5	1	.1	0

Food										
—1 c, romaine (30 gm)	0	0	0	0	0	0	0	6	1	.1	0
—1 c, mixed greens (40 gm)	0	0	0	0	0	0	0	7	1	.1	0
Escarole, cooked, fat not added in cooking (inc endive)—½ c (65 gm)	0	0	0	0	0	0	0	14	2	.2	0
Escarole, creamed—½ c (100 gm)	7	6	5	5	4	3	3	78	45	13.5	5
Fern shoots, cooked, fat not added in cooking—½ c (71 gm)	0	0	0	0	0	0	0	28	0	.1	0
Flowers or blossoms of sesbania, squash, or lily, fat not added in cooking—½ c (52 gm)	0	0	0	0	0	0	0	8	0	.2	0
Garlic, cooked—1 clove (2 gm)	0	0	0	0	0	0	0	3	0	0	0
Garlic, raw—1 clove (3 gm)	0	0	0	0	0	0	0	4	0	0	0
Gingerroot, crystallized, candied—1 oz (28 gm)	0	0	0	0	0	0	0	95	1	0	0
Gingerroot, raw—5 sl (1" dia) (11 gm)	0	0	0	0	0	0	0	8	1	.2	0
Gourd, dishcloth (towelgourd), boiled, drained, wo/salt—½ c 1" sl (89 gm)	0	0	0	0	0	0	0	50	3	.2	0
Gourd, white-flowered (calabash), boiled, drained, wo/salt—½ c 1" cubes (73 gm)	0	0	0	0	0	0	0	11	0	0	0
Greens, cooked, fat not added in cooking—½ c (83 gm)	0	0	0	0	0	0	0	18	2	.3	0
Hominy, cooked, fat not added in cooking—½ c (83 gm)	½	½	½	½	½	½	0	73	9	.9	0
Horseradish (marong-gay), pods, cooked, fat not added in cooking—½ c (59 gm)	0	0	0	0	0	0	0	21	1	.3	0
Horseradish, prepared—1 tsp (5 gm)	0	0	0	0	0	0	0	2	0	0	0
Hummus, raw—⅓ c (82 gm)	4	4	3	3	2	2	2	140	62	9.4	0
Jai, monk's food (mushrooms, lily roots, bean curd, water chestnuts)—½ c (94 gm)	1	1	1	1	1	1	1	88	19	3.0	0
Jicama, raw (inc yambean)—½ c (65 gm)	0	0	0	0	0	0	0	27	1	.4	0

	1200	1500	1800	2100	2400	2800	3200	3600	TOT CAL	FAT CAL	S/FAT CAL	CHOL MG
Vegetables (cont.)												
Kale, cooked, fat not added in cooking —1 c, fresh (130 gm)	½	0	0	0	0	0	0	0	41	5	.6	0
——1 c, frozen (130 gm)	½	0	0	0	0	0	0	0	41	5	.6	0
——1 c, canned (163 gm)	½	½	½	0	0	0	0	0	52	6	.8	0
Kale, frozen, boiled, drained, wo/salt —½ c chopped (65 gm)	0	0	0	0	0	0	0	0	19	3	.4	0
Kale, raw—½ c chopped (34 gm)	0	0	0	0	0	0	0	0	17	2	.3	0
Kale, Scotch, boiled, drained, wo/salt— ½ c chopped (65 gm)	0	0	0	0	0	0	0	0	18	3	.3	0
Kale, Scotch, raw—½ c chopped (34 gm)	0	0	0	0	0	0	0	0	14	2	.3	0
Kohlrabi, cooked, fat not added in cooking—½ c (83 gm)	0	0	0	0	0	0	0	0	24	1	.1	0
Kohlrabi, creamed—½ c (94 gm)	7	6	5	5	4	4	3	3	81	46	14.1	5
Lamb's-quarter, cooked, fat not added in cooking—½ c (90 gm)	0	0	0	0	0	0	0	0	29	6	.4	0
Leeks (bulb & lower-leaf portion), boiled, drained, wo/salt—¼ c chopped (26 gm)	0	0	0	0	0	0	0	0	8	1	.1	0
Leeks (bulb & lower-leaf portion), freeze-dried—1 tbsp (0 gm)	0	0	0	0	0	0	0	0	1	0	0	0
Leeks (bulb & lower-leaf portion), raw—¼ c chopped (26 gm)	0	0	0	0	0	0	0	0	16	1	.1	0
Lettuce, Boston (inc deer tongue lettuce, native lettuce, red leaf lettuce)—1 c shredded or chopped (55 gm)	0	0	0	0	0	0	0	0	7	1	.1	0
Lettuce, cos or romaine, raw—½ c shredded (28 gm)	0	0	0	0	0	0	0	0	4	1	.1	0

Food												
Lettuce, iceberg (inc crisp-head types, raw—1 leaf (20 gm)	0	0	0	0	0	0	0	0	3	0	0	0
Lettuce, loose-leaf, raw—½ c shredded (28 gm)	0	0	0	0	0	0	0	0	5	1	.1	0
Lettuce, manoa—1 c (50 gm)	0	0	0	0	0	0	0	0	8	1	.1	0
Lettuce, salad w/assorted vegetables, excluding tomatoes & carrots, no dressing—1 c (74 gm)	0	0	0	0	0	0	0	0	12	1	.2	0
Lettuce, salad w/assorted vegetables, inc tomatoes and/or carrots, no dressing—1 c (73 gm)	0	0	0	0	0	0	0	0	13	1	.2	0
Lettuce, salad w/avocado, tomato, and/or carrots, w/ or wo/other vegetables, no dressing—1 c (87 gm)	2	2	1	1	1	1	1	1	43	27	4.2	0
Lettuce, salad w/cheese, tomato and/or carrots, w/ or wo/other vegetables, no dressing—1 c (77 gm)	16	14	12	11	10	9	8	8	78	49	30.9	17
Lettuce, salad w/egg, tomato, and/or carrots, w/ or wo/other vegetables, no dressing—1 c (88 gm)	21	20	18	17	17	16	15	14	47	23	6.6	117
Lettuce, wilted, w/bacon dressing—½ c (63 gm)	8	7	6	5	5	4	4	4	53	42	15.6	6
Lotus root, cooked, fat not added in cooking—½ c (60 gm)	0	0	0	0	0	0	0	0	39	0	.1	0
Luffa (Chinese okra), cooked, fat not added in cooking—½ c (89 gm)	0	0	0	0	0	0	0	0	28	1	.4	0
Mixed vegetable juice (vegetables other than tomato)—½ c (123 gm)	0	0	0	0	0	0	0	0	32	2	.3	0
Mixed vegetables (corn, lima beans, peas, green beans & carrots), cooked, fat not added in cooking—½ c (91 gm)	0	0	0	0	0	0	0	0	53	1	.3	0
Mixed vegetables (corn, lima beans, peas, green beans & carrots), canned, low-sodium, fat not added in cooking—½ c (91 gm)	0	0	0	0	0	0	0	0	33	2	.4	0

Vegetables (cont.)

	1200	1500	1800	2100	2400	2800	3200	3600	TOT CAL	FAT CAL	S/FAT CAL	CHOL MG
Mountain yam, Hawaii, steamed, wo/salt—½ c cubes (72 gm)	0	0	0	0	0	0	0	0	59	1	.1	0
Mountain yam, Hawaii, raw—½ c cubes (68 gm)	0	0	0	0	0	0	0	0	46	1	.2	0
Mushroom, batter-dipped, fried—1 med (14 gm)	2	2	2	2	1	1	1	1	30	22	3.9	3
Mushroom, Oriental, dried (inc shiitake)—1 c (145 gm)	1	1	1	1	1	1	1	1	429	13	3.2	0
Mushrooms, canned, drained solids—½ c pieces (78 gm)	0	0	0	0	0	0	0	0	19	2	.3	0
Mushrooms, cooked, fat not added in cooking—½ c (78 gm)	0	0	0	0	0	0	0	0	21	3	.4	0
Mushrooms, creamed—½ c (109 gm)	9	7	6	6	5	5	4	4	95	58	17.5	6
Mushrooms, raw—½ c pieces (35 gm)	0	0	0	0	0	0	0	0	9	1	.2	0
Mushrooms, shiitake, cooked, wo/salt—4 mushrooms (72 gm)	0	0	0	0	0	0	0	0	40	2	.4	0
Mushrooms, stuffed—1 stuffed cap (24 gm)	9	8	7	7	6	6	5	5	54	28	8.0	32
Mustard cabbage, cooked, fat not added in cooking—½ c (85 gm)	0	0	0	0	0	0	0	0	10	1	.2	0
Mustard greens, cooked, fat not added in cooking—1 c, fresh (140 gm)	0	0	0	0	0	0	0	0	21	3	.2	0
—1 c, frozen (150 gm)	0	0	0	0	0	0	0	0	22	3	.2	0
—1 c, canned (153 gm)	0	0	0	0	0	0	0	0	23	3	.2	0
Mustard greens, raw—½ c, chopped (28 gm)	0	0	0	0	0	0	0	0	7	1	0	0
Mustard spinach (tendergreen), boiled, drained, wo/salt—½ c chopped (90 gm)	0	0	0	0	0	0	0	0	14	2	0	0

Food												
Okra, batter-dipped, fried (inc cornmeal-dipped, fried)—10 pieces (75 gm)	9	8	7	7	6	5	5	5	143	99	15.6	16
Okra, cooked, fat not added in cooking —½ c, frozen (92 gm)	0	0	0	0	0	0	0	0	32	2	.5	0
—1 c, fresh (160 gm)	½	½	½	½	0	0	0	0	55	3	.9	0
Okra, frozen, boiled, drained, wo/salt—½ c slices (92 gm)	½	0	0	0	0	0	0	0	34	3	.6	0
Onion rings, batter-dipped, fried—10 med rings (2"–3" dia) (60 gm)	7	6	5	5	5	4	4	3	106	74	11.6	12
Onion rings, breaded, pan-fried, frozen, prep heated in oven—7 rings (70 gm)	24	20	18	15	14	12	11	10	285	168	54.1	0
Onions, canned, solid & liquid—½ c chopped (112 gm)	0	0	0	0	0	0	0	0	21	1	.2	0
Onions, creamed—½ c (114 gm)	7	6	5	5	4	4	4	3	89	49	14.9	5
Onions, frozen, chopped, boiled, drained, wo/salt—½ c chopped (105 gm)	0	0	0	0	0	0	0	0	30	1	.2	0
Onions, mature, cooked, fat not added in cooking (inc pearl, or young)—1 med (94 gm)	0	0	0	0	0	0	0	0	26	1	.2	0
Onions, mature, cooked or sauteed, fat added in cooking (inc pearl, or young)—1 med (96 gm)	2	1	1	1	1	1	1	1	42	17	3.4	0
Onions, mature, raw (inc red onions)—1 med sl (⅛" thick) (14 gm)	0	0	0	0	0	0	0	0	5	0	.1	0
—1 c, chopped (160 gm)	½	0	0	0	0	0	0	0	54	4	.6	0
Onions, pearl, cooked (inc pickled or cocktail onions)—½ c (93 gm)	0	0	0	0	0	0	0	0	26	1	.2	0
Onions, spring (inc tops & bulb), raw—½ c chopped (50 gm)	0	0	0	0	0	0	0	0	13	1	.1	0
Onions, young green, cooked, fat not added in cooking—2 med (4⅛" long) (28 gm)	0	0	0	0	0	0	0	0	7	0	.1	0

	1200	1500	1800	2100	2400	2800	3200	3600	TOT CAL	FAT CAL	S/FAT CAL	CHOL MG
Vegetables (cont.)												
Onions, young green, raw—1 med (4⅛" long) (15 gm)	0	0	0	0	0	0	0	0	4	0	0	0
Onions, young green, cooked, fat added in cooking—2 med (4⅛" long) (30 gm)	1	1	½	½	½	½	½	½	15	7	1.4	0
Palm hearts, cooked, fat not added in cooking—1 heart (3¼"long) (38 gm)	0	0	0	0	0	0	0	0	39	1	.2	0
Parsley, cooked, fat not added in cooking—10 sprigs (12 gm)	0	0	0	0	0	0	0	0	4	0	.1	0
Parsley, freeze-dried—1 tbsp (0 gm)	—	—	—	—	—	—	—	—	1	0	—	0
Parsley, raw—10 sprigs (10 gm)	0	0	0	0	0	0	0	0	3	0	0	0
Parsnips, cooked, fat not added in cooking—½ c, pieces (78 gm)	0	0	0	0	0	0	0	0	63	2	.4	0
—1 med (98 gm)	0	0	0	0	0	0	0	0	79	3	.4	0
Parsnips, creamed—½ c (114 gm)	9	8	7	6	5	5	4	4	134	60	18.0	6
Peas & carrots, canned, low-sodium, fat not added in cooking—½ c (80 gm)	0	0	0	0	0	0	0	0	30	2	.4	0
Peas & carrots, canned, reg pack, solid & liquid—½ c (128 gm)	0	0	0	0	0	0	0	0	48	4	.5	0
Peas & carrots, canned, special dietary pack, solid & liquid—½ c (128 gm)	0	0	0	0	0	0	0	0	48	4	.5	0
Peas & carrots, cooked, fat not added in cooking—½ c (80 gm)	½	0	0	0	0	0	0	0	38	3	.6	0
Peas & carrots, creamed—½ c (122 gm)	9	8	7	6	6	5	4	4	119	62	18.7	6
Peas & carrots, frozen, boiled, drained, wo/salt—½ c (80 gm)	0	0	0	0	0	0	0	0	38	3	.5	0
Peas & corn, cooked, fat not added in cooking—½ c (81 gm)	½	½	½	0	0	0	0	0	77	5	.8	0

Food											
Peas & onions, canned, solid & liquid—½ c (60 gm)	0	0	0	0	0	0	0	30	2	.4	0
Peas & onions, cooked, fat not added in cooking—½ c (90 gm)	0	0	0	0	0	0	0	40	2	.3	0
Peas & onions, frozen, boiled, drained, wo/salt—½ c (90 gm)	0	0	0	0	0	0	0	40	2	.3	0
Peas & potatoes, cooked, fat not added in cooking—½ c (79 gm)	0	0	0	0	0	0	0	67	1	.2	0
Peas, cowpeas, field peas, or black-eyed peas (not dried), cooked, fat not added in cooking (inc crowder peas, pink-eyed peas, purple hull peas)—½ c, frozen (85 gm)	1	1	½	½	½	½	½	92	6	1.6	0
—1 c, fresh (165 gm)	1	1	1	1	1	1	1	179	12	3.1	0
Peas, creamed—½ c (122 gm)	9	8	7	6	5	5	4	144	62	18.8	6
Peas, edible-podded, frozen, boiled, drained, wo/salt—½ c (80 gm)	0	0	0	0	0	0	0	42	3	.5	0
Peas, green, Alaska, early or june, canned, diet pack, drained, low-sodium—1 oz (28 gm)	0	0	0	0	0	0	0	6	0	0	0
Peas, green, Alaska, early or june, canned, reg pack, drained—1 oz (28 gm)	0	0	0	0	0	0	0	7	0	0	0
Peas, green, canned, low-sodium, fat not added in cooking—½ c (85 gm)	0	0	0	0	0	0	0	42	2	.4	0
Peas, green, canned, reg pack, solid & liquid—½ c (124 gm)	0	0	0	0	0	0	0	61	4	.5	0
Peas, green, canned, special dietary pack, drained solids—½ c (85 gm)	0	0	0	0	0	0	0	59	3	.4	0
Peas, green, canned, special dietary pack, solid & liquid—½ c (124 gm)	0	0	0	0	0	0	0	61	4	.5	0
Peas, green, cooked, fat not added in cooking—½ c, fresh or frozen (80 gm)	0	0	0	0	0	0	0	67	2	.3	0

Vegetables (cont.)

	1200	1500	1800	2100	2400	2800	3200	3600	TOT CAL	FAT CAL	S/FAT CAL	CHOL MG
Peas, green, frozen, boiled, drained, wo/salt—½ c (80 gm)	0	0	0	0	0	0	0	0	63	2	.4	0
Peas, green, raw—½ c (73 gm)	0	0	0	0	0	0	0	0	59	3	.5	0
Peas, green, sweet, wrinkled (sugar peas), diet pack, drained, low-sodium—1 oz (28 gm)	0	0	0	0	0	0	0	0	5	0	0	0
Peas, green, sweet, wrinkled (sugar peas), reg pack, canned, drained—1 oz (28 gm)	0	0	0	0	0	0	0	0	6	0	0	0
Peas w/mushroom sauce—½ c (122 gm)	5	4	4	3	3	3	2	2	108	34	10.0	2
Peas w/mushrooms, cooked, fat not added in cooking—½ c (80 gm)	0	0	0	0	0	0	0	0	54	2	.4	0
Peppers, hot chili, raw (inc jalapeño)—1 pepper (45 gm)	0	0	0	0	0	0	0	0	18	1	.1	0
Peppers, green, cooked, fat not added in cooking—½ c (68 gm)	0	0	0	0	0	0	0	0	12	2	.3	0
—1 med pepper (2¾" long, 2½" dia) (114 gm)	0	0	0	0	0	0	0	0	20	3	.5	0
Peppers, hot chili, green, raw—1 pepper (45 gm)	0	0	0	0	0	0	0	0	18	1	.1	0
Peppers, hot chili, immature green, canned, chili sauce—1 c (245 gm)	0	0	0	0	0	0	0	0	49	2	0	0
Peppers, hot chili, mature red, canned, chili sauce—1 c (245 gm)	0	0	0	0	0	0	0	0	51	14	0	0
Peppers, hot chili, red, canned, excluding seeds, solid & liquid—1 pepper (73 gm)	0	0	0	0	0	0	0	0	18	1	.1	0

Peppers, hot, cooked, fat not added in cooking (inc Tuscan peppers, jalapeño, chili peppers)—1 c chopped (136 gm)	0	0	0	0	0	0	0	0	54	2	.3	0
Peppers, jalapeño, canned, solid & liquid—½ c chopped (68 gm)	0	0	0	0	0	0	0	0	17	4	.4	0
Peppers, red, cooked, fat not added in cooking—½ c (68 gm)	0	0	0	0	0	0	0	0	12	2	.3	0
———1 med pepper (2¾″ long, 2½″ dia) (114 gm)	0	0	0	0	0	0	0	0	20	3	.5	0
Peppers, sweet, green, canned, solid & liquid—1 c halves (140 gm)	0	0	0	0	0	0	0	0	26	4	.5	0
Peppers, sweet, immature green, cooked, stuffed w/beef and crumbs—1 pepper (2¾″ long) (185 gm)	30	27	24	22	20	18	17	16	315	92	43.5	70
Peppers, sweet, red, boiled, drained, wo/salt—1 pepper (73 gm)	0	0	0	0	0	0	0	0	13	2	.4	0
Peppers, sweet, red, canned, solid & liquid—½ c halves (70 gm)	0	0	0	0	0	0	0	0	13	2	.3	0
Pepper, sweet, green, raw—1 sm (5/lb) (74 gm)	0	0	0	0	0	0	0	0	19	3	.4	0
———1 med (approx 2¾″ long, 2½″ dia) (119 gm)	½	½	0	0	0	0	0	0	30	5	.7	0
———1 lg (2¼/lb, approx 3¾″ long, 3″ dia) (164 gm)	½	½	½	½	½	0	0	0	41	7	1.0	0
Pepper, sweet, red, raw—1 sm (5/lb) (74 gm)	0	0	0	0	0	0	0	0	19	3	.4	0
———1 med (approx 2¾″ long, 2½″ dia) (119 gm)	½	½	0	0	0	0	0	0	30	5	.7	0
———1 lg (2¼/lb, approx 3¾″ long, 3″ dia) (164 gm)	½	½	½	½	½	0	0	0	41	7	1.0	0
Pigeon peas, cooked, fat not added in cooking—½ c (77 gm)	1	1	1	1	1	½	½	½	91	10	2.1	0
Pimientos, cooked, fat not added in cooking—1 c (184 gm)	1	1	½	½	½	½	½	½	61	10	1.4	0

	1200	1500	1800	2100	2400	2800	3200	3600	TOT CAL	FAT CAL	S/FAT CAL	CHOL MG
Vegetables (cont.)												
Pinacbet (eggplant w/tomatoes, bitter melon, etc)—½ c (107 gm)	3	3	2	2	2	2	1	1	73	50	6.8	0
Poke greens, cooked, fat not added in cooking—½ c (78 gm)	0	0	0	0	0	0	0	0	16	3	.4	0
Potato, au gratin, prep fr home recipe using butter—½ c (122 gm)	28	24	21	19	17	15	14	13	160	84	51.9	29
Potato, au gratin, prep fr home recipe using margarine—½ c (122 gm)	20	17	15	13	12	11	10	9	160	84	38.8	18
Potato, baked, flesh & skin, wo/salt—1 potato (2⅓" x 4¾") (202 gm)	0	0	0	0	0	0	0	0	220	2	.4	0
Potato, baked, peel eaten, fat not added before serving—1 med (122 gm)	0	0	0	0	0	0	0	0	132	1	.3	0
Potato, baked, peel not eaten, fat not added before serving—1 med, raw (2¼"-3" dia) (93 gm)	0	0	0	0	0	0	0	0	86	1	.2	0
Potato, boiled, wo/peel, canned, low-sodium, fat not added in cooking—1 canned potato (approx 1" dia) (35 gm)	0	0	0	0	0	0	0	0	21	1	.2	0
Potato, boiled, wo/peel, fat not added in cooking (inc mashed, no fat or milk added)—1 med, raw (2¼"-3" dia) (122 gm)	0	0	0	0	0	0	0	0	104	1	.3	0
Potato, boiled, w/peel, fat not added in cooking—1 med, raw (2¼"-3" dia) (142 gm)	0	0	0	0	0	0	0	0	123	1	.3	0
Potato, cooked, w/cheese (inc au gratin)—½ c (122 gm)	20	17	15	13	12	11	10	9	177	84	38.2	19

Food											
Potato, french-fried, from fresh, deep fried (inc fried w/skins)—1 c (57 gm)	9	8	7	6	5	5	4	4	126	45	21.2	0
Potato, frozen, french-fried, cottage cut, heated in oven, wo/salt—10 strips (50 gm)	8	7	6	5	4	4	4	3	109	37	17.5	0
Potato, frozen, french-fried, extruded, heated in oven, wo/salt—10 strips (50 gm)	15	13	11	10	9	8	7	6	163	85	34.0	0
Potato, frozen, french-fried, home-prep, heated in oven, wo/salt—10 strips (50 gm)	8	7	6	5	5	4	4	3	111	40	18.7	0
Potato, frozen, french-fried, restaurant-prep, fried in vegetable oil—10 strips (50 gm)	10	8	7	6	6	5	5	4	158	75	22.5	0
Potato, frozen, french-fried, restaurant-prep, fried in animal fat & vegetable oil—10 strips (50 gm)	15	12	11	10	9	8	7	6	158	75	30.6	7
Potato, frozen, whole, unprepared—½ c (91 gm)	0	0	0	0	0	0	0	0	71	1	.4	0
Potato, hash brown, fr dry mix—½ c (78 gm)	15	13	11	10	9	8	7	7	169	80	31.4	9
Potato, hash brown, w/cheese (inc Ore Ida Cheddar Browns)—½ c (73 gm)	23	20	17	16	14	13	11	10	167	104	46.4	17
Potato, hash brown, frozen, plain, prep—½ c (78 gm)	15	13	11	10	9	8	7	7	170	81	31.6	9
Potato, hash brown, home-prep—½ c (78 gm)	19	16	14	12	11	10	9	8	119	98	38.2	10
Potato, home fries (inc fried raw, cottage style, German-fried, potatoes O'Brien)—½ c (97 gm)	16	13	12	10	9	8	7	7	214	76	36.1	0
Potato, mashed, dehydrated, prep fr flakes wo/milk, whole milk & butter added—½ c (105 gm)	17	14	13	11	10	9	8	8	119	53	32.5	15

	1200	1500	1800	2100	2400	2800	3200	3600	TOT CAL	FAT CAL	S/FAT CAL	CHOL MG
Vegetables (cont.)												
Potato, mashed, dehydrated, prep fr granules w/milk, water & margarine added—½ c (105 gm)	3	3	2	2	2	2	2	1	83	21	6.4	2
Potato, mashed, dehydrated, prep from granules wo/milk, whole milk & butter added—½ c (105 gm)	8	7	6	5	5	5	4	4	137	59	10.6	18
Potato, mashed, home-prep, whole milk added—½ c (105 gm)	2	1	1	1	1	1	1	1	81	5	3.2	2
Potato, mashed, home-prep, whole milk & margarine added—½ c (105 gm)	5	4	3	3	3	2	2	2	111	40	9.8	2
Potato, microwaved, cooked in skin, flesh & skin, wo/salt—1 potato (2⅓" x 4¾") (202 gm)	0	0	0	0	0	0	0	0	212	2	.4	0
Potato, microwaved, cooked in skin, flesh, wo/salt—1 potato (2⅓" x 4¾") (156 gm)	0	0	0	0	0	0	0	0	157	2	.4	0
Potato, microwaved, cooked in skin, skin, wo/salt—skin fr 1 potato (58 gm)	0	0	0	0	0	0	0	0	77	1	.2	0
Potato, patty (inc potato croquettes)— 2 patties (3"–3½" dia) (138 gm)	16	13	12	10	9	8	7	7	251	137	34.7	3
Potato puffs, cheese-filled—1 c (70 gm)	40	34	30	26	24	21	19	17	343	243	86.9	12
Potato, puffs (inc Ore Ida Crispy Crowns)—1 c (128 gm)	26	22	19	17	15	13	12	11	284	124	58.7	0
Potato, raw, flesh—1 potato (2½" dia) (112 gm)	0	0	0	0	0	0	0	0	89	1	.3	0

Food												
Potato, raw, skin—skin fr 1 potato (38 gm)	0	0	0	0	0	0	0	0	22	1	.1	0
Potato, roasted—1 med, raw, (2¼"–3" dia) (93 gm)	0	0	0	0	0	0	0	0	132	1	.4	0
Potato; scalloped, home-prep w/margarine—½ c (122 gm)	8	7	6	5	5	4	4	4	105	41	15.1	7
Potato, scalloped, w/ham (inc w/cheese)—½ c (116 gm)	15	13	11	10	9	8	8	7	117	46	26.5	19
Potato skins, chips (inc Tato Skins baked potato, sour cream n' chives, & cheese n' bacon flavor)—1 c, loose (26 gm)	15	12	11	9	8	7	7	6	150	108	32.7	0
Potato skins, w/adhering flesh, baked—skin fr 1 med, raw (2¼–3" dia) (29 gm)	0	0	0	0	0	0	0	0	39	0	.1	0
Potato skins, w/adhering flesh, fried—skin from 1 med, raw (2¼–3" dia) (34 gm)	4	4	3	3	2	2	2	2	137	38	9.5	0
Potato skins, w/adhering flesh, fried, w/cheese—skin fr 1 med, raw (28 gm)	9	8	7	6	6	5	5	4	113	43	18.0	7
Potato, stewed Puerto Rican style (papas guisadas)—1 med (163 gm)	8	7	6	6	5	5	5	4	183	47	15.6	8
Potato, sticks (inc french fry-shaped)—1 c (38 gm)	13	11	10	9	8	7	6	6	198	118	30.2	0
Potato, stuffed, baked, peel eaten, stuffed w/bacon & cheese—1 long type, raw (2⅓" dia, 4¾" long) (264 gm)	32	27	24	21	19	17	16	14	419	140	61.9	26
Potato, stuffed, baked, peel eaten, stuffed w/broccoli & cheese sauce—1 long type, raw (2⅓" dia, 4¾" long) (290 gm)	12	10	9	8	7	6	6	5	342	76	24.3	8
Potato, stuffed, baked, peel eaten, stuffed w/cheese—1 long type, raw (2⅓" dia, 4¾" long) (254 gm)	25	22	19	17	15	14	13	11	373	106	50.1	20

	1200	1500	1800	2100	2400	2800	3200	3600	TOT CAL	FAT CAL	S/FAT CAL	CHOL MG
Vegetables (cont.)												
Potato, stuffed, baked, peel eaten, stuffed w/chili—1 long type, raw (2⅓" dia, 4¾" long) (350 gm)	14	12	11	9	9	8	7	6	439	94	26.6	14
Potato, stuffed, baked, peel eaten, stuffed w/meat in cream sauce—1 long type, raw (2⅓" dia, 4¾" long) (274 gm)	17	15	13	12	11	9	9	8	360	97	31.6	19
Potato, stuffed, baked, peel eaten, stuffed w/sour cream—1 long type, raw (2⅓" dia, 4¾" long) (284 gm)	40	34	30	27	24	21	19	18	405	157	80.3	27
Potato, stuffed, baked, peel not eaten, stuffed w/broccoli & cheese sauce—1 long type, raw (2⅓" dia, 4¾" long) (241 gm)	12	10	9	8	7	7	6	5	261	76	24.4	8
Potato, stuffed, baked, peel not eaten, stuffed w/cheese—1 long type, raw (2⅓" dia, 4¾" long) (206 gm)	20	17	15	14	12	11	10	9	275	87	40.5	16
Potato, stuffed, baked, peel not eaten, stuffed w/chili—1 long type, raw (2⅓" dia, 4¾" long) (295 gm)	14	12	11	10	9	8	7	7	348	95	26.9	14
Potato, stuffed, baked, peel not eaten, stuffed w/meat in cream sauce—1 long type, raw (2⅓" dia, 4¾" long) (225 gm)	17	14	13	11	10	9	9	8	281	96	31.3	19
Potato, stuffed, baked, peel not eaten, stuffed w/sour cream—1 long type, raw (2⅓" dia, 4¾" long) (236 gm)	33	28	25	22	20	18	16	15	307	128	66.4	23
Pumpkin, canned, wo/salt—½ c (122 gm)	1	1	1	½	½	½	½	½	41	3	1.6	0

Food										
Pumpkin, cooked, fat not added in cooking—½ c (123 gm)	½	0	0	0	0	0	32	1	.6	0
Pumpkin, raw—½ c 1" cubes (58 gm)	0	0	½	0	0	0	15	1	.3	0
Radishes, Oriental, dried—½ c (58 gm)	1	½	0	½	½	0	157	4	1.2	0
Radishes, Oriental, raw—½ c slices (44 gm)	0	0	0	0	0	0	8	1	.1	0
Radishes, white icicle, raw—½ c slices (50 gm)	0	0	0	0	0	0	7	1	.2	0
Radishes, Japanese (daikon), cooked, fat not added in cooking—½ c (74 gm)	0	0	0	0	0	0	14	1	.2	0
Radishes, raw—½ c sl (58 gm)	0	0	0	0	0	0	10	3	.2	0
—1 med (¾"-1" dia) (5 gm)	0	0	0	0	0	0	1	0	0	0
Ratatouille—½ c (107 gm)	3	2	2	2	2	1	73	53	7.3	1
Rhubarb, frozen, cooked, w/sugar—½ c (120 gm)	0	3	2	2	0	0	139	1	.2	0
Rhubarb, raw—½ c diced pieces (61 gm)	—	—	—	—	—	—	13	1	—	—
Rutabaga, cooked, fat not added in cooking—½ c (85 gm)	0	0	0	0	0	0	29	1	.2	0
Rutabaga, raw—½ c (70 gm)	0	0	0	0	0	0	25	1	.2	0
Salsify (vegetable oyster), cooked, fat not added in cooking—½ c (68 gm)	0	0	0	0	0	0	46	1	.2	0
Sauerkraut, canned, low-sodium—½ c (71 gm)	0	0	0	0	0	0	13	1	.3	0
Sauerkraut, cooked, fat not added in cooking—½ c (71 gm)	0	0	0	0	0	0	13	1	.2	0
Sauerkraut juice, canned—1 c (242 gm)	0	0	0	0	0	0	24	0	0	0
Seaweed, dried (inc sea moss, kelp)—1 c (15 gm)	1	½	½	½	½	½	45	5	1.8	0
Seaweed, prep w/soy sauce—½ c (48 gm)	0	0	0	0	0	0	21	2	.5	0
Seaweed, raw (inc blanched)—½ c (40 gm)	½	0	0	0	0	0	12	3	.7	0

Vegetables (cont.)

	1200	1500	1800	2100	2400	2800	3200	3600	TOT CAL	FAT CAL	S/FAT CAL	CHOL MG
Sequin (Portuguese squash), cooked, fat not added in cooking—½ c (90 gm)	0	0	0	0	0	0	0	0	18	2	.5	0
Shallots, freeze-dried—1 tbsp (1 gm)	0	0	0	0	0	0	0	0	3	0	0	0
Snow peas (pea pods), cooked, fat not added in cooking—½ c (80 gm)	0	0	0	0	0	0	0	0	33	2	.3	0
Snow peas, raw—½ c (73 gm)	0	0	0	0	0	0	0	0	31	1	.3	0
Spinach & cheese casserole—½ c (100 gm)	31	28	26	24	23	21	20	19	138	67	29.6	117
Spinach, canned, drained solids—½ c (107 gm)	½	½	½	0	0	0	0	0	25	5	.8	0
Spinach, cooked, fat not added in cooking—1 c, fresh (180 gm)	½	½	0	0	0	0	0	0	41	4	.7	0
—1 c, frozen, leaf (190 gm)	½	½	0	0	0	0	0	0	43	4	.7	0
—1 c, frozen, chopped (205 gm)	½	½	½	0	0	0	0	0	47	5	.8	0
—1 c, canned (214 gm)	½	½	½	0	0	0	0	0	49	5	.8	0
Spinach, creamed—½ c (100 gm)	7	6	5	5	4	4	3	3	80	47	14.3	5
Spinach, frozen, chopped or leaf, boiled, drained, wo/salt—½ c (95 gm)	0	0	0	0	0	0	0	0	27	2	.3	0
Spinach, raw—1 c (30 gm)	0	0	0	0	0	0	0	0	7	1	.2	0
Spinach soufflé—½ c (50 gm)	15	14	13	12	11	10	10	9	65	41	11.8	64
Spinach w/cheese sauce—½ c (100 gm)	12	10	9	8	7	6	6	5	89	54	23.0	10
Squash fritter or cake—1 fritter (24 gm)	8	7	6	6	5	5	5	4	82	45	11.7	19
Squash, summer & onions, cooked, fat not added in cooking—½ c (93 gm)	0	0	0	0	0	0	0	0	21	2	.4	0

Food												
Squash, summer & onions, cooked, fat added in cooking—½ c (95 gm)	2	2	1	1	1	1	1	1	39	21	4.1	0
Squash, summer & tomato-cheese casserole—½ c (109 gm)	10	9	8	7	6	6	5	5	74	50	20.3	8
Squash, summer, breaded or battered, fried (inc zucchini pancakes)—½ c (110 gm)	13	11	10	9	8	7	7	6	165	122	21.7	19
Squash, summer, casserole, w/cheese sauce—½ c (109 gm)	10	8	7	6	6	5	5	4	92	46	18.8	8
Squash, summer, casserole, w/rice & tomato sauce—½ c (117 gm)	0	0	0	0	0	0	0	0	52	2	.5	0
Squash, summer, cooked, fat not added in cooking (inc yellow squash, zucchini)—½ c sl (93 gm)	0	0	0	0	0	0	0	0	18	3	.5	0
—1 c mashed (245 gm)	1	1	½	½	½	½	½	½	49	7	1.4	0
Squash, summer, creamed—½ c (109 gm)	7	6	5	5	4	4	4	3	80	48	14.7	5
Squash, summer, crookneck & straightneck, canned, drained, solids, wo/salt—½ c sl (108 gm)	0	0	0	0	0	0	0	0	14	1	.2	0
Squash, summer, crookneck & straightneck, frozen, boiled, drained, wo/salt—½ c sl (96 gm)	0	0	0	0	0	0	0	0	24	2	.4	0
Squash, summer, green, raw (inc zucchini)—½ c sl (57 gm)	0	0	0	0	0	0	0	0	8	1	.1	0
Squash, summer, soufflé—½ c (68 gm)	18	16	15	14	13	12	11	11	83	55	14.7	72
Squash, summer, yellow, raw—½ c sl (57 gm)	0	0	0	0	0	0	0	0	11	1	.2	0
Squash, summer, zucchini, Italian style, canned—½ c (114 gm)	0	0	0	0	0	0	0	0	33	1	.3	0
Squash, winter, baked w/cheese—½ c (112 gm)	14	12	10	9	8	7	7	6	165	94	27.8	8
Squash, winter, soufflé—½ c (79 gm)	15	14	13	12	11	11	10	10	67	19	6.2	77
Squash, winter, spaghetti, boiled, drained, or baked, wo/salt—½ c (78 gm)	0	0	0	0	0	0	0	0	23	2	.4	0

	THE 2-IN-1 SYSTEM											
	1200	1500	1800	2100	2400	2800	3200	3600	TOT CAL	FAT CAL	S/FAT CAL	CHOL MG
Vegetables *(cont.)*												
Squash, winter type, baked, fat & sugar added in cooking (inc acorn, butternut, hubbard)—½ c cubes, all varieties (107 gm)	2	2	2	1	1	1	1	1	65	24	4.7	0
Squash, winter type, baked, no fat or sugar added in cooking (inc acorn, butternut, hubbard)—½ c cubes, all varieties (103 gm)	1	½	½	½	½	½	0	0	40	6	1.2	0
Squash, winter type, mashed, fat & sugar added in cooking (inc acorn, butternut, hubbard)—½ c butternut (129 gm)	2	2	2	2	1	1	1	1	98	27	5.4	0
Squash, winter type, mashed, no fat or sugar added in cooking (inc acorn, butternut, hubbard)—½ c butternut (120 gm)	1	1	½	½	½	½	½	½	47	7	1.4	0
Squash, winter type, raw (inc pumpkin, acorn, butternut, hubbard)—1 c acorn cubes (140 gm)	½	0	0	0	0	0	0	0	52	3	.6	0
Succotash, (corn & limas), boiled, drained, wo/salt—½ c (96 gm)	1	½	½	½	½	½	½	0	111	7	1.3	0
Succotash, (corn & limas), canned, w/cream-style corn—½ c (133 gm)	1	½	½	½	½	½	0	0	102	6	1.2	0
Succotash, (corn & limas), canned, w/whole kernel corn, solid & liquid—½ c (128 gm)	½	½	½	½	½	0	0	0	81	5	1.1	0
Succotash, (corn & limas), frozen, boiled, drained, wo/salt—½ c (85 gm)	1	½	½	½	½	½	½	0	79	7	1.3	0

Food	35	31	27	25	23	21	19	18	337	99	52.3	76
Sweet potato & pumpkin casserole, Puerto Rican style—½ c (133 gm)	35	31	27	25	23	21	19	18	337	99	52.3	76
Sweet potato, baked, peel eaten, fat added before serving—1 med, raw (2" dia, 5" long) (151 gm)	3	3	2	2	2	2	1	1	215	37	7.3	0
Sweet potato, baked, peel eaten, fat not added before serving—1 med, raw (2" dia, 5" long) (146 gm)	0	0	0	0	0	0	0	0	180	1	.3	0
Sweet potato, baked, peel not eaten, fat added before serving—1 med, raw (2" dia, 5" long) (119 gm)	3	3	2	2	2	2	1	1	152	36	7.0	0
Sweet potato, baked, peel not eaten, fat not added before serving—1 med, raw (2" dia, 5" long) (114 gm)	0	0	0	0	0	0	0	0	117	1	.2	0
Sweet potato, boiled, wo/peel, fat not added in cooking—1 med, raw (2" dia, 5" long) (151 gm)	½	½	½	½	0	0	0	0	159	4	.9	0
Sweet potato, candied—½ c (98 gm)	3	2	2	2	2	1	1	1	147	32	6.4	0
Sweet potato, canned in syrup—½ c (98 gm)	0	0	0	0	0	0	0	0	87	2	.4	0
Sweet potato, canned wo/syrup—½ c (100 gm)	0	0	0	0	0	0	0	0	91	2	.4	0
Sweet potato, casserole or mashed, fr dried flakes—½ c (128 gm)	5	4	4	3	3	3	3	2	157	38	10.7	3
Sweet potato, casserole or mashed (inc sweet potato pudding)—½ c (124 gm)	5	4	3	3	3	2	2	2	141	33	9.3	3
Sweet potato, cooked, candied—1 piece (2½" x 2") (105 gm)	7	6	5	5	4	4	4	3	144	31	12.8	8
Sweet potato, dehydrated, flaked, prep w/water—1 c (255 gm)	0	0	0	0	0	0	0	0	242	3	0	0
Sweet potato, fried—½ c (103 gm)	9	8	7	6	5	5	4	4	209	83	20.5	0
Sweet potato, frozen, baked, wo/salt—½ c cubes (88 gm)	0	0	0	0	0	0	0	0	88	1	.2	0

THE 2-IN-1 SYSTEM

	1200	1500	1800	2100	2400	2800	3200	3600	TOT CAL	FAT CAL	S/FAT CAL	CHOL MG
Vegetables (cont.)												
Sweet potato leaves, squash leaves, pumpkin leaves, chrysanthemum leaves, bean leaves, or swamp cabbage, cooked, fat not added in cooking—½ c sweet potato leaves (32 gm)	0	0	0	0	0	0	0	0	8	1	.2	0
Sweet potato, yellow, Puerto Rican, cooked—½ c (70 gm)	0	0	0	0	0	0	0	0	73	2	.4	0
Taro, cooked, wo/salt—½ c sl (66 gm)	0	½	½	0	0	0	0	0	94	1	.2	0
Taro leaves, cooked, fat not added in cooking—½ c (73 gm)	1	½	½	½	½	0	0	0	38	6	1.2	0
Taro shoots, cooked, wo/salt—½ c sl (70 gm)	0	0	0	0	0	0	0	0	10	1	.1	0
Thistle leaves, cooked, fat not added in cooking—½ c (90 gm)	0	0	0	0	0	0	0	0	21	5	.2	0
Tomato & celery, cooked, fat not added —½ c (122 gm)	0	0	0	0	0	0	0	0	23	2	.4	0
Tomato & corn, cooked, fat not added —½ c (121 gm)	½	½	½	½	0	0	0	0	54	6	.9	0
Tomato & lima beans, cooked, fat not added—½ c (121 gm)	0	0	0	0	0	0	0	0	74	3	.5	0
Tomato & okra, cooked, fat not added —½ c (109 gm)	0	0	0	0	0	0	0	0	27	2	.4	0
Tomato & onion, cooked, fat not added—½ c (119 gm)	0	0	0	0	0	0	0	0	30	3	.4	0
Tomato & vegetable juice, mostly tomato (inc V-8)—¾ c (182 gm)	0	0	0	0	0	0	0	0	35	1	.2	0

Food												
Tomato & vegetable juice, mostly tomato, low-sodium—¾ c (182 gm)	0	.2	2	35	0	0	0	0	0	0	0	0
Tomato aspic—½ c (114 gm)	0	.1	1	32	0	0	0	0	0	0	0	0
Tomatoes, broiled—½ c (120 gm)	0	.4	3	29	0	0	0	0	0	0	0	0
Tomatoes, green, cooked (inc fried)—½ c (90 gm)	26	26.3	100	149	8	8	9	10	11	12	14	16
Tomatoes, green, raw—1 med (123 gm)	0	.3	2	30	0	0	0	0	0	0	0	0
Tomatoes, raw (inc plum & Italian tomatoes)—1 cherry (17 gm)	0	0	0	3	0	0	0	0	0	0	0	0
—1 sl or wedge (44 gm)	0	.1	1	8	0	0	0	0	0	0	0	0
—1 plum tomato (62 gm)	0	.2	1	12	0	0	0	0	0	0	0	0
—1 med (2⅗" dia) (123 gm)	0	.3	2	23	0	0	0	0	0	0	0	0
Tomatoes, red, fried—½ c (90 gm)	26	26.3	100	145	8	8	9	10	11	12	14	16
Tomatoes, red, ripe, stewed—1 c (101 gm)	0	3.6	20	59	1	1	1	1	1	1	1	2
Tomatoes, red, ripe, canned, wedges in tomato juice—½ c (131 gm)	0	.3	2	34	0	0	0	0	0	0	0	0
Tomatoes, red, ripe, canned, whole, reg pack—½ c (120 gm)	0	.4	3	24	0	0	0	0	0	0	0	0
Tomatoes, red, ripe, canned, whole, special dietary pack—½ c (120 gm)	0	.4	3	24	0	0	0	0	0	0	0	0
Tomatoes, red, ripe, canned, w/green chilies—½ c (120 gm)	0	.1	1	18	0	0	0	0	0	0	0	0
Tomatoes, scalloped—½ c (118 gm)	0	12.3	63	109	2	2	3	3	4	4	5	5
Tomatoes, stewed—½ c (128 gm)	0	5.3	28	88	1	1	1	1	2	2	2	2
Tomatoes, green, pickled—1 tomato (30 gm)	0	.1	1	11	0	0	0	0	0	0	0	0
Tomato juice—¾ c (182 gm)	0	.1	1	31	0	0	0	0	0	0	0	0
Tomato juice, canned, wo/salt added—½ c (122 gm)	0	.1	1	21	0	0	0	0	0	0	0	0
Tomato juice cocktail—¾ c (182 gm)	0	.2	1	35	0	0	0	0	0	0	0	0
Tomato juice, low-sodium—¾ c (182 gm)	0	.1	1	31	0	0	0	0	0	0	0	0

	1200	1500	1800	2100	2400	2800	3200	3600	TOT CAL	FAT CAL	S/FAT CAL	CHOL MG
Vegetables (cont.)												
Tomato juice, w/clam or beef juice—¾ c (182 gm)	0	0	0	0	0	0	0	0	26	2	.5	0
Tomato w/corn & okra, cooked, fat not added—½ c (106 gm)	½	½	0	0	0	0	0	0	44	4	.7	0
Tree fern, cooked, wo/salt—½ c chopped (71 gm)	—	—	—	—	—	—	—	—	28	1	—	0
Turnip, cooked, fat not added in cooking—½ c pieces (78 gm)	0	0	0	0	0	0	0	0	14	1	.1	0
—1 med (116 gm)	0	0	0	0	0	0	0	0	21	1	.1	0
Turnip greens, canned, low-sodium, cooked, fat not added in cooking—½ c (78 gm)	0	0	0	0	0	0	0	0	11	11	.5	0
Turnip greens, boiled, drained, wo/salt—½ c chopped (72 gm)	0	0	0	0	0	0	0	0	15	2	.4	0
Turnip greens, canned, solid & liquid—½ c (117 gm)	½	½	0	0	0	0	0	0	17	4	.7	0
Turnip greens, cooked, fat not added in cooking—1 c, fresh (144 gm)	½	½	0	0	0	0	0	0	29	3	.7	0
—1 c, canned (159 gm)	½	½	½	0	0	0	0	0	32	3	.8	0
—1 c, frozen (165 gm)	½	½	½	0	0	0	0	0	33	3	.8	0
Turnip greens, frozen, boiled, drained, wo/salt—½ c (82 gm)	½	½	0	0	0	0	0	0	24	4	.7	0
Turnip greens w/roots, cooked, fat not added in cooking—½ c (81 gm)	0	0	0	0	0	0	0	0	14	1	.2	0
Turnip, raw—½ c (65 gm)	0	0	0	0	0	0	0	0	18	1	.1	0
Turnips, creamed—½ c (118 gm)	9	8	7	6	5	5	4	4	97	60	18.2	6
Vegetable combination (excluding carrots, broccoli & dark-green leafy), cooked, w/soy-base sauce—½ c (93 gm)	3	3	2	2	2	2	1	1	76	37	7.2	0

Food											
Vegetable combination (excluding carrots, broccoli & dark-green leafy), cooked, w/cream sauce—½ c (114 gm)	9	7	6	6	5	5	4	4	124	57	17.3	6
Vegetable combination (inc carrots, broccoli, and/or dark-green leafy), cooked, w/soy-base sauce (inc stir-fry vegetables)—½ c (93 gm)	0	0	0	0	0	0	0	0	47	2	.4	0
Vegetable combination (inc carrots, broccoli, and/or dark-green leafy), cooked, w/cream sauce—½ c (114 gm)	8	7	6	6	5	5	4	4	95	55	16.9	6
Vegetable combination (broccoli, carrots, corn, cauliflower), cooked, fat not added in cooking—½ c (70 gm)	0	0	0	0	0	0	0	0	28	1	.1	0
Vegetable combination (excluding carrots, broccoli & dark-green leafy), cooked, w/cheese sauce—½ c (114 gm)	3	3	2	2	2	2	2	1	80	13	5.7	3
Vegetable combination (excluding carrots, broccoli & dark-green leafy), cooked, w/pasta—½ c (68 gm)	0	0	0	0	0	0	0	0	54	3	.4	0
Vegetable combination (excluding carrots, broccoli & dark-green leafy), cooked, w/sauce & pasta—½ c (99 gm)	3	2	2	2	1	1	1	1	96	31	5.9	0
Vegetable combination (inc carrots, broccoli, and/or dark-green leafy), cooked, w/cheese sauce—½ c (114 gm)	3	2	2	2	1	2	1	1	61	11	5.0	3
Vegetable combination (inc carrots, broccoli, and/or dark-green leafy), cooked, w/pasta—½ c (68 gm)	0	0	0	0	0	0	0	0	41	2	.3	0

	1200	1500	1800	2100	2400	2800	3200	3600	TOT CAL	FAT CAL	S/FAT CAL	CHOL MG
Vegetables (cont.)												
Vegetable combination (inc carrots, broccoli, and/or dark-green leafy), cooked, w/sauce and pasta (inc Birdseye New England Style Mixed —½ c (99 gm)	3	2	2	2	1	1	1	1	83	30	5.7	0
Vegetable juice cocktail, canned—½ c (121 gm)	0	0	0	0	0	0	0	0	22	1	.2	0
Vegetable mixture, dried (inc Salad Crunchies)—1 tbsp (6 gm)	½	½	0	0	0	0	0	0	22	6	.7	0
Vegetables, mixed, frozen, boiled, drained, wo/salt—½ c (91 gm)	0	0	0	0	0	0	0	0	54	1	.3	0
Vegetable stew, wo/meat—1 c (239 gm)	5	4	4	3	3	3	2	2	134	21	9.8	3
Vegetable tempura—1 c (63 gm)	13	12	11	10	10	9	8	8	102	57	11.3	53
Water chestnut—2 med (20 gm)	0	0	0	0	0	0	0	0	10	0	0	0
Watercress, cooked, fat not added in cooking—½ c (69 gm)	0	0	0	0	0	0	0	0	6	0	.1	0
Watercress, raw—½ c chopped (17 gm)	0	0	0	0	0	0	0	0	2	0	0	0
Wax gourd (Chinese preserving melon), boiled, drained, wo/salt—½ c cubes (87 gm)	0	0	0	0	0	0	0	0	11	2	.1	0
Winter melon, cooked (inc Chinese melon, togan)—½ c (88 gm)	0	0	0	0	0	0	0	0	11	2	.1	0
Yam, boiled, drained, or baked, wo/salt—½ c cubes (68 gm)	0	0	0	0	0	0	0	0	79	1	.2	0
Yam, raw—½ c cubes (75 gm)	0	0	0	0	0	0	0	0	89	1	.3	0
Zucchini w/tomato sauce, cooked, fat not added in cooking (inc summer squash)—½ c (117 gm)	0	0	0	0	0	0	0	0	21	1	.2	0

APPENDIXES

APPENDIX A

Measurement Equivalents

Most measurements in the 2-in-1 tables are given in a standard American serving size—1 ounce, 4 ounces (¼ pound), 1 tablespoon, 2 fluid ounces, ½ cup, 1 cup—as well as in specific units such as 1 drumstick, 1 wing, 1 chop, 1 rib, 1 can, 1 jar, and so forth.

Besides the standard American weights and measures, every item also has its weight given in grams. This was done to facilitate quick and accurate comparisons between various products.

It might help to remember that there are two "ounces" in American measurements. The regular ounce, used in weighing, is called avoirdupois, while the ounce used for liquid measurement is called a fluid ounce.

1	regular ounce	28.35	grams
3½	regular ounces	100	grams
4	regular ounces	113	grams
16	regular ounces	454	grams, 1 pound
1	teaspoon	⅓	tablespoon
3	teaspoons	1	tablespoon
1	tablespoon	½	fluid ounce
8	tablespoons	½	cup

16	tablespoons	1	cup
½	cup	4	fluid ounces
1	cup	8	fluid ounces, ½ pint
2	cups	1	pint
2	pints	1	quart
4	quarts	1	gallon

APPENDIX B

The Mathematics
of the 2-in-1 System

This appendix is a technical discussion on the mathematical makeup of the 2-in-1 System. It is included for those readers who want to know exactly how the 2-in-1 scores were calculated.

The National Heart, Lung, and Blood Institute, in its suggestions for the dietary treatment for reducing high serum cholesterol, has recommended that the intake of saturated fat be limited to a maximum of 10 percent of daily calories, and that cholesterol consumption be limited to a maximum of 300 milligrams per day. The 2-in-1 System is based on those recommendations.

In order to find a mathematical formula from which the system could be constructed, we first had to define the serum cholesterol influence from both dietary cholesterol and saturated fat consumption. After much research, it was decided to base the system on formulas devised originally by Dr. Ancel Keys of the University of Minnesota. His work on dietary cholesterol, saturated and unsaturated fats, and their impact on serum cholesterol has been published in numerous scientific and technical publications and has been widely accepted by the medical, nutritional, and scientific communities.

One formula used in designing the 2-in-1 System states that the percentage of saturated fat intake (as compared to total caloric intake), multiplied by 2.7, equals the production of that amount of serum cholesterol in milligrams. Thus, if you have a daily consumption of 1,800 calories, of which 180 calories are saturated fat, you will consume 10 percent of your total calories in saturated fat (180 divided by 1,800 = 0.1). Multiplying this result (10%) by 2.7 yields 27 milligrams of serum cholesterol produced by the saturated fat. With the recommended maximum quantity of saturated fat intake of 10 percent at all caloric levels, this figure of 27 is the same for every amount of caloric consumption, as long as 10 percent of that consumption consists of saturated fat calories. As 1 gram of fat always equals 9 calories, a person with an intake of 1,800 daily calories could, therefore, consume 20 grams of saturated fat (10% of 1,800 = 180, divided by 9 = 20), while an individual with an intake of 3,200 calories could consume 35.56 grams of saturated fat (10% of 3,200 = 320, divided by 9 = 35.56), while both would produce the same amount, 27 milligrams, of serum cholesterol.

The formula for converting dietary cholesterol to serum cholesterol is somewhat more complicated. The number of milligrams of cholesterol consumed is first divided by the multiple that the caloric intake bears to 1,000. Thus, if your consumption is 1,800 calories, and you consume 300 milligrams of cholesterol, you first divide 300 by 1.8. From this figure you derive the square root, and then multiply by 1.5. In this example then, you would divide 300 by 1.8, giving you 166.67. The square root of this number is 12.91, which, multiplied by 1.5, equals the production of 19.36 milligrams of serum cholesterol. A caloric intake of 3,200, with the same dietary cholesterol consumption of 300 milligrams, would result in the following equation: 300 divided by 3.2 = 93.75; the square root of 93.75 = 9.68, which, multiplied by 1.5, results in the production of 14.52 milligrams of serum cholesterol. Therefore, the same consumption of 300 milligrams of dietary cholesterol produces 19.4 milligrams of serum cholesterol if your total consumption is 1,800 calories, but only 14.5 milligrams if you consume a total of 3,200 calories daily.

To complete the formula for the 2-in-1 System, the dietary cholesterol and the saturated fat were then related as follows: In the first example (with 1,800 daily calorie consumption), 180 calories of saturated fat produced 27 milligrams of serum cholesterol, and 300 milligrams of dietary cholesterol produced 19.36 milligrams of serum cholesterol. Therefore, in the Institute-suggested diet of a maximum of 10 percent saturated fat plus 300 milligrams of dietary cholesterol, the 1,800-calorie diet would produce an allowable maximum total of 46.36 milligrams of serum cholesterol (27 milligrams plus 19.36 milligrams).

We then applied these formulas, with eight variations for each of the caloric levels used in the 2-in-1 tables, to arrive at the specific score for each individual food item.

Here is an example: If you consume 1,800 calories daily, and if you have a food item with 10 grams of saturated fat (90 calories) and 60 milligrams of dietary cholesterol, the equation would be that of 90 calories of saturated fat equal 50 percent of the 180 allowable calories of saturated fat. This translates to a production of 13.5 milligrams of serum cholesterol (50 percent of the maximum 27 milligrams). The 60 milligrams of dietary cholesterol are equal to 20 percent of the allowable 300 milligrams of dietary cholesterol, which translates to a production of 3.87 milligrams of serum cholesterol (20 percent of the maximum 19.36 milligrams). Adding these two numbers, 13.5 and 3.87, shows that the item produced 17.37 milligrams of serum cholesterol. The total allowed for 1,800 calories was 46.36 milligrams. Therefore, dividing 17.37 milligrams into 46.36 milligrams, we find that the item produced 37.47 percent of the allowable serum cholesterol, giving it a value of 37 in the 2-in-1 tables.

In the second example above, a daily consumption of 3,200 calories would give the following equation: The 27 milligrams of serum cholesterol produced by 10 percent saturated fat, added to the 14.52 milligrams of serum cholesterol produced by 300 milligrams of dietary cholesterol, results in a total of 41.52 milligrams of serum cholesterol allowed under the Institute guidelines. In the prior food example, with 90 calories of saturated fat and 60 milligrams of dietary cholesterol, the result would be as follows: The 90 saturated fat calories are equal to 28 percent of the total 320 allowable saturated fat calories, translating to a production of 7.56 milligrams of dietary cholesterol (28 percent of 27 milligrams). The 60 milligrams of dietary cholesterol equal 20 percent of the allowable 300 milligrams, and therefore produced 2.904 milligrams of serum cholesterol (20 percent of 14.52 milligrams). Adding these two numbers (7.56 and 2.904) equals 10.464 milligrams of serum cholesterol produced by the item. Dividing this number (10.464 milligrams) into 41.52 milligrams, the maximum allowed for 3,200 calories, we find that it equals 25.20 percent, thus giving the item a 2-in-1 score of 25.

Therefore, the same food item represents 37 of the 100 points allowed by the 2-in-1 System if you eat 1,800 calories daily, but only 25 of 100 points if your daily intake is 3,200 calories.

The explanation of the mathematical calculations and formulas used to formulate the 2-in-1 tables may be difficult to understand, but we have included them for those readers who might be interested in the manner in which we arrived at the equations necessary in designing the tables.

Of far greater importance than your understanding of the mathematical calculations, however, is your growing familiarity, on a daily basis, with the 2-in-1 scores for the food items you consume, and a daily intake that is strictly limited to 100 points.

Index